Fictional Matter

Fictional Matter

Empiricism, Corpuscles, and the Novel

Helen Thompson

PENN

UNIVERSITY OF PENNSYLVANIA PRESS

PHILADELPHIA

Published by
University of Pennsylvania Press
Philadelphia, Pennsylvania 19104-4112
www.upenn.edu/pennpress

Printed in the United States of America on acid-free paper
1 3 5 7 9 10 8 6 4 2

Cataloging-in-Publication Data is available from the Library of Congress.
ISBN 978-0-8122-4872-2

For Jeff

CONTENTS

Introduction

> I could not see any impossibility in the Nature of the Thing, that one kind of Metal should be transmuted into another.
>
> —Robert Boyle, *The Origin of Forms and Qualities* (1666–67)

> What Sartre said of humans—that their existence precedes their essence—has to be said of all the actants: of the air's spring as well as society, of matter as well as consciousness.
>
> —Bruno Latour, *We Have Never Been Modern* (1991)

This book brings eighteenth-century novels into contact with the science of chemistry. The historically transformative aspect of that science is communicated in the moniker for it coined by the seventeenth-century experimentalist Robert Boyle: "Corpuscularian Philosophy."[1] Corpuscles, or chemical units of matter, are neither classical atoms nor modern-day elements. Nor are they imaginary entities banished from the scene of laboratory work. Corpuscles are durable but sometimes divisible, mechanical but attracting, insensible but real. The corpuscle's complexities, I argue, compel us to rethink perception, ontology, and realism in the emergent genre of the novel.

Reflecting my biggest stake in literary history, this book argues against a "realist" regime of transparently apprehended and transparently rendered facts, both in science and fiction. I route this literary-historical claim through chemistry's experimental involvement with sensed qualities like sourness or acidity, which do not isolate sensational knowledge from insensible particles. Corpuscular chemistry, I argue, defines empirical understanding that does not segregate worldly things from imperceptible causes. Drawing on recent revisionist histories of science, which restore agency to early modern micromatter beyond mere mechanical reaction, my book traces a path from chemistry through empiricism that transforms our portrayal of realist representation.

This is not the transcription of passive, distantly witnessed objects, but sensory knowledge placed in inextricable relation to particles, sensory knowledge that particles *produce.*

Fictional Matter refutes literary history's dominant account of early modern science, Steven Shapin and Simon Schaffer's argument for the social construction of experimental fact. The pretense to truth of collectively witnessed fact, Shapin and Schaffer argue, is forged by a science that refuses to speculate on imperceptible causes. Theirs is a strenuously literal portrait of empirical knowledge, which places science and representation in a phenomenal enclosure delimited by visible things. But early modern chemistry—indeed, chemistry from the ninth century forward—does empirically vindicate the truth of matter that cannot be sensed. Processes like distillation, precipitation, and sublimation make reactants vanish and reappear again. In this history of experiment, imperceptibility is not radical nescience and perceptibility is not self-evident fact. Together, imperceptibility and perceptibility enable access to minuscule constituents of things conjured from or back to human touch, smell, taste, or sight part by infinitesimal part. In Boyle's science, the physicality of matter is wedded to its sensory evanescence. In John Locke's empiricism, sensational understanding summons matter that stimulates ideas. My book argues for empirical knowledge whose epistemology, whose metaphysics, and whose representations do not divide perceived from imperceptible things.

Corpuscular chemistry is driven by a deep refusal of scholastic or neo-Aristotelian essence, a refusal that propels the new science from the English Civil War forward. The corpuscle instantiates this denial of essence by refusing to embody humanly known qualities. Such qualities—for example, warmth or sweetness—do not flow from a substance locked inside or a supervening influence visited upon matter. Rather, non-elemental corpuscles provoke human sensation because they amass into changeable structures, which Boyle names textures. In eighteenth-century science and empiricism, it is neither essence nor elements but changeable texture that engenders the perceived identity of things. By denying ontology a secure home in matter, chemistry defines an empirical world whose sensed endowments compel articulation of the textures, forms, and relations that produce them.

Throughout this book, I engage chemical science to motivate my departure from reigning assumptions about Lockean empiricism in eighteenth-century literary scholarship. Michael McKeon's *Origins of the English Novel* makes the influential formulation: "Locke's indivisible 'truth' is overseen by a standard of objective truth whose first premise is the empiricist credo that the instrument of verification can and must be separated from the object verified."[2] Locke's treatment of perceptual knowledge, I argue, never affirms a static "standard of objective truth," because Locke adopts Boyle's corpuscular doctrine of qualities. Empirical knowledge is not predicated, for Locke, on the break between verifier and "object verified," between transparently instrumental sensory input and the impervious facticity of its "objective" referent. My claim for empiricism runs counter to McKeon's: as I argue of the physical and ideational entities Locke calls secondary qualities, Locke does not divorce percipients from the bodies they know, whether minuscule or macroscopic. As the ground for truth, empiricism institutes not separation but relation. The realism that empirical knowledge bequeaths literary history is not mimetic reflection of objects in the world. This realism entails the figures, forms, and experiences through which novels think the contingently produced event of qualitative understanding.

Corpuscles, Secondary Qualities, and Empirical Knowledge

In *Fictional Matter*, I do not take science as factual backdrop for the novel's fictional play. Both scientific discourse and prose fiction, I argue, elaborate modes of figural, formal, and experimental access to imperceptible things, crossing the epistemological pretenses of empirical knowledge with the awareness that such knowledge is *produced*. But I begin my introduction with Boyle's air-pump trials to tackle the question of the corpuscle's reality. Against theorists of science like Ian Hacking and Karen Barad, discussed at further length below, I affirm that early modern chemical particles are variably qualified, divisible, and real.

Since the ninth century, chemical processes like sublimation—passage directly from a solid to a gaseous state—demonstrate that reactants are comprised of minuscule parts. As one exposition attests of subliming sulfur, "The Islamic authors were fascinated with the process because, as far as they knew, the sulfur started out in one place (the bottom of the flask), disappeared, and then reappeared in another place (the sides of the flask)."[3] Its disappearance does not make sulfur an imaginary entity. On the contrary, it concretizes the physicality of tiny parts which, for a time, the chemist cannot see. Boyle invokes the newly discovered corrosives hydrochloric acid, sulfuric acid, and nitric acid, which disclose the texture of previously impenetrable metals: "by the Solution of Gold in *Aqua regis* [mixed nitric and hydrochloric acid], it appears, that the ponderousest of Bodies, if it be reduc'd to parts minute enough, may be kept from sinking in a Liquor much lighter then it self."[4] Aqua regia proves that even gold, "the ponderousest of Bodies," is comprised of "parts minute enough" to float in acidic solution. This proof does not require that Boyle see the minute parts. On the contrary, it requires that, for a while, he *not* see them. A base added to the solution neutralizes the acid, returning the metal to experimental perception. Sublimation, dissolution, and other chemical processes vindicate the reality of particles through the event of their vanishing and return to sense.

Another proof of the existence of particles was published among Boyle's trials of his air-pump and partially evacuated receiver, *New Experiments Physico-Mechanicall, Touching The Spring of the Air* (1660). Boyle was fascinated by one of the air-pump's effects: as pressure descends in the glass receiver, bubbles appear in a vessel of clear water placed inside. To grasp Boyle's account of what happens when ambient pressure on the water falls, we must revisit the overall mandate for his air-pump experiments: Boyle aims to debunk an occult variant of scholastic agency, "the supposed Aversation of Nature to a *Vacuum*."[5] Boyle replaces this "supposed Aversation" with an account of air's most crucially intuited physical attribute. Air expands to fill void space—or, as Boyle writes, there is an "Elastical power in the Air"—because "Aerial Corpuscles" possess a particular *shape*. Comparing these amassed corpuscles to "a Fleece of Wooll," Boyle likens each one to a "little Spring, [which can] be easily bent or rouled up; but will also, like a Spring, be still endeavouring to stretch it self out again."[6] Boyle's air-pump trials summon "the springy Texture of the Corpuscles of the Air"[7] to explain what happens in the exhausted receiver. Springy texture exemplifies the logic of corpuscular philosophy's rejection of "natures abhorrency of a *Vacuum*."[8] The core physicality of each corpuscle—what Locke will call its primary qualities—includes the attribute of shape. The mechanical force of coiled shape effects the perceptible springiness of "endeavouring" air, justifying the double name Boyle assigns the new chemistry, "*Corpuscular or* Mechanical *Philosophy*."[9]

We can now return to the apparition of bubbles in what had seemed a homogeneous fluid. Boyle suggests that water, itself a particulate body, contains "Pores or invisible little recesses."[10] When external pressure descends, equally invisible aerial corpuscles exert their elastic power from within the fluid: "there may lurk undiscernable parcels of Air, capable, upon the removal of the pressure of the ambient Air (though but in part) and that of the Liquor wherein it lurks, to produce conspicuous bubbles."[11] Lurking in liquor's pores, "undiscernable parcels of Air" become manifest to empirical sense as a result of the corpuscle's springy shape. Hidden air becomes "conspicuous" when it no longer meets compressive force on the fluid in whose recesses it is contained. Rendered empirically vivid by Boyle's analogy to wool, the endowment of springiness limns each individual corpuscle. The physicality of an imperceptible corpuscle which, like wool, may be "easily bent or rouled up," is obviously figural. But as a primary mechanical attribute that summons bubbles made of compressible parts from water in the emptied receiver, the corpuscle's physicality is also real.

Boyle lists the attributes native to imperceptible corpuscles: "Bigness, Shape, Motion (or Rest,) Scituation and Contexture."[12] These features operate to produce qualities the chemist can sense, like the springiness which, at low pressure, makes bubbles out of formerly invisible parcels of air. Corpuscular philosophy, which replaces nature's horror of a vacuum with the mechanical force of minuscule physical parts, is thus wedded to a theory of perception. The chemist never perceives a single corpuscular spring: instead she apprehends something quite different, like the elasticity of environing air that holds a column of mercury upright in an inverted tube. From the side of experience, corpuscular physicality entails sensed effects not always resident in matter as such. To cite an instance dear to Locke, who repeats it in *An Essay Concerning Human Understanding* (1690), Boyle ascribes humanly felt heat or cold to the corpuscular attribute of "Motion (or Rest)":

> *Heat* consists either onely or chiefly in the local motion of the small parts of a body Mechanically modified by certain conditions, of which the principal is the vehemency of the various agitations of those insensible parts; and if it be also true, as Experience witnesses it to be, that, when the minute parts of a body are in or arrive at such a state, that they are more slowly or faintly agitated than those of our fingers or other organs of feeling, we judge them cold: These two things laid together seem plainly enough to argue, that a Privation or Negation of that Local Motion . . . may suffice for the denominating a body Cold, as Coldness is a quality of the Object, (which as 'tis perceiv'd by the mind, is also an affection of the Sentient:) . . . I say Coldness as to sense; because as 'tis a Tactile Quality, in the popular acception of it, 'tis relative to our Organs of Feeling; as we see that the same luke-warm water will appear hot and cold to the same mans hands, if, when both are plung'd into it,

> one of them shall have been newly held to the fire, and the other be benummed with frost.[13]

In Boyle's account, the felt attribute heat proceeds from the "vehemency" of "insensible parts," while "more slowly or faintly agitated" parts effect cold. But persons are comprised of imperceptible corpuscles too. When apprehended as a "Tactile Quality," temperature is determined respective to the motion of the "minute parts" making up the "Organs of Feeling" in contact with the object. To stress the relational status of empirically known qualities, Boyle stipulates that heat or cold exist only "*relative* to our Organs of Feeling." As proof, he invokes the "*same* luke-warm water" whose qualitative discrimination by the "*same*" man hinges on the corpuscular motion of the right or left hand that senses it.

As "a quality of the Object"—gauged relative to the insensible parts of its percipient—temperature is corpuscular motion. As "perceiv'd by the mind," temperature is an "affection of the Sentient," a feeling or an idea. To deploy terminology invented by Locke, the corpuscular motion of water or hands are primary qualities, while the heat and cold experienced by persons are secondary qualities. Although they are caused by the primary attributes size, shape, motion, and texture, secondary qualities do not mimetically reflect those attributes: the relation between primary vehemence and secondary heat is not mimetic but *productive*. We can thus isolate an illuminating difference between eighteenth-century corpuscles and the variant of primary qualities recently championed by the speculative realist philosopher Quentin Meillassoux. For Meillassoux, the key virtue of primary qualities is their capacity to drive "the mathematical de-subjectivation of the world" and instate an extrasubjective reality "henceforth stripped of its sensible qualities."[14] But as Meillassoux concedes, Locke never distinguishes primary from secondary qualities to institute so stark a break.[15] Intuited from the vantage of experiment or experience, the primary reality of shape does not subsist extraneous to the chemist's qualitative understanding. Boyle's spring illustrates primary physicality that interfaces with human sense, Boyle writes, as "Elastical power." Ascribed to air as its texture's *power*, elasticity perceived by persons sutures the shape of corpuscles to human perceptual experience.

An overarching goal of my book is to rethink the significance of secondary qualities, which are not free-floating ideas but productive powers rooted in matter. Indeed, Locke adopts the keyword *power* to revise the location of ideas engendered by tiny bodies. Perceptual ideas seem like they belong inside minds, but Locke goes so far as to insert them into objects, because he aims to underscore matter's acquisition of powers known only after the event of human perception. A long-standing scholarly conundrum, Locke's location of secondary qualities like heat in both mind and matter may initially strike the *Essay*'s reader as an inadvertent lapse from Platonic or Cartesian dualism. But latent in matter as its capacity to produce qualitative understanding, secondary qualities *are* neither strictly physical attributes nor strictly intangible ideas. At a crux that, I argue, defines empirical knowledge in science and the novel, "secondary qualities" signify throughout this book as both sensory perceptions and corpuscular texture's power to effect them. (In addition, I sometimes deploy a cognate term, "secondary power.")

For the deepest feminist investment of this book, I am indebted to Judith Butler's philosophical genealogy of the category of matter. Against her Platonic account of how matter has been constructed as an irreducible entity, I argue in Chapter 1, chemistry offers a history of body that is divisible, only contingently unified even at the level of the particle. As matter's power to produce sensed attributes like heat, secondary qualities sustain a metaphysics of objects and ideas that cannot rigorously divide them. This metaphysics resonates with the feminist theorist Elizabeth Grosz's recent appeal to modalities of mind-body difference that do not hew to a Platonic template.[16] Rather than insulating ideality as a category hostile to matter, Boyle's chemistry—and Locke's rendition of it as empirical understanding—posit human ideas summoned into being as perceptions. Neither antagonistic nor antecedent to matter, empirical ideas are effected by corpuscular stimuli. This is not to deny the ideality of ideas but to rethink their provenance and affiliation with matter.

Beyond Mechanism: Chymistry and the Divisible Corpuscle

As Boyle's spring suggests, corpuscles do not facilitate the particle's mathematical or pointillist reduction. Air's elastic force may be quantified, as we will see with eighteenth-century pneumatic scientists like Stephen Hales. But rendered by analogy with a fleece of wool, the aerial corpuscle's secondary power entails the attribute of shape. Granted figural lucidity, mechanical causes displace forces like nature's abhorrence of a vacuum, as Boyle writes of temperature: "an Imminution of such a degree of former motion . . . which is sufficient to the Production of sensible Coldness, may be Mechanically made, since Slowness as well as Swiftness being a Mode of Local motion is a Mechanical thing."[17] Rather than occult or essential endowments, sensible qualities are "Mechanically made." Yet the mechanical likeness Boyle claims for macro- and submicroscopic bodies does not exhaust the corpuscle's causal repertoire. I now survey recent revisions of the historical status of mechanism. With matter's secondary power, this vastly enriched account of particulate agency undoes the static facticity of things, from particles to species.

To address the history of mechanism, I return to Boyle's bubbles. At normal pressure, bubbleless water may now be viewed as also an experimental artifact. But bubbles disclosed at low pressure do more than expose imperceptibility as also an effect of compressed air's spring. To pursue the anti-elemental—and, as we will see, the transmutational—mandate of corpuscular chemistry, Boyle wonders whether the difference between air and water can be breached: "whether or no Air be a Primogenial Body (if I may so speak) that cannot now be generated or turn'd either into Water or any other Body."[18] As he proceeds to argue, aerial corpuscles refuse to incarnate elemental ontology because air's spring can be made from other things:

> And if we will have the Air to be a *congeries* of little slender Springs, it seems not impossible, though it be difficult, that the small parts of divers Bodies may be [*sic*] a lucky concourse of causes be so connected as to constitute such little Springs, since . . . even the bare alter'd position and connexion of the parts of a Body may suffice to give it a Spring that it had not before, as may be seen in a thin and flexible Plate of Silver; unto which, by some stroaks of a Hammer, you may give a Spring, and by onely heating it red hot you may make it again flexible as before.[19]

For Boyle, elastic power is as adventitious to the identity of the aerial particle as the "Spring" lent silver by a hammer. Just as flexibility or resistance are effected in silver by fire or forge, spring may proceed from "small parts" that recombine to instantiate air. But while silver remains silver in this instance, air may be produced from other things. Boyle's "little Springs," which are contrived from "small parts of divers Bodies," reveal the existence of parts even smaller than the corpuscle of air they compose. Air's identity is not fixed at the level of the single corpuscle because this corpuscle, we discover, is not indivisible. Boyle's aerial corpuscle may be engendered by the "bare alter'd position and connexion" of subcorpuscular parts that render its spring reproducible by the "lucky concourse" of other, even tinier, bodies.[20]

No individual particulate body entrenches the identity of air. On the contrary, air subsists as "*congeries* of little slender Springs" each of whose shape may be comprised of smaller parts. We thus encounter a corpuscle which, unlike the units proposed by Epicurus or Descartes, may be divided. The divisible corpuscle heralds one deeply consequential difference, for the history of chemistry, between corpuscular science and classical or reductionist atomism. As we have seen, Boyle deploys the word "mechanical" to flag the polemical thrust of antischolastic science, which replaces mystified essence with clearly schematized physical causes. But corpuscular mechanism does not presume an indivisible particle. Eighteenth-century literary scholarship tends to assume that particulate bodies are equivalent to Lucretius's hard atoms or Descartes's abstracted units, but Boyle posits a corpuscle that refuses irreducibly to incarnate air.

Chemistry's next departure from reductionist atomism also recalls Boyle's denial that aerial corpuscles incarnate "Primogenial" elemental bodies. As we have seen, the spring whose shape contains air's elastic power may—however luckily—be composed of smaller discrepant parts. But *Spring* does not always deny air's claim to some kind of substantive integrity. Writing of "that deservedly Famous Mechanician and Chymist" Cornelius Drebell (1572–1634), a Dutch alchemist, dyer, and mariner who in 1620 built a submarine for James I which was tested in the Thames, Boyle cites Drebell's rationale for the longevity of the men inside the craft: "*Drebell* conceiv'd, that 'tis not the whole body of the Air, but a certain Quintessence (as Chymists speake) or spirituous part of it, that makes it fit for respiration, which being spent, the remaining grosser body, or carcase (if I may so call it) of the Air, is unable to cherish the vitall flame residing in the heart: So that (for ought I could gather) besides the Mechanicall contrivance of his vessell he had a Chymicall liquor, which he accounted the chiefe Secret of his submarine Navigation."[21] Medieval alchemists—or, in Boyle's parlance, "Chymists"—appropriated the Aristotelian appellation "a fifth element, a fifth essence, a quinta essence [*essentia*], a quintessence"[22] to refer to purified products of distillation like the alcohol or aqua vitae yielded by wine.[23] But however warily Boyle cites the source of air's proposed "Quintessence," it appears to contradict his claim for the smaller parts that may comprise corpuscular spring. While air's finer or "spirituous part" is embodied by corpuscles—its separation reflects the insensible nonhomogeneity of starting air—it cannot be alienated from the property that "makes it fit for respiration." That is, while spring may be reconstituted from unspecified subcorpuscular parts, the "Chymicall liquor" Drebell dispenses to the inhabitants of his submarine heralds his success in isolating air's vivifying virtue.

Spring does not resolve these contradictory accounts of air composed of random things and air whose quintessence remains intact.

Air's quintessence persists into a corpuscular episteme that debunks elemental ontology. The coexistence of quintessence and mechanically replicable spring is illustrated by the submarine itself, whose sailors survive by means of both "Mechanicall contrivance" and "a Chymicall liquor." Up to this point, I have stressed corpuscular philosophy's anti-essentialist and mechanical mandate. Yet air's quintessence does not signal Boyle's anachronistic relapse into a logic his science discards. Rather, vivifying liquor marks one of numerous modalities of extramechanical agency central to the explanatory and experimental success of early modern chemical science. Until the final decades of the twentieth century, this science was historicized as rigidly anti-occult—and anti-alchemical—reductionist mechanism. Since then, historians including Antonio Clericuzio, William R. Newman, and Lawrence M. Principe have launched a reappraisal that illuminates Boyle's submerged debt to a long, not unilaterally mechanistic genealogy of particulate matter theory. Clericuzio, Newman, and Principe definitively distinguish corpuscular chemistry from systems like "the Cartesians' science," which, as Clericuzio writes, "may be described as reductionist."[24]

This break with reductionist mechanism has led historians to reassess the importance of alchemy. No longer regarded as an error soon to be cleared by mechanical automatism, transmutation is now seen as an experimental enterprise indistinguishable from corpuscular chemistry. As Clericuzio remarks of the ensuing terminological reduplication in the history of science, "Chemistry, or chemical works also mean alchemy and alchemical works."[25] Newman and Principe argue that historians should deploy the more inclusive "archaic" term "chymistry," Newman writes, "to refer to early modern alchemy-chemistry."[26] My Chapter 1 pursues the overlay of chemistry and alchemy sustained by non-elemental corpuscles, an overlay vindicated, Boyle claims, by his experimental success at transmuting gold into silver. Following Newman and Principe, I employ "chymistry" as an expansive cognate for chemistry, using the former word to emphasize the historically productive confluence of alchemy and corpuscular science.

In a crucial departure from mechanical reductionism, we find a divisible particle. Indeed, Boyle scorns the notion of absolute indivisibility as an idealist abstraction, irrelevant to laboratory reactions in which corpuscular unity is only relationally sustained. Placed in proximity to other things, corpuscles may stay intact, divide into smaller parts, or become something else whose own durability will, in turn, be determined in relation. What enables such outcomes is the extramechanical force Boyle dubs stickiness and Isaac Newton names attraction. Among other corpuscular capacities like seminal influence, attraction is critical to the renovated history of chymistry. Mechanism facilitates Boyle's denial of neoAristotelian ontology without dictating the reductionist impoverishment of matter's powers. As we will see in Chapter 1, the corpuscle's primary physicality admits extramechanical augmentation.

Experimenters employ quantitative methods like gravimetric analysis, but these are not the hallmark of chemistry's purchase on matter, as Newman and Principe remark: "how sophisticated did mathematical methods need to be even for the striking advances of the eighteenth-century chemical revolution?"[27] As Clericuzio suggests, it is time to appreciate a historical impetus for science that is "not mechanical but qualitative."[28] The novel's engagement with empirical knowledge, I argue, is likewise qualitative. Rather than aspiring to a truth standard based on mathematical reduction, novels project the forms, relations, and powers through which empirical apprehension of reality happens. It is this reality—qualitative reality—which eighteenth-century chemistry, empiricism, and novels think about knowing.

Corpuscles, Scientific Realism, and New Materialism

Advanced by Shapin and Schaffer's *Leviathan and the Air-Pump* in 1985, the claim that "[h]istorians have paid relatively little attention to Boyle's experimental researches after 1660"[29] is no longer accurate. The history of chemistry now accommodates Boyle's career-long investment in transmutation, a revision advanced by Principe, who, refuting Shapin and Schaffer, shows the experimental coherence of Boyle's alchemy and pursuits like the air-pump trials.[30] Most critical to Boyle's revised legacy, historians of chymistry repudiate Shapin and Schaffer's claim that Boyle "abominated anti-mechanical notions."[31] As Newman suggests, this claim recurs to a "preexisting history of ideas"[32] which does not interrogate the mechanistic bias characterizing, Principe writes, an "older positivist historiography."[33] My own difference from Shapin and Schaffer's unilaterally mechanistic science focuses on the ensuing account of perceptual knowledge. Despite Boyle's ascription of macroscopic effects—like air's "Power of self-Dilatation"[34]—to the springiness of insensible parts, Shapin and Schaffer argue that as the condition of his pretense to experimental fact, Boyle represses any intimation of invisible causes. Their claim for the constitutive exclusion of imperceptible causality from the new science is affirmed by scholars from Bruno Latour to John Bender. I now review how, Shapin and Schaffer argue, Boyle divides witnesses from objects to institute what Ian Hacking calls a spectator theory of knowledge.[35]

Shapin and Schaffer recruit Boyle's air-pump experiments to argue that "experimental philosophy, empiricist and inductivist, depended upon the generation of matters of fact that were objects of perceptual experience."[36] The keywords "matters of fact," Shapin and Schaffer emphasize, divorce scientific data from human interference: "matters of fact . . . were not of one's own making: they were, in the empiricist language-game, discovered rather than invented."[37] The scientist's initial "perceptual experience" models his role in the ensuing "empiricist language-game": perception records preexisting objects in the world, which garner labels that do not perturb the resultant experiential data. This language-game conserves the transparency of words that operate as counters for sensory input, which likewise does not disrupt the integrity of objects the witness observes from a distance. To enlist corpuscular philosophy in the apparition of data impervious to both perception and language, however, Shapin and Schaffer deny its status as philosophy. Boyle's *Spring of the Air* instead consists of "piecemeal reporting of experimental trials"[38] that meets Shapin and Schaffer's key criterion for facts whose social construction is masked by their sensory self-evidence: "the quest for real causes ought to be carefully segregated from the factual enterprise."[39] It is, in particular, *corpuscular* causes that must be exiled from Shapin and Schaffer's fact-making scene: "Since the world of corpuscles is inaccessible to our senses . . . causal inquiry was to be tactically segregated from the main tasks of the natural philosopher; hypotheses about causes were . . . distal to fact production."[40] Shapin and Schaffer restrict the "main" ambit of Boyle's science to what is empirically perceived. Most important for my purposes, their witness of experimental fact cannot pursue "causal inquiry" that recurs to the power of corpuscular spring.

To claim that corpuscles are "segregated" from Boyle's experimental history of air is to repress the bulk of his *Spring*, whose defining empirical conceit cannot be detached from Boyle's account of insensible causes: "Air is confessedly endow'd with an Elastical power that probably proceeds from its Texture."[41] The secondary power of texture—which belongs to amassed corpuscles Boyle likens to a fleece of wool—proceeds, as we have seen, from "the Shape of the Springy Particles of the Air."[42] By arguing that corpuscular science isolates air's perceptible effects from its "Springy" shape, Shapin and Schaffer ignore the warrant for Boyle's extension of mechanical agency to insensible bodies. Shapin and Schaffer deny empirical salience to, they write, "entities that would not be made sensible,"[43] but this denial overlooks sublimation, dissolution in aqua regia, and other reactions that prove the physicality of tiny parts as the very result of their sensory disappearance. Only by virtue of such proof can Boyle claim aerial corpuscles as "a heap of little Bodies"[44]—that is, as *body* at all.

The history of chemistry conjoins vanishing and sensory recovery to prove that things are made of insensible parts. With Simon and Schaffer, recent appeals to "scientific realism"[45] repress this long-standing chemical proof. My introduction's epigraphs excerpt Boyle's suggestion "that one kind of Metal should be transmuted into another" alongside Latour's application of the existentialist dictum "existence precedes essence" to bodies like springs as well as to persons. This juxtaposition aims to illuminate their agreement: as Jean-Paul Sartre will suggest of men, the existence of transmutable metal far precedes its belated ascension to essence. Latour, however, relies on Shapin and Schaffer's portrait of Boyle: as Latour writes while crediting their history of science, "the silent behaviour of objects" or "inert bodies" invite transcription into experimental fact by "signifying" to privileged witnesses.[46] Boyle relays phenomena whose experiential impact cannot be narrowed to "signifying": he sublimes, calcines, distills, acidifies, precipitates, transmutes, and reduces. At the same time, he inhales, smells, touches, tastes, and sees (poorly: Boyle's weak eyesight led him to valorize the acuity of his other senses).[47] Finally, he posits the secondary powers of insensible texture. Boyle, Latour's avatar of modern science, is not a modest witness but a chymist.

In reaction to the postmodern credo, Latour writes, "that matter is immaterial,"[48] he and other theorists of science including Hacking, Donna Haraway, and Karen Barad promote an atomic reality whose experimental specification interconnects with the situated reality of the scientist.[49] For Hacking, premodern particles fail to achieve this reality: rather, they inaugurate what Hacking calls "representationalism" as soon as the classical philosopher Democritus advances his atomistic hypothesis. Once Democritus posits atoms, Hacking argues, the world turns into "mere appearance"[50] and perceptible qualities break away from the atomic object's unknowable "inner constitution."[51] Hacking credits the demise of this representationalist impasse to experimental technology which, "by 1910," makes particulate truth experientially known: "Anti-realism about atoms was very sensible . . . a century ago. Anti-realism about *any* sub-microscopic entities was a sound doctrine in those days. Things are different now."[52] Hacking does not consider laboratory processes like sublimation or distillation, practiced a millennium earlier, during which reactants vanish and reappear as micro-thin deposits coat the upper walls of the flask or condensing apparatus. He dates the reality of particles to 1910 by overlooking chemical processes that predicate the physicality of minuscule parts on the revelation of their vanishing and return to sense.

But as it turns out, things are not so different now. Barad indicts an "Enlightenment culture of objectivism" that divides objects "with their own independent sets of determinate properties" from disembodied scientific spectators.[53] Yet when Barad, who approves Hacking's claim for the unreality of particles from Democritus till the early twentieth century, gives a first-person account of the existence of the atom, her syntax is telling. As she and others "watch[]" a scanning tunneling microscope (STM) trained "on a sample of graphite," Barad writes, "individual carbon atoms were imaged before our very eyes."[54] With this rendition of her empirical experience, Barad links atomic reality to a humanly "imaged" idea. But she does not suggest, in her words, that "our image of matter" or "our imaginations and . . . possibilities for imaging"[55] claim any affinity with the inner constitutions Hacking shields from human knowledge before 1910. The reality of Barad's carbon atom seems to be buttressed by a certain reductionism at the level of the image, as she writes: "humans do not possess a perceptual apparatus that can directly detect atomic events, and we therefore depend on pointers and other macroscopic devices to help us discern the results of experiments."[56] Instruments like pointers produce "results" or "measurements" exemplified for Barad, in the quantum physicist Niels Bohr's words, by "marks—such as a spot on a photographic plate, caused by the impact of an electron."[57] Although STM images, results, "marks," and "spot[s]" are not atoms, Barad grants them a reality she denies particles proven, for example, by the emergence of bubbles from clear water at low pressure.

Barad's STM-imaged carbon atom is real, while minuscule parts disclosed, in part, by vanishing are terminally unknowable. Is matter real when imaged by an STM, but unreal when imaged by a human retina? Is the bitterness Boyle tastes to identify silver corpuscles in a mixture unreal, although Barad does not doubt the reality of images she sees? Barad seems to predicate the superior reality of modern marks on their resistance to figuration. Spots repress the imagistic dimension of the image, fusing Barad's claim for a technologically novel particulate reality to an anti-figural vindication of its presence to human understanding. The atom is real today because "the impact of an electron" observes an implicitly anti-figural criterion flouted by the spring of an aerial corpuscle.

Jane Bennett's definition of the theoretical remit of new materialism, which reclaims the vibrant agentive capacities of nonhuman bodies, may help clarify the issue. To reject models of action based on the political individual, Bennett calls for a decentered understanding of perception: new materialism "attempts a more radical displacement of the human subject than phenomenology has done."[58] But the mandate that phenomenology displace persons exists in tension with Bennett's account of the objects she encounters in a storm drain grate, an encounter that conjures the vitality of these assembled objects' inanimate presence. Bennett's appeal to these things' more than "passive 'intractability' " is leveraged, she writes, by the "tableau that they formed with each other . . . [and] with me."[59] Bennett calls for an extrasubjective phenomenology while her embrace of "sensory attentiveness to the qualitative singularities of the object"[60] recruits both the storm drain grate and a perceptually attentive "me." A "me" difficult to extricate from Bennett's realist phenomenology extends to the scale of the imperceptible. Recapitulating a scientist's description of the crystalline "microstructure" of metals, she concludes that "metallic *vitality* . . . can be seen in the quivering of these free atoms at the edges between the grains of the polycrystalline edifice."[61] This depiction of texture is almost as vivid as Boyle's. But Bennett elides the fact that she has *not* "seen" it. Her attempted removal of subjectivity from the event of human sensory understanding only knits her more deeply into the figural elaboration of insensible matter's powers.

To image metallic crystals and carbon atoms is neither to reinstate Platonic ideality nor to insist, as Latour claims of postmodernism, that these things are really words. It is to affirm, for realism today no less than for early modern corpuscular science, a form of knowledge that implicates insensible matter's power to make itself humanly known. I use the word "form" to signify not an ideal blueprint imposed *on* matter but modes of experimental and representational access *to* matter. These modes of knowledge insinuate perceptual understanding back into particles not to mitigate their reality but to figure their capacity to effect sensory experience.

Chemists like Boyle and Newton, as well as dyers, painters, gem and glassmakers, miners, metallurgists, brewers, tanners, submarine and shipbuilders, and other artificers, really *made* things with corpuscles (they also suffered from ingesting, breathing, and touching them[62]). This suffices to meet Hacking's standard: "We shall count as real what we can use to intervene in the world to affect something else, or what the world can use to affect us."[63] In the pages that follow, I do not question the corpuscle's reality. The issue my book pursues, highlighted by Barad's acknowledgement and repression of human imaging of atoms, is empirical understanding engendered as the relational effect of matter's power.

Science, Perception, and the Novel

Ian Watt's *Rise of the Novel* commences: "Modern realism, of course, begins from the position that truth can be discovered by the individual through his senses."[64] Despite groundbreaking new scholarship on materiality and the novel, the implicit support for narrative realism remains the sensory accessibility—even if it is enhanced by microscopes—of what novelists describe.[65] Claims for the role of atomistic materialism in eighteenth-century literature must confront Watt's standard, which defines "truth" that can be delivered only by perception. *Rise of the Novel* concludes with a return to sensory knowledge: the divergence between "the novel with an emphasis on society" and "novelists who explore the inner lives of their characters" solicits, Watt proposes, "a close epistemological analogue—dualism."[66] Crediting Descartes, Watt suggests that just as Cartesian skepticism dictates an "ego" alienated from the "external world," so "although different novelists have given different degrees of importance to the internal and external objects of consciousness, they have never completely rejected either; on the contrary, the basic terms of their inquiry have been dictated by the narrative equivalent of dualism—the problematic nature of the relation between the individual and his environment."[67] Watt's overarching concern is an individual whose nascent perceptual autonomy propels his or her break from traditional social order. Watt aligns this individual's "problematic" separation from the collective with "dualism" that, according to his Cartesian rubric, insulates things "discovered . . . through the senses" from "internal . . . consciousness" of the self. For Watt, realist reference to external objects and realist reference to inner selves run on parallel but alienated narrative tracks.

Watt assumes "the ultimate realist premise of a one-to-one correspondence between literature and reality."[68] Whether trained on inner or outer objects, narrative realism's "premise" is unilaterally and mimetically referential. I depart from Watt by interrogating the critical givens of mimetic reference and mind's isolation from matter. The novel, I argue, evokes things that cannot be sensed, things whose power to produce sensation is not mimetically transmitted by sensation itself. As a chymical genealogy of knowledge and representation, my account is neither dualist nor Cartesian. Indeed, although *Rise of the Novel* precedes Shapin and Schaffer's exposition of experimental science, a key feature of Watt's realism anticipates their claim for the literal visibility of empirical fact: the "one-to-one correspondence" of referent to "reality" bars insensible causes from the novel as well as from experiment. My book disputes this "ultimate realist premise" by challenging the dualist presumption that segregates references to mind from references to body. Instead, I argue, eighteenth-century novels make explicit the *production* of sensational understanding as the reader's encounter with forms and powers that enable empirical knowledge.

Bender's recent study of science and representational realism, *Ends of Enlightenment*, plays out the literary-historical consequences of Shapin and Schaffer's account of empirical fact. Bender posits a novel defined by observational detachment and quantitative abstraction or, Bender writes, "experimental objects and mathematical symbols."[69] This science subsists on praxis "derived from mathematics" and, as Bender affirms with Shapin and Schaffer, "the derivation of 'facts' from nature directly via the senses."[70] His quotation marks endorse the social construction of an "edifice of scientific factuality"[71] whose refusal of insensible causes shores up its pretense to incontestable self-evidence. Shapin and Schaffer's science limns an epistemological expanse restricted to phenomenal things, a limit Bender ascribes to the chymist Newton: "experiment could make fact—the real thing—available to him through sensory observation."[72] What are the literary-historical stakes of this denial of modern science's involvement with the contingent, the relational, and the imperceptible?

Bypassing groundbreaking scholarship on Newton's alchemy and the forces, such as attraction, that he posits in his influential *Opticks* (1704), Bender enlists Newton in a hard break between occultism and mechanism that operates to "suspend or erase the speculative function"[73] in science.[74] For Bender, novels mimic science by promising ocular access to—by making readers "virtual witnesses"[75] of—anti-"speculative" fact. Science and novels alike refuse to concede that the self-evidence of such fact is socially enabled: "a certain denial of fictionality marks both the earlier eighteenth-century novel and early science."[76] Perhaps in response to the constructivist conundrum which would render scientific and novelistic truth claims indistinguishable, Bender argues that they diverge as the century advances. By "midcentury," Bender's "empiricist novel" assumes a more "manifest fictionality," "a manifest yet verisimilar fictionality" demarcated from "the factual truth of demonstration."[77] Scientific fact precludes "speculative" recourse to micromatter, but its pretense to truth is assisted, Bender argues, by contrast with the more flagrantly fictional representations found in novels. Yet the constructivist dilemma endures. How does science secure its own claim to truth? What operates to prevent even occasional disclosures of an "edifice of . . . factuality" sustained, Bender argues with Shapin and Schaffer, by science's refusal to postulate insensible causes?

The threat of such disclosure is redundant because, taking Shapin and Schaffer at their word, Bender assumes the total scientific adequacy of what is seen. A corollary to this fact-making containment of sight becomes evident: Bender treats reading, like experimental witnessing, not as formal, connotative, or stylistic experience but as a strictly instrumental mimetic encounter that delivers "one-to-one" or "signifying" information in Watt's and Latour's sense. While at midcentury, novels fall away from fact, experimental truth claims are shored up by the veridical force, Bender writes, of "enough empirical data": "empirical, sense-based facts"[78] sustain the preemptive stipulation of a science whose refusal to limn corpuscular causes demarcates its break from "fictionality." Bender's "sense-based" science enforces the transparently referential operation of reading, whether the reader peruses Boyle's *Spring of the Air* or a novel.

To pursue the repercussions of this emphatically literal account of empirical perception, I turn to Bender's exposition of a visual artifact that makes manifest the formal infrastructure of realist representation. By reconsidering William Hogarth's visual argument, I aim to reclaim scientific and literary knowledge that exposes its formal conditions of possibility *as* the reader's encounter with both sensed qualities and the powers that produce them. An exemplar of "hyperrealism," Bender writes, Hogarth's *Satire on False Perspective* (1754) (Fig. 1), "as Shapin and Schaffer say, constituted 'new perceptual objects' and generated 'matters of fact that were objects of perceptual experience.'"[79] Drawing on art historical argument, Bender suggests that Hogarth promotes his viewer's exercise of experimental witnessing: "the objects Hogarth describes exist in a closed world possessing distinct features of the 'absorptive' mode in pictorial representation that Michael Fried has described in French painting from the 1760s onward."[80] But Fried's *Absorption and Theatricality* specifies aspects of painterly absorption that operate, precisely, to exclude Hogarth's satire. In part a reaction against Rococo mannerism, in part an effort to neutralize the emergent awareness, Fried writes, "that paintings

Figure 1. After William Hogarth (British, 1697–1764 London). Satire on False Perspective: Frontispiece to "Kirby's Perspective," Feb. 1754. Engraver: Luke Sullivan (Irish, 1705–1771 London). Etching and engraving; second state of two, image: 8 1/8 × 6 13/16 in. (20.6 × 17.3 cm) plate: 8 7/8 × 7 1/4 in. (22.6 × 18.4 cm) sheet: 9 3/8 × 8 in. (23.8 × 20.3 cm). Gift of Sarah Lazarus, 1891 (91.1.33). The Metropolitan Museum of Art, New York, N.Y., USA. Image copyright © The Metropolitan Museum of Art. Art Resource, N.Y.

are made to be beheld," absorption is an anti-theatrical visual stance sustained by paintings which internalize "the supreme fiction that the beholder did not exist, that he was not really there, standing before the canvas."[81] An anti-dramatic "realism" that squelches any taint of theatrical contrivance, the absorptive tableau posits a viewer defined by "the conviction of his absence from the scene of representation."[82] As a compositional mode, absorptive paintings deliver aesthetic experience conveyed by an image that denies it operates as "representation." Hogarth's *Satire* does the opposite. A botched deployment of compositional rules that enable the viewer to perceive depth, *Satire* systematically and aggressively disables the latent conditions of visual realism. That is, Hogarth does not let the realist witness look in peace. He denatures figural conventions that dictate not just a better or worse rendition of space, but the representational experience of space itself.

As satire that reveals *how* a realist beholder overlooks representational forms, Hogarth's print is anti-absorptive—that is, theatrical—to its core. By interrupting that reflexive perceptual operation, Hogarth shows how lines on a two-dimensional surface stimulate the viewer to perceive depth. His foreground works visually until a fisherman's line crosses beneath the line of a man half his size; a bridge over a river functions until smoke from a rifle on the far side annuls the depth it projects; the background stays in place until an inn placard hangs behind miniature trees deep in the further distance. Hogarth's print solicits and frustrates the viewer's attribution of space to the two-dimensional image, offering in its stepwise triggering and refusal of depth something like a phenomenology of realist perception, which is, in other words, an anti-phenomenology of the qualities realist viewers attribute to things rather than to their perceptual interaction with them. To deploy Locke's terms, the secondary quality of space implicates primary perspectival conventions that Hogarth forces the viewer to "see."[83] With Hogarth, I argue, eighteenth-century novelists like Henry Fielding activate the structure of primary-secondary difference to make evident the forms that produce empirical—or aesthetic or readerly—experience. This is to suggest that "realist" reference in eighteenth-century literary representation is not the mimetic imitation of transparently encountered things in the world. Eighteenth-century novels make explicit the *production* of empirical reality as the reader's encounter with forms and powers that enable sensational knowledge.

Hogarth's argument, as I apprehend it, precludes the dichotomization of transparently known big things and terminally fictional tiny ones, a split that denies the insinuation of causal power in the first instance and equates it with unscientific romance in the latter. More to the point, the dichotomy of big fact and little fiction curtails the epistemological reach of science just as surely as it renders novels irrelevant to the production of modern knowledge. Eighteenth-century chemical science does not abandon its engagement with reality when it figures insensible bodies. Nor does it pretend that such things—as well as bigger things—can be transparently known. In *Air's Appearance*, Jayne Elizabeth Lewis gestures to a critically kindred appreciation of empirical understanding when she evokes corpuscular elasticity that is also the figural spring of Boyle's prose: "Why is it so hard to see that Boyle's quantifiable 'parts' of air can also harbor more mysterious, and liberating, 'Forms'?"[84]

In this capacity, my book's title, *Fictional Matter*, aims to rethink fiction as well as matter. If corpuscles are real but not impervious to imaging, sensation, and even ideas, then eighteenth-century novels likewise participate in a project of qualitative understanding driven by the powers of insensible texture. Qualities effected in relation, qualities that do not mimetically mirror the forces that effect them, constitute knowledge in science and the novel, which do not divide dissimulated as opposed to open interest in what cannot be directly sensed between them. The novels I read below deploy different parts to motivate the sensible qualities of persons that other persons perceive—a minuscule plate of epidermal skin in the case of race (Chapter 4), a typographical character in the case of status (Chapter 5), or what *Spring* calls air's "carcase," a unit of expired breath, in the case of sex (Chapter 6). With chemical science, novels explore the experience and ontology of qualities not locked in irreducible units of body. The outcome for literary history is a novel that partakes in eighteenth-century thinking about the perception and ontology of qualities, including qualities of persons, not essentially or mimetically resident in matter.

It is a critical irony that provocative scholarship in eighteenth-century literary history has adopted an old story about mechanism, while new materialist appeals to agentive matter bypass the chemical corpuscle. In literary studies, Jonathan Kramnick declares of British interest in particles: "The point of looking at the smallest units out of which things are made is thus to suggest that persons are no more (or less) than material objects and material objects no more (or less) than their constituent parts."[85] This may be the case for Epicurean materialism's denial of the category of soul.[86] But in eighteenth-century corpuscular science and empiricism, whose relation to classical and Cartesian atomism I discuss in Chapter 1, corpuscles do not propel a unilaterally reductionist drive. Nor do corpuscles instantiate the antagonism to ideas, or ends, or extramechanical agency whose obduracy would refute a humanizing history of the novel. The counter to literary scholarship that enshrines the potency of individual volition is not a history of science that banishes ideas from empirical knowledge of matter.

The old reductionist mechanism, one warrant for a novel relegated to the domain of obtrusive fiction, can no longer explain some obtrusive oddities in the history of science. For example, as Newman writes of Newton's manuscript *Of Natures obvious laws & processes in vegetation*, "Newton believed that metals and minerals 'vegetated' (i.e. grew) within the earth, the subterranean mineral veins corresponding to the branches of terrestrial trees."[87] Citing "som further cause" beyond "the slight mutation of the textures of bodys in common chymistry & such like experiments," Newton invokes "a more subtile secret & noble way of working in all vegetation."[88] Showing his debt to Boyle as well as the chymical theorist of putrefactive agency Joan Baptista Van Helmont (1579–1644)[89] and the Helmontian theorist of vegetative growth Kenelm Digby (1603–1665),[90] Newton writes of the emergence of metals from putrefying matter: "In ye same Oare severall metalls are found all wch vegetate distinctly . . . sometimes metalls grow like trees in ye earth."[91] Ultimately, "minerall dissolutions & fermentations" may account, Newton suggests, for the agency of gravity.[92] At the same time, metallic vegetation sustains the chymical promise of growing gold.[93] Boyle notes the corresponding referential range he accords the word "life": "For 'tis ascrib'd not onely to all sorts of Animals and Plants, but by many Chymists and Mineralists to Stones and Metals growing in the bowels of the Earth."[94] In the new science, vegetable growth is as prevalent a model as mechanism in motivating the corpuscle's "way of working."

Sounding a cautionary note, Hacking writes: "realism has, historically, been mixed up with materialism . . . Such a materialism will be realistic about atoms, but may then be anti-realistic about 'immaterial' fields of force."[95] The cost of a materialism secured by reciprocal anti-realism about "immaterial" agents is intimated by Barad: "realism," she suggests, must be "reformulated in terms of . . . that reality of which we are a part . . . rather than some imagined and idealized human-independent reality."[96] The reality Barad defends is perceived by humans. As it enters the frame of the eighteenth-century novel, this reality does not possess facticity extraneous to the event of its apprehension. Secondary qualities are enmeshed with the scene of their perception, their powers neither finally instantiated nor finally known. In the chapters that follow, I track the formal, experiential, and ontological repercussions of corpuscles that harbor neither essence nor sensed qualities as such. In the novels I read below, the qualitative endowments of things or persons proceed from the forms and relations through which matter's powers are felt.

Chapter 1 explores Boyle's doctrine of qualities, his denial that qualitative understanding proceeds from scholastic essence or chymical elements. As Boyle writes of temperature, "a Privation or Negation of that Local Motion . . . may suffice for the denominating a body Cold." The exercise of "denominating" is no transparent imposition of language but a relational adjudication of what name this particular perceptual encounter—which, as we recall, involves the corpuscles of the sentient hand—might solicit. Boyle offers an account of chemical denomination proceeding from the particulars of sometimes violent sensory contact. This chapter pursues the complex dialectic of substance and structure propelled by Boyle's divisible corpuscle, his claim to transmute silver, and the experiment that weds the particle's reality to its vanishing. I depart from Butler's claim that body since Plato has been constructed as irreducible, suggesting that Boyle offers a radically penetrable person as the avatar of empirical understanding. Chapter 1 concludes with Robert Hooke's influential book of microscopic images, *Micrographia* (1665), to explore indistinguishable figurations of mechanical and vegetable motion.

By locating Locke's corpuscularianism in literary history, I aim to extend Newman's claim for the legacy of Boyle's science: "It was Boyle's corpuscular theory that contributed, through Locke, to the movement usually labeled 'British empiricism.'"[97] Chapter 2 situates Locke's *Essay* in the context of Civil War and puritan systems of sense-based pedagogy to foreground a paradox: human understanding proceeds from sensory ideas, but the insensible particles that stimulate perception must be represented by words. This chapter elucidates a corpuscular Locke, whose equivocal treatment of texture entails the monitory confluence of figure and fiction. Corpuscles propel Locke's ambivalence over the legitimacy of figuring matter in language, which, for him, confuses empirical sensation and synesthetic receptivity extending to the practice of reading novels. But as I suggest above, Locke's secondary qualities recombine sensory ideas and corpuscular texture. Locke's ambivalence tips into his subversion of dualistic metaphysics and transparent linguistic access to things, because matter's secondary power precludes the realist enclosure of the entities to which words refer. I close with plots by Eliza Haywood that posit objects—women—who change as a function of the discrepant ideas they elicit over space and time. Haywood's novelistic persons contain no inner guarantor of character's fixity but are known and named by the men who experience them.

Chapters 3 through 6 affirm that some novelists—Daniel Defoe, Jonathan Swift, Henry Fielding, and Samuel Richardson—directly engage chemical science and the corpuscle. While others, like Haywood, Penelope Aubin, and William Rufus Chetwood, innovate ontology, form, and qualitative knowledge without representing particulate science, all of these authors represent perception that recurs not to essence but to the relational effects of secondary powers. Chapter 3 places Defoe's fictional first-person history *A Journal of the Plague Year* in the context of the ascendancy of chemical medicine during England's Great Plague (1665–66). Following chemical physicians' etiology of seeds of disease, Defoe posits a porous person whose body admits and exudes matter too tiny to be perceived. My chapter argues that character in the *Journal* is shaped by persons who cannot sense the particles they contain. Character's innerness proceeds not from psychology but from contents that cannot be perceived, affiliating hidden interiority and matter's still latent powers. In Defoe's novel, depth is motivated by corpuscles and by boundaries whose perviousness cannot be staunched.

Chapters 4, 5, and 6 turn to sensible attributes that qualify persons: race, status, and sex. Corpuscles, I argue, secure none of these qualities. Chapter 4 delineates Boyle's and Newton's conflicting treatments of color. The chapter examines two mid eighteenth-century corpuscular theorists of race, John Arbuthnot and John Mitchell, who deploy color science to affirm a critical historical outcome: the political inhumanity of southern persons is sustained not by essence but by particles—of air, of skin—that refuse to entrench essence. The chapter turns to two novelists, Chetwood and Aubin, whose travel narratives stage climate's capacity to undo attributes that are only locally secured. In so doing, Aubin's travel plots span the options for color's relation to matter mapped by Boyle and Newton. Aubin enlists North African masters and African slaves to secure the Anglo-patriarchal value of white feminine chastity, but she cannot contain a field of ontological possibilities that includes reproductive or seminal influence.

Chapter 5 examines Fielding's treatment of status. Fielding satirizes alchemical science, but he is indebted to it for the intervention that defines his early novelistic output. In *Shamela*, *Joseph Andrews*, and *Jonathan Wild*, Fielding attacks the specter of literary hypocrisy by recruiting Lockean epistemology and what I designate alchemical form. Extending back to medieval translations of Arabic sources, alchemy's transmutational procedures embed qualities inside opposing qualities to articulate chemical reactivity as mutually qualitative and formal structure. Fielding, I argue, devises an alchemical model of character whose primary endowments are disclosed by the phenomenal impact of the printed page. In *Shamela*, *Joseph Andrews*, and *Jonathan Wild*, Fielding conjures an ontology of status from the reader's explicitly formal experience of the printed artifact.

Chapter 6 addresses Richardson's novel *Clarissa*, which attempts to specify the difference between the paragon Clarissa and the prostitutes who closely surround her. Richardson seems to observe a binary metaphysics to define Clarissa's superiority, but because he employs eighteenth-century respiratory theory to characterize women, his novel's account of used air extends to Clarissa. Accounts of breathing, sweat, and secretion by corpuscular scientists James Keill and Stephen Hales place *Clarissa*, I suggest, in a relay of expiring bodies that includes both prostitutes and African slaves. The chapter moves from the failure of a metaphysics of breath to differentiate Clarissa to other, still elusive determinants of her sexed virtue. As Newman remarks, the success of Boyle's "attack on theories of Aristotelian inspiration linking qualities to [essence]" confronts Boyle "with that most Aristotelian of questions—what is it that truly identifies a thing as itself and distinguishes it from others?"[98] The rigor with which Richardson interrogates the qualities that distinguish Clarissa leaves his novel, I argue, unable to determine what "truly identifies" sexed virtue effected as relational feeling, virtue that partakes in a metaphysics of frowzy corpuscles propounded by Swift as well as by early modern respiratory science.

In my Epilogue, I complete the arc *Fictional Matter* spans with two canonically modernizing moments in the histories of science and literature. First is Antoine-Laurent Lavoisier's claim, against corpuscular science, that oxygen is an element. Second is the resolution of Watt's dualist account of the eighteenth-century novel, Jane Austen's narrative technique of free indirect discourse. I appraise both moments from the vantage of my book's reconception of empirical knowledge. Rather than decisive ruptures, Lavoisier's new system of elemental nomenclature and Austen's third-person view of selves and things, I argue, manifest continuity with the corpuscular episteme. Lavoisier's moniker "oxygen" refers reactive tendencies to insensible particulate figure; Austen's most famous marriage plot recounts not love at first sight but sensory adjustment achieved in relation. By locating the impetus of Lavoisier's and Austen's modernizing representations in corpuscular science, I affirm their deep engagement with a realism—and a reality—where perceptible qualities implicate the forms, figures, and powers through which empirical understanding is produced.

CHAPTER 1

Boyle's Doctrine of Qualities

[W]e may learn what to think of the opinion of some Eminent Modern Philosophers who teach, that a fluid body is always divisible into bodies equally fluid, as Quantity into quantities, as if the particles of fluid Bodies must also be fluid themselves.
—Robert Boyle, "The History of Fluidity and Firmness" (1669)

Robert Boyle's manifesto *The Sceptical Chymist* (1661) is easily enlisted to emblematize reductionist science. *Sceptical Chymist* deploys mechanism to advance a profound denial of essence. To refute elemental substance in chemistry—in particular, Aristotle's earth, air, fire, and water as well as Paracelsus's mercury, sulfur, and salt—Boyle extirpates these ingredients from humanly sensed objects. Attributes chemists perceive, like color, do not reflect the presence of any given element or, as it was called, principle: "bodies exhibite colours, not upon the Account of the Predominancy of this or that Principle in them, but upon that of their Texture, and especially the Disposition of their superficial parts, whereby the Light rebounding thence to the Eye is so modifi'd, as by differing Impressions variously to affect the Organs of Sight."[1] Color does not flow from the mystified influence of "this or that Principle in" the object. Instead, color is a perceptual effect produced by "Texture": not elemental constituents, but the structural arrangement of the object's minuscule parts. At its most radical, *Sceptical Chymist* promotes the wholesale displacement of essence by texture. No single particle instantiates the source of sensed attributes like color, because color proceeds from structural permutations of indifferent units of matter. Distinguished solely by size, shape, motion, and cumulative texture, these units embody mechanism impervious to essence.

Yet *Sceptical Chymist* does not consistently uphold this reductionist impetus. With new scholarship on Boyle, I paint a more complex portrait of Boyle's unit of matter, the corpuscle. Corpuscles are neither the indivisible bodies of classical materialism nor are they, as Karen Barad suggests, "simply representative fictions, useful heuristics, or mere bookkeeping devices."[2] Summoned into—and out of—experimental awareness by processes like sublimation and distillation, singly imperceptible parts do not occupy the history of chemistry as inert, pointillist, or abstract fragments. But matter not reducible to mechanical units still refuses to conserve the sensed identity of things—even things like gold, which, Boyle affirms, may be transmuted into silver. The chemical corpuscle is neither an Epicurean atom nor a modern-day element. Its capacities are not exhausted by reductionist automatism. This corpuscle, whose endowments Boyle gleans from a long history of chymical science, does not enable transparently realist reference or self-evident fact.

Perceived qualities are produced by texture, like the superficial corpuscles which stimulate color in the viewer's eye. But attributes like color do not inhabit corpuscular texture as such. In eighteenth-century science and empiricism, sensed qualities engender a mode of knowledge distinct from the mimetic directive which, Ian Watt argues, motivates the rise the novel. Watt identifies "more dispassionate and scientific scrutiny of life" as grounds for a generic stance, formal realism, whose formal as opposed to personal or factional representative pretenses Watt aligns with "scientific objectivity."[3] Watt never acknowledges the ensuing tension between "scientific objectivity" and his claim for an "individualist . . . reorientation" of narrative veracity defined as "truth to individual experience—individual experience which is always unique and therefore new."[4] As I aim to show in this chapter, the encroaching discrepancy between "unique," often sensational novelistic "experience" and "dispassionate" "scrutiny" is not due to fiction but to the presumed "objectivity" of the science Watt cites. Early modern science does not, I argue, uphold a depopulated standard of mimetic truth to which the novel's rendition of individual experience would somewhat paradoxically aspire. Experimental knowledge implicates neither realist mimesis nor private feeling but corpuscular texture that effects sensed qualities in relation.

Chemical scientists experiment with neither bookkeeping devices nor irreducible atoms. Isaac Newton, like Boyle, invests corpuscles—and, as we will see, smaller parts of corpuscles—with the extramechanical power to attract. In recent eighteenth-century literary scholarship, reductionism and materialist mechanism have enabled a host of anti-idealist outcomes. For

Wolfram Schmidgen, mechanical atoms allow mixture's triumph over the purity of species. For Sandra Macpherson, impenetrable materiality shores up the integrity of harm, especially harm wreaked on women. For Macpherson and Jonathan Kramnick, causal mechanism banishes volition from plots that preclude the insinuation of ends.[5] But the recovery of matter's extramechanical endowments does not dictate a return to Platonic ideality or essence, because corpuscles disclose an alternative mind-body relation. Ideas are not insulated from bodies but provoked by corpuscles, as when, Boyle writes, "differing Impressions variously . . . affect the Organs of Sight." Empirical ideas engendered as sensory impressions are neither antecedent nor hostile to the corpuscular matter that effects them.

This chapter pursues aspects of ontology and experience that shape empirical knowledge and, through the influence of John Locke, the novel. The first section elucidates Boyle's treatment of sensed qualities as relational effects of texture. Section 2 examines the extramechanical attributes of Boyle's corpuscle, which enable both identity and transmutation. Section 3 argues for the critical significance of science's sometimes reducible unit of matter. The final section takes up Robert Hooke's *Micrographia* (1665), which illustrates the indifference, even to amplified sight, of mechanical motion and vegetable growth.

Sensible Qualities: Relations

In *The Origin of Forms and Qualities* (1666–67), the companion manifesto to *Sceptical Chymist*, Boyle delivers an "*EXCURSION about the* Relative Nature *of* Physical Qualities" mobilized, he writes, to correct "the Grand Mistake, that hath hitherto obtain'd about the Nature of Qualities."[6] Following the lead of Francis Bacon, Boyle refutes a "Mistake" that encapsulates the premise Shapin and Schaffer assign Enlightenment science: the experimenter's belief that qualities perceived in relation to an object are resident as such in the isolated thing.[7] To attack this presumption, Boyle argues that the invention of the first lock, followed by the first key, demonstrates the ineluctably "Relative Nature" of their defining endowments: "it became a Main part of the Notion and Description of a Lock, that it was capable of being made to Lock or Unlock by that other Piece of Iron we call a Key, and it was Lookd upon as a Peculiar Faculty and Power in the Key, that it was Fitted to Open and Shut the Lock, and yet by these new Attributes there was not added any Real or Physical Entity, either to the Locke, or to the Key, each of them remaining indeed nothing, but the same Piece of Iron, just so Shap'd as it was before."[8] One "Piece of Iron" acquires its "Peculiar Faculty and Power" as a function of the other: the key's most instrumental feature proceeds from extrinsic developments like the apparition of another piece of metal. But Boyle's point is not just that external things elicit qualities from other things, because he stipulates properties restricted to "just so Shap'd" masses of iron. Manifestly, the lock's shape stays the same with or without its complement. Whatever "new Attributes" locks and keys summon from each other, there is not, Boyle insists, "added any Real or Physical Entity" to either. Metallic shape distills Boyle's mandate for the reality instantiated by corpuscles: the autonomous physicality of shape is adequate to justify other, extrinsically summoned qualities like "being made to Lock or Unlock by that other Piece of Iron." Boyle's account of perceptual knowledge is thus tightly bound to his account of corpuscular matter. Shape that cannot be extricated from the isolated piece of metal models the zero-degree reality of Boyle's corpuscle. With shape, a restricted roster of core physical attributes—attributes inherent in a body before it is placed in proximity to others—are designated by Boyle *mechanical affections.*

Anchored in and as corpuscular body, mechanical affections debunk physicality in excess of what already realizes the faculties of a key. In this capacity, Boyle justifies his selective deployment of classical materialism. As he relates, ancient atomists explicate the perceived variety of things without recourse to incorporeal agency or essence: "as *Leucippus, Democritus, Epicurus,* and others . . . have taught, the difference of Bodies proceeds but from the various Magnitudes, Figures, Motions, and Textures of the small parts they consist of, (all the qualities that make them differ, being deducible from thence)."[9] Boyle rejects the atheistic denial of soul and the polarized metaphysics of Epicurean materialism; he deploys the moniker "mechanical" to flag the intelligibility of corpuscular features, like shape, that cause perceptible "difference[s] of Bodies."[10] Mechanical attributes invested in corpuscles as such lend them features, like figure, motion, and cumulative texture, whose agentive repertoire is lucid and familiar.

By denying that the key's power to open the lock augments its physical existence, Boyle does not render that quality unreal. Like a chemical reagent, the key still *does* something. Rather, by refusing to grant supplemental reality to the key's power to open, Boyle counters scholastic ontology. To preclude the influence of neoAristotelian essences or, in scholastic terminology, "Incomprehensible Substantiall Formes,"[11] Boyle roots the object's distinguishing traits in an anterior reality delimited by corpuscles.[12] A key's capacities can be ascribed to neither materiality in excess of metal nor "Incomprehensible Substantiall Formes" also alien to the iron they invest, "as if," Boyle writes, "this Forme were some distinct and Operative substance that were put into the Body as a Boy into a Pageant."[13] Boyle borrows from Epicurean materialism to disprove an "Operative substance" that would detach the source of sensed qualities from the core physicality of corpuscles.

Citing Epicurus, Boyle ascribes humanly perceived qualities to the "various Magnitudes, Figures, Motions and Textures" of minuscule particles—that is, their size, shape, motion, and cumulative grouping into an array or structure. Another example from Boyle's "excursion" demonstrates the capacity of texture to manifest new attributes even though, over the course of chemical history, the physical object—gold—remains the same:

> [Qualities] are not in the Bodies that are Endow'd with them any Real or Distinct Entities, or differing from the Matter its self, furnish'd with such a Determinate Bigness, Shape, or other Mechanical Modifications. Thus though the modern Gold-Smiths and Refiners reckon amongst the most distinguishing Qualities of Gold, by which men may be certain of its being True, and not Sophisticated, that it is easily dissoluble in *Aqua Regis*, and that *Aqua Fortis* will not work upon it; yet these Attributes are not in the Gold any thing distinct from its peculiar Texture, nor is the Gold we have now of any other Nature, then it was in *Pliny*'s time, when *Aqua Fortis* and *Aqua Regis* had not been Found out.[14]

William R. Newman recounts: by "the seventeenth century, the mineral acids—sulfuric, hydrochloric, nitric, and the mixture of the latter two, called *aqua regia*, had been discovered. These remarkable chemicals were able to dissolve and separate the metals with a speed and activity"[15] unknown to earlier experimenters. Nitric acid (aqua fortis), which dissolves silver, and mixed hydrochloric and nitric acid (aqua regia), which dissolves gold, permit the chymist to analyze apparently homogeneous alloys, proving that they are composed of discrete, durable, retrievable metallic corpuscles. To claim the verdict of the latter test as a relative attribute, Boyle argues that recent history elicits gold's "most distinguishing Qualit[y]." Only "modern" science conjures the authenticating standard of dissolubility in aqua regia. The experimental contingency of gold's proof by mineral acids, Boyle argues, underscores the persistence of its "peculiar Texture." Gold's corpuscular physicality is impervious to the indefinite sequence of attributes other bodies may draw from it.

Boyle limits the "Real" attributes of bodies to the "Determinate Bigness, Shape, or other Mechanical Modifications" of their constituent corpuscles. In a decisive move for his place in the history of science, he justifies the latter modifier: "which Attributes I call the Mechanicall Affections of Matter, because to Them men willingly Referre the various Operations of Mechanical Engines."[16] We turn in the following section to corpuscularianism that is *not* exclusively "Mechanical," most emphatically at the level of the individual corpuscle. *Forms and Qualities* schematizes particulate physicality by recourse to analogy with "Engines" to counter the mystified force of scholastic substance. In the crucial case of qualitative difference, Boyle restricts its cause to "Corporeall Agents"[17] whose hermeneutic emphasis is less unilateral reductionism than insistent and literal physicality, as when he credits the qualities of a key to "*nothing*" but shaped iron. Like gold that dissolves when the proper corrosive is found, the key's endowments proceed from mechanical affections in relation to other bodies, not supervening informing influence.

In corpuscular chemistry and empiricism, the body that elicits qualities from another is most often human. Mineral acids dissolve metal to verify it in the eyes of the goldsmith, but some things act directly on persons. As he turns to the sensory acumen incarnated by human percipients, Boyle refutes Barad, who, glossing Shapin and Schaffer, invokes "the Enlightenment ideal of the detached observer, the modest witness, who . . . when all is said and done simply stands back and watches."[18] Bracketing the flagrant immodesty of the transmutational enterprise, I instead underscore the violence with which Boyle defines the phenomenon of perception as texture's interaction with texture.[19] To show the corpuscular antecedents of the humanly sensed quality of poisonousness, Boyle instances "Nuns" in a cloister who consume "beaten Glass" "mixt" with their peas by a "distracted Woman":

> Now though the powers of Poisons be not onely look'd upon as *real* Qualities, but are reckoned among the *Abstrusest* ones: yet this Deleterious Faculty, which is suppos'd to be a Peculiar and Superadded Entitie in the beaten Glasse, is really nothing distinct from the Glass its self, (which though a Concrete made up of those Innocent Ingredients, Salt and Ashes, is yet a hard and stiffe Body,) as it is furnish'd with that determinate Bigness, and Figure of Parts, which have been acquir'd by Comminution. For these Glassy

> Fragments being many, and Rigid, and somewhat Small, (without yet being so small as Dust,) and endow'd with sharp Points and cutting Edges, are enabled by these Mechanical Affections to Pierce or Wound the tender Membranes of the Stomach and Guts, and cut the slender Vessels that they meet with there . . . And it agrees very well with this Conjecture, that beaten Glass hath diverse times been observ'd to have done no Mischief to Animals that have swallowed it: For there is no Reason it should, in case the Corpuscles of the Powder either chance to be so small, as not to be fit to wound the Guts, which are usually lin'd with a slimy substance, wherein very minute Powders may be as it were sheath'd . . . especially in some Individuals, whose Guts and Stomach too may be of a much stronger Texture, and better Lin'd or Stuff'd with Gross and Slimy Matter, then those of others.[20]

The quality of poisonousness is not "Superadded" to the bare physicality of ground glass. Indeed, as Boyle notes, glass is a chemical "Concrete" issuing from the mixture of "Innocent Ingredients." These innocuous—and, as "Salt and Ashes," incongruous—constituents do not harbor a perdurable poisonous "Entitie" whose irrelevance Boyle spoofs with the scholastic jargon "Deleterious Faculty." Instead, because they are not "so small as Dust," Boyle's "Glassy Fragments" offer a blown-up etiology of felt effects proceeding from the mechanical affections of even smaller particles. Since Boyle's macrofragments possess "sharp Points and cutting Edges," the mechanical affection of shape or "Figure" makes them fatal when they come in contact with "the tender Membranes of the Stomach and Guts." In this scenario, the body that manifestly acts on another poisons, but the body that suffers models agency too: as their receptivity to "Corpuscles" that "Pierce or Wound," Boyle's nuns sustain the enabling condition of empirical knowledge.

Boyle's avatars of the relation that enables humanly sensed qualities, pierced nuns do not model modest witnesses. Rather, they suffer the physicality of particulate figure as it meets the flesh that takes it in. As encounters with things that enter—and may wound or "cut" along the way—the contingency of human perceptual experience is amplified by the contrasting fate of Boyle's "Animals." This difference may be ascribed to the mechanical affection of size, if the fragments "chance to be so small," but animal susceptibility depends just as much on the "Texture" of the receiving gut. In the ultimate proof that poisonousness occurs only in relation, the identical particles become anodyne when they encounter a perceiving surface "Lin'd or Stuff'd with Gross and Slimy Matter." While the impenetrable texture of slimy matter may sometimes extend to humans as well, Boyle recurs to the schematics of wounding when he reprises the terminological and scientific revision *Forms and Qualities* aims to make: "though, for shortness of speech, I shall not scruple to make use of the word *Qualities* . . . , yet I would be understood to mean them in a sense suitable to the Doctrine above deliver'd. As if I should say, that Roughnesse is apt to grate and offend the Skin, I should mean, that a File or other Body, by having upon its Surface a multitude of little hard and exstant Parts, and of an Angular or sharp Figure, is qualify'd to work the mentioned Effect."[21] A quality is really a "qualify'd" body whose corpuscular physicality is rendered in terms of the inverse vulnerability of percipient parts. *Sceptical Chymist* asserts the homologous operation of taste: "an Affection of matter that relates to our Tongue," this quality transpires in persons when corpuscles "pierce into and make a perceptible Impression upon the Nerves or Membranous parts of the Organs of Tast."[22] The quality of smelliness is likewise exemplified after sal ammoniac (ammonium chloride) reacts with salt of tartar (potassium carbonate): "you will finde your Nose invaded with a very strong smell of Urine, and perhaps too your Eyes forc'd to water, by the same subtle and piercing Body that produces the stink."[23] With the cognate verb "forc'd," Boyle defines empirical experience *as* the porosity of texture that admits more subtle or offensive entering things.

For Boyle, the word "qualities" really refers to mechanical affections of particles "qualif'd to work the mentioned Effect." Because "Angular," "sharp," or "piercing" corpuscles ultimately deliver "a perceptible Impression," their attributes are sustained by the persons they offend, invade, and pierce into. But the receptive sensitivities of gut, skin, tongue, nose, and eyes do not exhaust the repercussions of sensed qualities that add nothing to the physical texture of the object. While points and edges concretize the mechanical provenance of glass's impression on persons, *Forms and Qualities* claims the same relational status for perceived effects less obviously incurred by particles that grate, force, or pierce. In a reiterated scene of perceptual and referential genesis, Boyle enlists not only particles that cut but also particles that stimulate ideas:

> [T]he body of Man having several of its external parts, as the Eye, the Ear, &c. each of a distinct and peculiar Texture, whereby it is capable to receive Impressions from the Bodies about it . . . And to these Operations of the Objects on the Sensories, the Mind of

> Man, which upon the account of its Union with the Body perceives them, giveth distinct Names, calling the one Light or Colour, the other Sound, the other Odour, &c . . . Whence men have been induc'd to frame a long Catalogue of such Things as, for their relating to our Senses, we call Sensible Qualities; and because we have been conversant with them, before we had the use of Reason . . . we have been from our Infancy apt to imagine, that these Sensible Qualities are Real Beings, in the Objects they denominate . . . whereas indeed . . . there is in the Body . . . nothing of Real and Physical, but the Size, Shape, and Motion, or Rest of its component Particles, together with that Texture of the whole . . . nor is it necessary they should have in them any thing more, like to the Ideas they occasion in us.[24]

Boyle desists from explicating the mind's "Union with the Body," but this union entails corpuscles that stimulate "Ideas" rather than wounds. Neither Boyle nor Locke, who repeats Boyle to affirm the sensory origin of words acquired "before we had the use of Reason," queries the integrity of collectively felt "Impressions" instituted as shared language. For Boyle as for Locke, the automaticity of perceptual reflex grounds a linguistic consensus inaugurated by the "Catalogue" of descriptors whose contents Locke claims as the simplest units of human understanding. The scientific intervention that galvanizes empiricism is not the split of sensed qualities from objects but, on the contrary, the provenance of ideas as "Impressions" stimulated by environing bodies. This is not a dualistic metaphysics which insulates the transparency of thought—or, indeed, language—but a corpuscular genealogy of perception that Locke will seize as the starting point for human reason.

Boyle makes the same claim about ideas that he makes about ground glass. Sensible qualities are not unreal, unmoored from physical objects to hover in a skeptical netherworld; but neither do human perceptions exist *in* the physical object as such. Like poisonousness, "Colour" or "Odour" is "*nothing*" other than the mechanical affections of particles in relation to the texture of "Sensories." Scent cannot be designated a "true Entitie" not because it is divorced from the object but because it does not augment the productive physicality of corpuscles. Boyle thereby distinguishes between corpuscular reality "*in them*" and sensible qualities produced "*in us*," a divide that does not jeopardize the infallibility of sensation schematized by fragments that enter porous flesh. He does, however, add a caveat that applies to effects like color rather than effects like wounded guts: as the sum of mechanical affections that enforce an Epicurean criterion for what is physical, particles do not incarnate attributes "like to the Ideas they occasion in us." Ideas are not mimetically "*like*" the insensible texture that stimulates them. Boyle thus assails a grand mistake promulgated, his linguistic creation story suggests, when sensible qualities receive "distinct Names" like sweet, soft, or sour. Such names elide the relational occasion of qualities by collapsing perceived effects back into the substance of the object. By refuting substantial forms, Boyle challenges the false ontology of qualities perpetrated by descriptive words: "several Names, as *Heat, Colour, Sound, Odour* . . . are commonly imagin'd to proceed from certain distinct and peculiar Qualities in the External Object, which have some resemblance to the Ideas, their action on the Senses excites in the Mind."[25] Sometimes, "some resemblance" can be sustained: piercing fragments *are* piercing at the level of human sense. But with "Ideas" like warmth, the resemblance between "Qualities in the External Object" and human impressions does not hold. This marks a critical distinction: ideas reflect the power of corpuscles to act on receptive senses, but they do not mimetically represent the texture of the perceived object. No imperceptible mechanical affections—no particulate attributes amenable to figural amplification like that illustrated by shards of glass—mimetically anticipate the sensory effect of heat.

While poisonousness proceeds from corpuscular shape, the physicality of some particulate attributes is enhanced by its non-resemblance to the ideas it effects. In "The History of Fluidity and Firmness" (1661), Boyle argues that water's "seeming continuity" belies the submicroscopic attributes that produce this macroscopic outcome: "the particles whereof the Liquor consists, being too small to be visible, and being not only voluble, but in actual motion, the pores or vacant spaces intercepted between them, must also be too little to be discern'd by the Eye, and consequently the body must appear an uninterrupted or continu'd one: not to mention, that, were the parts of the Liquor less minute, their shifting of places would hardly be perceiv'd by the Eye, each displac'd Corpuscle being immediately succeeded by another like it."[26] Because human ideas of "uninterrupted" fluidity proceed from "pores," "particles," and "shifting" that cannot be seen or felt, this example departs from Boyle's ground glass. Rather than sharpness that acts directly on flesh, the defining attribute of water's porous body is effected by human sensory incapacity to register corpuscular detail. The humanly perceived continuity of water fails to mirror—and, indeed, utterly misrepresents—the mechanical affections of porousness and rapid motion. But Boyle enlists this failure to vindicate the liaison between particulate materiality and the way the object "must appear." That is, Boyle justifies a relational effect not as invasion or piercing but as the viewer's inability to apprehend corpuscular attributes which "would hardly be perceiv'd." In the case of all continuous but really interrupted bodies—that is to say, all humanly sensed bodies—the continuity of the object proceeds from imperceptibly "too little" parts and pores. This is not an appearance severed from terminally inscrutable atoms, because Boyle denies the mimetic resemblance of particles and ideas to vindicate the *productive* relation between them. Bodies like water are empirically accessed not through mimetic realism but through the relational form of experimental knowledge.

Sensations like fluidity or smelliness are not instantiated by corpuscles. The ensuing dilemma, for Locke as well as Boyle, concerns not the reliability of human sense but the capacity of qualities to delimit stable designations of things. When Boyle adopts classical atomism to stipulate the bedrock physicality of particles, he claims that this restricted set of features engenders "all the qualities that make [sensed objects] differ." But how is such difference secured if the defining attributes of things may be transformed by new relations to other things, like the gut that renders glass toxic or the aqua regia that makes gold dissolve? *Forms and Qualities* poses this question in chymical terms. Paracelsian chymists, also known as spagyrists for the analytic operation of *spagyria* or separation, augment the antique elemental duo of mercury and sulfur with a third so-called principle or element, salt. (Of these names, none designates a modern chemical referent. The alchemical ingredients "mercury" and "sulfur" were primordial constituents of the seven historically known metals gold, silver, lead, tin, copper, iron, and mercury.) As the newest principle to round out the Paracelsian *tria prima*, the category "salt," Boyle complains, evinces discrepant chemical attributes: "although some sorts of salts be very fugitive, (as the volatile salts of Harts-horne, Urine *&c.*) other very fix'd (as that drawn from the Calx of Tartar . . .) and though some saline Liquors as vinegar and Juice of Lemmons are acid, and dissolve Pearle, Coral, *&c.* which Lixiviate Salts, whose Tast is fiery, will precipitate what the others have dissolv'd, yet all these are numbred among *Salts*, because they agree in the Accounts upon which we allow bodies that Denomination; namely their being very sapid and readily dissoluble in water."[27] The problem with Paracelsian "*Salts*" is emphatically one of difference, because this salt contradicts itself in every register but two, "being very sapid and readily dissoluble in water." What I will call this object's *qualitative identity*, its experientially adjudicated claim to the qualities that establish what the chemist will call it, does not regulate its possession of antipathetic traits like volatility and fixity or acidity and alkalinity ("Lixiviate Salts" are basic). As the "Denomination" granted by chymists who "allow" any flavorful, water-soluble substance the title, salt collapses reactants as discrepant as "vinegar" or "volatile salts of Hartshorne." If modifiers like poisonous and sweet tend to install humanly sensed attributes in particles that do not bear them as such, then "Denomination" reflects the novel challenge incurred by Boyle's ontology of qualities: at what level of qualitative specificity should the chemist name a body? If chemical qualities may be revised by new relations, how are anything's denominating attributes secured?

In *Sceptical Chymist*, the exercise of denomination affords a chymical genealogy of revolutionary puritan demands for linguistic reform. Galvanized as a millennial embrace of Baconian natural history and a backlash against obscurantist scholastic learning, puritan appeals, Charles Webster relates, to "naked truth"[28] are echoed when Boyle applauds "naked Knowledge of the Truth."[29] But Boyle embraces "naked Knowledge" neither as the historical state of chemistry nor, indeed, the stylistic limit of his own prose. *Sceptical Chymist*'s account of word use sits at a far remove from Barad's gloss of the instrumentality of referential language to empirical science: "language is a transparent medium that transmits a homologous picture of reality to the knowing mind."[30] Far from transparency, for Boyle chymical denomination reflects the ineptitude of Paracelsian chymists' "gross Way of judging of the Nature of Bodies."[31] *Sceptical Chymist* champions more refined analytic techniques to refute "the Contrariety of Bodies which according to the Chymists must be huddl'd up together under one Denomination."[32] But the science Boyle deploys to separate "huddl'd up" contrarieties entails its own linguistic conundrum: over time and in relation, anti-elemental texture may not undergird the fixity of the same denomination.

Boyle accuses scholastics, spagyrists, and "vulgar"[33] chymists of two failings: the failure to break huddled bodies into differently qualified parts, and the failure to grasp that identifying qualities can be conjured by reaction.[34] Both failures of elemental denomination—denomination that is too gross and denomination that is too easy—invalidate "the specious Titles"[35] vulgar chymists grant substances like earth or salt. Concerning the arbitrariness of criteria that fail to unify the Aristotelian element earth, Boyle notes that "notwithstanding this Resemblance in some one Quality, there may be such a Disparity in others, as may be more fit to give them Differing Appellations."[36] While discrete "Appellations" are enjoined by disparities that belie elemental coherence, multiple names may also be solicited by the same body over experimental time: "the Principles of Chymists may receive their Denominations from Qualities, which it often exceeds not the power of Art, nor alwayes that of the Fire to produce."[37] A refutation of scholastics and spagyrists who "play with words," *Sceptical Chymist* attacks the falsely ontologizing "Liberty they take in using names."[38] Boyle's corrective is the experimental redress of chemical reference, chemistry's more stringent arbitration of what is "more worthy to be called an Elementary Principle" and what "seem[s] to have a better title . . . to the name."[39] A mandate for linguistic renewal affiliated with interregnum systems of sensory pedagogy serves, for Boyle, to restrict elemental designation to objects that "well merit the name of Principles."[40] Analytic agents like fire promise to isolate "worthy" contenders from grossly or easily qualified bodies. But when Boyle attests to the experimental replication of chemical attributes—"do but consider what sleight and easily producible qualities they are that suffice . . . to Denominate a Chymical Principle or an Element"[41]—he specifies the conundrum engendered by corpuscles that do not entrench an ontology of elemental substance. What anchors some matter's "better title" if corpuscular texture can be made to change in relation?

Because substances like salt or mercury are, Boyle contends, made of corpuscles, they lose their title to any perdurable denomination:

> [T]hese differing substances that are call'd Elements or Principles, differ not from each other as Metals, Plants, and Animals . . . but these [former] are only Various Schemes of matter or Substances that differ . . . but in consistence (as Running Mercury and the same Metal congeal'd by the Vapor of Lead) and some very few other accidents, as Tast, or Smel, or Inflamability . . . So that by a change of Texture not impossible to be wrought by the Fire and other Agents that have the Faculty not only to dissociate the smal parts of Bodies, but afterwards to connect them after a new manner, the same parcell of matter may acquire or lose such accidents as may suffice to Denominate it Salt, or Sulphur, or Earth.[42]

Boyle distinguishes "Metals" as well as "Plants, and Animals" from would-be elements whose attributes recur solely to "Texture." (The difficulty of qualitative identification undergirded by corpuscles is anticipated by mercury, a metal whose insulation from the anti-ontological momentum of texture Boyle immediately ignores.) Again Boyle specifies that sensible attributes like "Running" as opposed to "congeal'd" consistency demarcate how bodies "differ": it is thereby a *differential* system of naming that corpuscular texture solicits. Denomination proceeds not from preexisting substance that authorizes names in advance but from palpable differences that institute breaks between things on the side of sensation. But such differences are not entrenched in corpuscles that permanently stimulate human feelings of "Tast, or Smel, or Inflamability." On the contrary, the "small parts" whose texture effects the object's denominating qualities are not impervious to "change." Among the "differing substances" Boyle treats, no denominating variable resists undoing. No single "parcell of matter" can anchor the differential institution of chemical names, because "Fire and other Agents" can undo any parcel's texture. If denomination is propelled by qualitatively felt difference, reactions able to "dissociate" corpuscles and "connect them after a new manner" divest names of security. Only experiential affirmation that it continues to elicit the same differentiating sensations can perpetuate the object's denomination over time.

Names hinge on perceived difference, and difference proceeds, Boyle repeats, from the corpuscular attributes "Motion or Rest, . . . Bigness, Figure, Texture and the thence resulting Qualities of the small parts which are necessary to intitle the Body whereto they belong to this or that Peculiar Denomination."[43] "Small parts" are comprised of mechanical affections that result in sensed qualities; the latter qualifications "*intitle*" the matter to "this or that Peculiar Denomination." Provoked by imperceptible parts, names recur to a physical referent that *is* corpuscular texture. Boyle's doctrine of qualities is equally a doctrine of denomination: names are solicited by differential sensations effected—in Boyle's anomalous application of a word he reserves for discrepancies of phenomenal experience—by "Qualities of the small parts." As doctrine whose influence on Locke is profound, *Sceptical Chymist* does not promote the transparency of reference that, according to Shapin and Schaffer, ensures the pretended "capacity of knowledge to mirror reality"[44] unperturbed by experiential intervention. Boyle instead implicates the chemist's gut, nose, eyes, mind, skin, and tongue in the denomination to which sensation entitles its engendering parcel of matter for the present experimental moment. Not only the temporal persistence of denominating difference but difference itself may be reconstituted by new relations. These relations produce effects delineated by names that do not entrench qualitative identity at the level of the object's component parts:

> how sleight a variation of Textures without addition of new ingredients may procure a parcel of matter divers names, and make it be Lookt upon as Different Things;
>
> I shall invite you to observe with me, That Clouds, Rain, Hail, Snow, Froth, and Ice, may be but water, having its parts varyed as to their size and distance in respect of each other, and as to motion and rest.[45]

In his most unassailable assertion that no single particle fixes linguistic difference, Boyle invokes water's susceptibility to be "Lookt upon as Different Things." Because it subsists in "variations of Textures" that operate on the sensory apparatus of persons, difference is extrinsic to parts that stay the same. But names like "Clouds, Rain, Hail, Snow, Froth, and Ice" obfuscate their reference to identical corpuscles. Effectuated in relation, these denominations codify discrepancy while referring to the selfsame matter. Texture, and no innately qualified particle, stimulates the semantic departure of rain from ice.

Yet despite *Sceptical Chymist*'s anti-elemental warrant, difference unanchored by particles fails to license the indefinite transformational license seemingly implied in Boyle's famous claim: "almost All things may be made of All things, not immediately, but by intervention of successive Changes and Dispositions."[46] Boyle's cautionary gesture to the tedium of alchemical cookery, the often expurgated proviso "not immediately, but by intervention of successive Changes" intimates that just as a differential system of signification does not devolve into linguistic free play, so rigorous chymical analysis does not admit one texture's easy conversion into another. In fact, some corpuscles conserve attributes extraneous to the mechanical affections Boyle delimits. As we will now see, Boyle's corpuscle retains extramechanical powers—and fails to realize the basic mechanical attribute of unity—precisely as Boyle defends the consummate triumph of corpuscular texture over elemental substance: transmutation.

Transmutation: Ontology

Before the new Boyle scholarship, *Sceptical Chymist* was lauded as the definitive demystification of matter henceforth abstracted into mathematical or mechanical units.[47] Yet that characterization is belied by corpuscles that break into smaller parts. While Boyle often employs the terms "corpuscles" and "particles" as synonyms, their referent differs from the entities he calls *prima naturalia* or *minima naturalia*, even smaller parts that comprise the body of the corpuscle:

> [T]here are also Multitudes of Corpuscles, which are made up of the Coalition of several of the former *Minima Naturalia*; and whose Bulk is so small, and their Adhæsion so close and strict, that each of these little Primitive Concretions or Clusters (if I may so call them) of Particles is singly below the discernment of Sense, and though not absolutely indivisible by Nature into the *Prima Naturalia* that compos'd it, or perhaps into other little Fragments, yet . . . they very rarely happen to be actually dissolv'd or broken, but remain entire in great variety of sensible Bodies.[48]

By positing a corpuscle "not absolutely indivisible" into smaller *minima*, Boyle concurs with prevailing theories of chymical matter which, Newman shows, use the terms atom, particle, and corpuscle "without the implication of absolute indivisibility, either mathematical (like a geometrical point) or physical (like the atoms of Lucretius)."[49] In Newman's words, Boyle's claim for "operationally indivisible particles" rather than "indivisible point atoms"[50] marks a critical departure from Descartes or Epicurus, because the corpuscle's irreducibility hinges upon its *experimental* resistance to being "actually dissolv'd or broken." Indivisibility is no static attribute because corpuscular coherence depends on resistance to analysis by acid or fire. An experimental and not notional variable, operational indivisibility is claimed by Newman as a turning point in the history of chemical causation: "it was precisely the relatively stable corpuscles initially revealed by processes such as sublimation, distillation, and calcination . . . that would provide the backbone of Boyle's mechanical philosophy, not the sterile and unattainable *prima naturalia* that have traditionally been seen as the sine qua non of this doctrine."[51] Boyle's "relatively stable corpuscle" crucially distinguishes chemical matter from indivisible bodies activated by accident. This revised history of early modern matter theory entails the insinuation of mechanical *and* extramechanical agencies into the heart of the particle: Boyle's selective recourse to Epicurean materialism may guarantee the core physicality of texture, but it does not enforce reductionism predicated on inert, experimentally "unattainable," impenetrable points.

Texture is composed, Boyle writes, of "uneasily dissoluble"[52] corpuscles. Reconstituted as a relational entity, the indivisible corpuscle subsists as experimental proof of the "Adhaesion" of its even smaller parts. The moniker "corpuscle" thus designates "onely those very little masses or clusters of particles, of what kind soever they be, that stick so firmly to one another, as not to be divisible and dissipable by that degree of fire in which the body is said to be fixt."[53] In this key statement of the relational adjudication of corpuscular unity, "that degree of fire" or acid determines the chymist's assessment of whether the particulate body can be "said to be fixt." Rather than the Lucretian or Cartesian atom—for Boyle, a placeholder for the speculative fetish of irreducible solidity—what occupies chymistry are "little masses or clusters" the test of whose intracorpuscular adhesion justifies both relative fixity and irreversible change. Of the critical capacity of the corpuscle to "remain entire" or, as Boyle writes of "*Alkalys* . . . destroy'd with Ebullition by *Aqua fortis*,"[54] to become something else, *Sceptical Chymist* elaborates:

> [T]here may be some Clusters of Particles, wherein the Particles are so minute, and the Coherence so strict, or both, that when Bodies of Differing Denominations, and consisting of such durable Cluster [*sic*], happen to be mingl'd . . . each of the little Masses or Clusters may so retain its own Nature, as to be again separable, such as it was before . . . But . . . there are other Clusters wherein the Particles stick not so close together, but that they meet with Corpuscles of another Denomination, which are dispos'd to be more closely United with some of them, than they were among themselves. And in such case, two thus combining Corpuscles losing that Shape, or Size, or Motion, or other Accident, upon whose Account they were endow'd with such a Determinate Quality or Nature, each of them really ceases to be a Corpuscle of the same Denomination it was before; and from the Coalition of these there may emerge a new Body, as really one, as either of the Corpuscles was before they were mingl'd, or, if you please, Confounded: Since this Concretion is really endow'd with its own Distinct qualities, and can no more by the Fire, or any other known way of *Analysis*, be divided again into the Corpuscles that at first concurr'd to make it.[55]

Irreducible unity is neither the ontological nor the physical sine qua non of Boyle's corpuscle. When placed in relation, this corpuscle may either "retain its own Nature" or "emerge a new Body." In a transformation that leaves no remainder, one corpuscle may "*really*" recombine to become "another." Just as the first body "*really* ceases to be," the second is "as *really* one, as either of the Corpuscles was before." By repeating the stressor "really"—again, the emergent corpuscle "is *really* endow'd with its own Distinct qualities"—Boyle underscores the warrant for chymical denomination: rather than peripatetic or spagyrical essence, this consists of not absolutely indivisible particles. What the unified particle really is subsists as the serial experimental trial of its uneasy dissolubility. (We may also observe that intracorpuscular adhesion violates Boyle's restriction of a single particle's mechanical affections to size, shape, and motion.)

Uneasy dissolubility, or "strict" "Coherence," cannot be determined in isolation. For both "again separable" and "combining" bodies, corpuscular integrity is not freestanding but rather the relational outcome of whether each unit's constituent particles "are dispos'd" to unite with another's. No chymist can denominate this corpuscle outside the frame of experimental time, since it may really be two distinct things, first the combining corpuscle and then the new body. But these temporally discrete entities do not defer the mechanical reduction of matter into featureless points, a resolution that "clusters" might seem only to forestall. Instead, one corpuscle that becomes another leverages uneasy divisibility to justify discrepant intensities of chemical combination. The quiddity of this new body is real but temporally bounded, "Distinct" but produced in relation. Unlike an atom, the corpuscle occupies experimental relations that render what it really is a function of other matter. From the vantage of the chymist, corpuscular unity is determined by the presence of other bodies just as much as the qualities of taste or smell.

Wolfram Schmidgen claims Boyle's irreversibly new "Concretion" as the hallmark of Boylean science's fundamental commitment to mixture.[56] Boyle's decisive rebuttal of peripatetic ontology—signified by the neoAristotelian term for essence, form or substantial form—pivots, however, on his experimental disproof of the scholastic claim that "Elements alwaies loose their Forms in the mix'd Bodies they constitute."[57] When he retrieves a starting reactant's denominating qualities from the solution in which they seem to have vanished, Boyle demonstrates the relative superficiality of some mixtures: "even a mix'd Body . . . doth, in a further mistion, retain its Form and Nature, and may be immediately so divorced from the Body, to which it was united, as to turn, in a trice, to the manifest Exercise of its former Qualities."[58] Most urgently, a rationale for corpuscular integrity, even in combination, is requisite to the existence of any relatively stable chemical entity; not all mixtures produce irreversibly new things. But a paradox is already evident in Boyle's reference to "Differing Denominations" of singly insensible corpuscles, which further illuminates the significance of the "durable Cluster" extricated from reaction with its prior qualitative identity intact. Boyle offers an example of this durably denominated corpuscle: "As when Gold and Silver being melted together in a Due Proportion . . . *Aqua Fortis* will dissolve the Silver, and leave the Gold untoucht; by which means . . . both Metalls may be recover'd from the mixed Mass."[59] An obvious question is begged by Boyle's denomination of the corpuscle: how can the chymist perceive qualities of an insensible body? One answer occupies the history of chemistry as the retrieval of discrete metals "recover'd from the mixed Mass" of alloyed silver and gold. Because nitric acid dissolves only silver, Boyle separates two starting metals from an apparently homogenous material. The separation of blended metals constitutes proof that *while they are mixed* their metallic identity must be conserved in the form of parts too tiny to perceive. Recovery *from* mixture vindicates the perdurable denomination of corpuscles as the very effect of their prior disappearance. Singly insensible corpuscles can be named because the aggregate metal they comprise vanishes into—and reappears from—a homogeneous mixture of two things.

Sceptical Chymist and *Forms and Qualities* attempt to specify the intensities of chemical change spanned by sometimes retrievable corpuscles, a range not exhaustively refined when Boyle contrasts work on "the *sensible* (not insensible) parts . . . as when a Joyner makes a Stool . . . or a Turner a Bowl" and "Productions, wherein the *insensible* parts of Matter are alter'd."[60] As palpable evidence against Aristotle's assignment of new essential identity to mixtures, Boyle argues that "since a Pint of wine, and a pint of water, amount to about a Quart of Liquor, it seems manifest to sense, that these Bodies doe not Totally Penetrate one another . . . but that each retains its own Dimensions; and Consequently, that they are by being Mingl'd only divided into minute Bodies."[61] The total volume of mixed fluid refutes Aristotelian substance.[62] Rather than a new essential entity whose constituents "Totally Penetrate," the combined liquid adds up to twice as much because its parts are "only divided into minute Bodies." A mixture that seems mingled but whose parts just as manifestly "*retain*" the mechanical affection of size, the water and wine herald modalities of reaction in which corpuscular identity is not clearly signaled by human sense.[63] Boyle deploys a critical keyword to affirm that apparently transformed mixtures may contain minuscule, durably denominated parts when he writes: "Gold, that tinged the Menstruum . . . was reduc'd or præcipitated out of it."[64] As a synonym for "precipitated" (or, concreting and falling out of the fluid in which its dissolved corpuscles were suspended), "reduc'd" does not mean that gold is broken into smaller, finally irreducible units. This terminological proviso marks the departure of chymistry from the reductionist mechanism to which it has been wrongly assimilated. When gold dissolved in an acid is "*reduc'd*," it is recovered from the fluid in its former metallic state. As evidenced by the Latin *reducere* or "to lead back,"[65] reduction indicates not breaking down but retrieval of an experimentally durable body. We can thus parse Boyle's claim that "mixtures of Metalline Bodies, with the Saline parts of *Aqua fortis* or other corrosive Liquors" do not permanently destroy the corpuscles comprising the metals, even though they are temporarily lost to ocular sense, "as is evident by their being reducible into Silver or Quicksilver [mercury], as they were before."[66]

Corpuscles "reducible into [what] they were before" brake the anti-elemental impetus of Boyle's skeptical chymistry. As Newman recapitulates, Boyle's predecessor, the German doctor and corpuscular chymist Daniel Sennert (1572–1637), canonizes the experiment that recuperates dissolved "Silver or Quicksilver": this process entails "dissolving precious metals in strong acids, and then precipitating them out, seemingly unchanged, by means of alkalies." Newman continues: "For [Sennert] and for Boyle, the reduction of dissolved metals into their original or 'pristine' state (*reductio in pristinum statum*) became a sort of crucial experiment."[67] Because the *reductio* experiment retrieves metallic corpuscles from the transparent acid mixture, the metal's prior "optical disappearance"[68] shows that it is made of singly insensible bodies. As Newman asserts, the *reductio* experiment represents "evidence for the reality of semipermanent corpuscles beneath the level of sense."[69] The ontological warrant of Boyle's chymistry is not unidirectional. What Newman designates "semipermanent" or "robust"[70] corpuscles are neither irreducible, mechanical units delimited solely by size, shape, and motion nor a license for miscibility in which prior identities are irretrievably extinguished. The science Boyle dubs both corpuscular and mechanical philosophy is mobilized by corpuscles the persistence of whose integrity is determined in relation over experimental time.

Imperceptible in the clear mixture, silver's corpuscles hover at the edge of sense as they are summoned back out of solution by a base or, as Boyle denominates this agent, a "Salt": "when I try'd whether the Particles of Silver, dissolv'd in *Aqua fortis*, would not . . . convene . . . into little Concretions of smooth and flat surfaces, I found, that having (to afford the Metalline Corpuscles scope to move in) diluted one part of the Solution with a great many parts of distill'd Rain water . . . a Plate of Copper being suspended in the

Liquor, and suffer'd to lie quiet there a while, . . . there would settle, all about it, swarms of little Metalline and Undiaphanous Bodies, shining in the water like the scales of small Fishes."[71] Recent studies of eighteenth-century science and literature align the advent of human sensory access to the microworld with the Royal Society publication *Micrographia*, authored by the first Royal Society curator of experiments Robert Hooke (1635–1703). Hooke replicates his experience of blown-up reality in plates that accommodate microdetail to the scale of the two-dimensional page. By contrast, Boyle's disclosure of "swarms" of singly insensible particles emerging from "Solution" takes place over experimental time. Boyle motivates silver's corpuscularity—from the metal's disappearance in clear fluid to its emergence as "Undiaphanous" "scales"—as the imminence of its particles' appeal to human sight. In so doing, this passage offers just as important an approximation of micromatter whose sensory impact humans must experimentally summon. What enables empirical awareness of both silver's reduction (tiny parts that re-"convene" as the starting metal) and its microfigure ("little Concretions of smooth and flat surfaces") are operationally indivisible corpuscles that traverse narrative time to solicit experimental sense.

A supplement to Boyle's materialist roster of mechanical affections, the variable tendency of subcorpuscular parts to stick dictates either recovery or a really new body. To instance the latter, *Forms and Qualities* experimentally vindicates a triumphal (if, Boyle notes, not very "Lucriferous") case of corpuscular redenomination: gold's transmutation into, he claims, "true Silver."[72] (Unfortunately, after melting the proof of his success and placing it on paper to dry, Boyle "being suddenly call'd out of my Chamber, an ignorant Maid, that in the mean time came to dress it up, unluckily swept this Paper, as a foul one, into the fire."[73]) Boyle's rationale for the transmutation of gold reveals the persistence of agential disparities not squelched by his recourse to classical atomism. To explicate his experimental success, Boyle does not entirely dismiss ancient alchemical notions of tinctures or even souls of gold:

> [H]owever the Chymists are wont to talke irrationally enough of what they call *Tinctura Auri*, and *Anima Auri*; yet, in a sober sense, *some such thing* may be admitted, I say, *some such thing*, because as on the one hand, I would not countenance their wild Fancies about these matters, some of them being as unintelligible, as the Peripateticks substantial Forms, so, on the other hand, I would not readily deny but that there may be some more noble and subtle Corpuscles,

> being duely conjoyn'd with the rest of the Matter, whereof Gold consists, may qualifie that Matter to look Yellow, to resist *Aqua fortis*, and to exhibit those other peculiar *Phænomena*, that discriminate Gold from Silver, and yet these Noble parts may either have their Texture destroy'd by a very piercing Menstruum, or by a greater congruity with its Corpuscles, then with those of the remaining part of the Gold, may stick more closer to the former, and by their means be extricated and drawn away from the latter.[74]

On the one hand, Boyle cites "wild Fancies" dating from the chymist Zosimos of Panopolis (300 A.D.), who, Principe explains, "calls transmutation the 'tingeing' of metals . . . [and] likewise calls a transmuting agent a 'tincture,' that is, something able to tint or color."[75] Arguing for the "continuity between sixteenth-century Paracelsian thought and the new science," Jole Shackelford explains that the concept of tincture imports "seminal"[76] powers into chymical matter theory: as a "dye, something that tints or colors something else," a tincture "transfers its form—its color—to a body. Alchemists latched onto this terminology to express the multiplication of species that is evident when a small amount of tincture or elixir is used to color or transmute a base material."[77] As Shackelford's collapse of the words "form" (again, in its neo-Aristotelian sense) and "color" suggests, Boyle's willingness to concede "some more noble and subtle Corpuscles" which "qualifie that Matter to look Yellow, to resist *Aqua fortis*" signals a remarkable inversion of the relation between qualitative identity and some anterior source of that identity's conservation or transferal. Boyle's "*some such thing*" makes gold gold on the basis of yellowness and resistance to nitric acid; but if "*some such thing*" can make other metal gold too, Boyle specifies the latter goldenness in terms of chymical qualities adjudicated by experiment. Some such thing is not "the Peripateticks substantial Forms," because it motivates qualities after their relational experience by the chymist who has, for example, discovered analytic agents like aqua regia. If some such thing conserves an ontology of gold in parts "duly conjoin'd" to the bulk of the metal, then these parts justify not essence but experimentally elicited qualities. As it resides in some such thing, gold's ontology implicates not incorporeal informing influence but rather the power to "qualifie that Matter" to provoke human denomination.

Some such thing serves not to fix an ontology of gold but to justify its undoing. Boyle repeats: "yet Gold it self is not absolutely indestructible by

Art."[78] While the alienability of goldenness compels Boyle to posit something whose removal exempts "the rest of the Matter" from that title, he mobilizes this divestiture in familiarly mechanical terms. In a striking effort to accommodate tincturing and mechanical modes of causation, gold's qualifying thing is incarnated as "more noble and subtle Corpuscles"—that is, as *also* matter. Embodied by subtler corpuscles, Boyle's thing may have its own "Texture destroy'd" by a superlatively "piercing" solvent. This tingeing thing, we find, *also* possesses corpuscular texture that may be undone. Or, the thing may retain its seminal agency and yet "be extricated" from gold because of a tendency to "stick instead" to other ambient parts. Because the qualifying power of tincture is susceptible to mechanical destruction or chymical adhesiveness, what seems like an ontology of gold preserved in "Noble parts" is inalienable neither at the level of the metal nor at the level of the parts themselves. Subtle corpuscles which recombine tingeing influence and texture's proneness to change do not conserve gold's qualitative identity. On the contrary, they facilitate the transference of gold's denominating attributes.

The corpuscle's mechanical affections fail to preclude such extramechanical properties as stickiness and tingeing agency. In the long history of chemistry, this accommodation reflects a dialectical "interplay," and not a dualistic antagonism, Newman writes, "between the two types of qualitative explanation . . . the structural and the substantial."[79] In *Forms and Qualities*, Boyle indicts Sennert for resorting to a threadbare scholastic expedient, subsidiary forms that essentialize a reactant's smaller parts: "besides the Specifick Form, as *Sennertus* calls it . . . there may reside in those Body's and especially in some determinate parts of them, certain other Formes proper to those parts." Boyle condemns "subordinate Forms"[80] as an attempt to entrench "subordinate" essence that cannot staunch the anti-ontological imperative of corpuscular texture. But however caustically he glosses Sennert, Boyle cannot obscure the persistence of his own reliance on "some such thing": "For as tis not absurd to think, that our Menstruum may have a particular Operation upon some Noble, and (if I may so call them) some Tinging parts of the Gold, so it is not impossible, but that the Yellowness of that rich Metal may proceed not from any particular Corpuscles of that Colour, but from the Texture of the Metal."[81] Boyle does not only restrict the source of gold's "Colour" to "Texture" while also admitting the action of "Tinging parts": he knits seminal potency into the body of the corpuscle. His repressed debt to Sennert's accommodation of structure and substance belies its ongoing centrality to the insight Newman specifies from the vantage of contemporary chemistry: "something approximating the modern idea of the chemical bond, which holds elemental particles together intimately and yet allows them to retain their identity and to be recaptured intact."[82] Eighteenth-century mechanism dictates neither featureless, irreducible atoms nor the radical miscibility of parts impervious to identity, but rather the variable stickiness of bonds which cement a really new body or conserve metals dissolved in a liquid. The chemical dialectic of structure and substance, seminal influence and mechanical affections, recombines—or divides—possibly ennobling ends and the uneasily divisible corpuscle.

Boyle's noble corpuscles and his texture, his tingeing parts and his more or less adhesive *minima*, allow for the prospective transmutation of qualitative identity by means of even smaller, acutely piercing, or stickier parts. Newman affirms: "Boyle's mechanical philosophy and chymistry were deeply integrated."[83] As the motor for both the corpuscle's recovery and its new denomination, Boyle incorporates texture and extramechanical attributes like tingeing and stickiness. We thus encounter particles whose uneven power to make gold operates not to secure gold's essence but rather to justify the relational alienation of its denominating attributes. I now ask how such particles revise the historical status of matter—and, more particularly, matter's putative irreducibility—in contemporary agential realism and feminism. This turn to recent appraisals of matter's past brings us to Boyle's poisoned nuns and the normative contours of the eighteenth-century empirical person.

Starting Matters: Metaphysics and Penetrability

With Boyle's pierced nuns and his "very piercing Menstruum," we can remark the confluence of Boyle's figures for chymical as well as sensory receptivity to relational impression. Metals and persons claim a proneness to chymical—and empirical—penetrability that contravenes recent claims for the abject passivity entailed by that attribute. First, I return to corpuscular mechanism to address critical claims for its especially vexing reliance on analogy. This return underscores, from the side of persons as well as particles, chemistry's dialectic of structure and substance.

In "About the Excellency and Grounds of the Mechanical Hypothesis" (1674), Boyle assigns mechanical principles to the "hidden Transactions" of submicroscopic parts.[84] Unlike peripatetics or vulgar chymists who recruit "Un-mechanical Principles and Agents"[85] like substantial forms to infer unobservable causes from observable effects, Boyle insists that mechanical laws govern particles whose likeness to gross bodies persists below the threshold of sense: "both the Mechanical affections of Matter are to be found, and the Laws of Motion take place, not onely in the great Masses, and the middle-size'd Lumps, but in the smallest Fragments of Matter; and a lesser portion of it, being as well a Body as a greater, must, as necessarily as it, have its determinate Bulk and Figure: And he that looks upon Sand in a good Microscope, will easily perceive, that each minute Grain of it has as well its own size and shape, as a Rock or Mountain."[86] Crucially for treatments of seventeenth-century mechanism that take the reductive thrust of "smallest Fragments" at its word, the explanatory reach of "Mechanical affections" and "Laws" is licensed by Boyle's deployment of analogy. The familiar physicality of "Lumps" delimits the materialist coordinates of "Bulk and Figure" claimed by a corpuscle that is "as well a Body as a greater." The likeness of large and insensible bodies is sealed by Boyle's "minute Grain" of sand, the microscopic revelation of whose resemblance to "great Masses" extends backwards to guarantee the physicality of indefinitely smaller parts. Scholars who read the analogical consistency of mechanically qualified rocks and submicroscopic *minima* as a terminal deferral of substance inevitably proclaim this analogy's tautological endpoint. As Daniel Tiffany insists, mechanism driven by the sameness of fragments and "great Masses" nests engines inside engines: "Thus all bodies, and indeed the very constitution of matter, are to be understood as machines."[87] Tiffany's reductionist approximation of mechanical philosophy and, most crucially, "the regime of analogy upon which philosophical atomism is founded"[88] refers to nothing but tinier and tinier machines, an analogical mise en abyme whose substantive impoverishment Catherine Wilson likewise affirms: "This promotion of analogy as a means to knowledge of nature has a strange unsatisfactoriness . . . Analogies invite a regress."[89] Bernadette Bensaude-Vincent and Isabelle Stengers strongly assert the fruitlessness of a science bereft of any qualifying resource save the reshuffling of indistinguishable component parts: "Boyle's atomism therefore implied the impossibility of chemistry ever becoming a science with a theory on which to base and explain its practices. The only possible chemical theory would in effect be mechanist and its subject would have to be textures."[90] But Boyle's chymistry does not devolve into a regressive appeal to smaller engines. Durable corpuscular denomination, as evidenced by silver recovered from acid solution in the *reductio* experiment, is enabled by extramechanical attributes like stickiness or some such thing. As Bensaude-Vincent and Stengers affirm without noting the anachronism of the charge, a historiography of science that equates particles and inert, geometrically qualified units dictates "the impossibility of chemistry."

New appraisals of seventeenth-century science have enriched the meanings of mechanism, a revision traceable from Boyle's endorsement of puritan and natural philosophical appeals to practical understanding. Following Bacon's *New Organon* (1620), which claims as the "true ends of knowledge . . . the uses and benefits of life,"[91] Boyle endorses the epistemological orientation of the person who marshals micro-understanding to fulfill Baconian purposes:

> I shall not deny but that the Atomical, the Cartesian, or some such Principles, are likely to afford the most of satisfaction to those speculative Wits that aim but at the knowledge of Causes; so I think that the other sort of men may very delightfully & successfully prosecute their ends, by collecting and making Variety of Experiments and Observations, since thereby learning the Qualities and Properties of those particular Bodies they desire to make use of, and observing the power that divers Chymical Operations . . . have of altering such Bodies, and varying their effects upon one another, they may by the help of Attention and Industry be able to do many Things . . . very Useful in humane life.[92]

Contrary to Shapin and Schaffer's claims, Boyle eschews only certain kinds of causal indulgence—namely, "speculative" systematizing of the kind that preoccupies "Wits." But Boyle does not thereby advertise his refusal to venture any essay at microcausation. Boyle does not impugn corpuscular agency tout court when he distinguishes "learning the Qualities and Properties of those particular Bodies" from wrangling over postulates deduced from "Atomical" or "Cartesian" systems, because his experimenter's "divers Chymical Operations" muster the minuscule constituents of precipitates, sublimates, and mixtures. Rather than renouncing corpuscular causes, Boyle affiliates himself with an "other sort of men," those who "prosecute . . . ends" without mounting abstract philosophical systems: "We see that the Artificers that never dream'd of the Epicurean Philosophy, have accommodated Mankind with a Multitude of useful Inventions."[93] Undeterred by speculative ambition, "Artificers" actually do things. Shackelford forges a link between Boyle's unphilosophical artisan and the agency of something like Boyle's tingeing part: "Artisan was socially synonymous with 'mechanic' . . . [T]he mechanics often cannot give a good explanation of what they do; they just do it . . . [H]e has a knowledge of the process that is implanted in him by years of craftsmanship, which enables him to carry out that process with predictable results." Craftsmanship that is motivated but mechanical, end-directed but rote, performed by a human but not always deliberative is encapsulated, Shackelford suggests, by "*scientia*—a knowledge of a process that enabled the artisan . . . to carry out a chemical (or mechanical) activity, whether in the workshop or at the atomic level."[94] Like artificers, tingeing parts facilitate ends somewhat in the manner of machines to cross *scientia* and mechanical texture.

Scientia is prosecuted by tingeing parts and mechanics, ennobling corpuscles and end-driven inventors. In this sense, Shackelford suggests, "the concepts of machine and agent are interleaved."[95] Also insinuating human practice into seventeenth-century deployments of the mechanism concept, Sylvia Berryman explicates its overlay of Greek and medieval connotations. While "mechanics, traditionally conceived, is a paradigm of intelligibility"[96] for Boyle, its medieval signification implicates the "manual arts" practiced by persons like artisans: "It is difficult to see how this second sense of the term 'mechanical' . . . could have given rise to a reductive materialism . . . Arts such as metallurgy, glass-making, and agriculture hardly tend to suggest the view that matter is passive, pure extension: these might equally be the arts inspiring chemical or hermetical philosophy."[97] Just as, Newman writes, the recovery of silver corpuscles from a mixture "allows Boyle to go from the observational world to the microlevel without relying on mere analogy,"[98] so *scientia* cannot be banished from seventeenth-century mechanism or even from mechanically delineated corpuscles. Chymical practice does not predicate the existence of microparts on "mere analogy," even though Boyle emphasizes materialist physicality by depicting figured lumps. Mechanism does not preclude human or corpuscular retention of ends—whether use or tingeing influence—even though it skirts idealisms like atomistic indivisibility, peripatetic essence, and philosophical consciousness.

In their claims for eighteenth- and twentieth-century atoms, recent feminist characterizations of matter do not acknowledge the chymical dialectic of structure and substance. Judith Butler's influential *Bodies That Matter* interrogates the feminist breakthrough launched by Simone de Beauvoir, who separates extrinsically imposed gender from corporeal sex, to query the political leverage granted by the former category if it fails to disturb the anatomical body. Addressing the integrity of sex still "presumed as the irreducible point of departure for the various cultural constructions it has come to bear,"

Butler questions the liaison between corporeality and "presumed" resistance to culture, language, and power: "I want to ask how and why 'materiality' has become a sign of irreducibility, that is, how it is that the materiality of sex is understood as that which only bears cultural constructions and, therefore, cannot be a construction? . . . Is materiality a site or surface that is excluded from the process of construction, as that through and on which construction works? . . . And what kinds of constructions are foreclosed through the figuring of this site as outside or beneath construction itself?" Butler also flags the query that dogs contemporary feminism from the other side: "Does anything *matter* in or for poststructuralism?"[99] Writing alongside post-poststructuralist appeals to scientific realism and new materialism, Butler navigates between "materiality" lost, critics of deconstruction claim, in the referential shuffle and "materiality" whose imperviousness to construction renders it the platform for an only superposed gender. It is the proneness of the latter materiality to guarantee "irreducibility"—whether the irreducibility of postmodernism's "lost matter"[100] or the irreducibility of sex—that Butler challenges. She contests the self-evidence of sex by puncturing matter's imperviousness to culture and, in particular, the cultural imposition of heteronormative order: "matter has a history (indeed, more than one) and . . . the history of matter is in part determined by the negotiation of sexual difference . . . [M]atter is fully sedimented with discourses on sex and sexuality."[101]

Butler's claim "that matter is fully sedimented with discourses" has provoked strong feminist and realist rebuttal. According to Vicki Kirby, "If we are only ever dealing with the signification of matter rather than with the stuff of matter as such," then the sex Butler makes permeable to discourse is vulnerable to the charge of "ideality."[102] This charge seems to repeat the attack on poststructuralism Butler attempts to parry, but Kirby takes a different tack. Rather than irreducibility extraneous to culture, matter's own power to signify shields it from the totalizing encroachment of constructivism: for Kirby, Butler "denies the possibility that nature scribbles or that flesh reads."[103] Kirby recruits "quantum physics" to animate uncanny responsiveness to human measurement that "ruled out any sense of an enclosed identity of the particle": "It appeared that subatomic particles did not behave as objects."[104] Comprised of phenomena that do not sit out in the world awaiting referential summons, Kirby concludes, "perhaps matter is considerably more articulate than Butler has imagined."[105] Barad underscores Kirby's claim: "Butler's theory ultimately reinscribes matter as a passive product of discursive practices rather than as an active agent participating in the very process of materialization."[106] Unlike the subatomic particle, which frustrates discursive delimitation in advance, Butler's matter, Barad claims, is always only a linguistic effect: "while Butler correctly calls for the recognition of matter's historicity, ironically, she seems to assume that it is ultimately derived (yet again) from the agency of language or culture."[107]

Barad's own appeal to "historicity" does not enlist matter as "an active agent" prior to the quantum revolution authored by the early twentieth-century physicist Niels Bohr. Indeed, Barad negatively characterizes matter "up through the nineteenth century" as "invisible, indivisible, immutable, impenetrable corpuscles running aimlessly in the void, constituting the sum total of existence."[108] Barad endorses matter's activity only, it must be stressed, after the turn of the twentieth century. To specify science's objects before that date, she cites the hard, imaginary, abstract units postulated by Shapin and Schaffer. "Immutable" and "aimless[]" points, these are not corpuscles in any historical or scientific sense, but a reductionist shibboleth which forecloses the chemical particle's relational powers. Barad rebukes Butler for denying agency to matter, but she reenacts a historical foreclosure at least as profound.

Barad and Butler thus agree on two of matter's historical attributes: it is "indivisible" and "impenetrable." As cognates for Butler's adjective—*irreducible*—these endowments support Butler's claim for matter's capacity to obfuscate the fact of its own construction. To isolate the motive for this denial, Butler draws upon Luce Irigaray's feminist treatment of one of Western philosophy's founding visions of mind-body difference, Plato's creation story *Timaeus*. Irigaray aligns what she dubs "the feminine"[109] with this cosmogony's inscrutable third term, a womb-like entity that assists the metaphysics of world genesis by serving as a bridge between divine forms and their earthly replicas. Plato's womb or receptacle is neither ideal template nor body, Irigaray claims, but an incoherent middle agent that sustains their difference. She reprises *Timaeus* to capture the tertiary "feminine" modeled by Plato's womb in her deconstructive enactment of "[p]assages, and hemorrhages, between sensible and intelligible."[110] The receptacle that receives intelligible imprints to engender worldly copies facilitates Plato's metaphysics while, Butler suggests after Irigaray, as "a matter that exceeds matter"[111] it makes a mockery of them. By exposing the "hemorrhages" that pervade this canonical inauguration of Western metaphysics, Irigaray shows that it is haunted by an anterior feminine which exposes philosophy's founding failure to isolate mind from body.

Butler glosses Irigaray's deconstructive treatment of Plato as a "penetrative textual strategy"[112] to suggest that Irigaray violates a metaphysical system most emphatically figured by his womb's passive receptivity to engendering impression. As Butler moves to her own reappraisal of matter, she conflates these metaphysical and figural coordinates to affirm "the exclusive allocation of penetration to the form, and penetrability to a feminized materiality."[113] This assertion of metaphysical and figural identity marks the pivotal historicizing move in Butler's account. The figural proscription's legibility is most graphically evidenced by enforcement of the corporeal schematics of "compulsory heterosexuality" sustained as the criminalization of "a masculine penetration of the masculine . . . a feminine penetration of the feminine, or a feminine penetration of the masculine."[114] The disciplinary coarticulation of passively materialized feminine penetrability and a masculine anteriority to embodiment it may elect to enter extends through a present tense in which the "morphogenesis" of straight sex respects the figural and metaphysical strictures informing Plato's "fantasy of heterosexual intercourse."[115] As a response to Butler's own framing questions, her history turns the irreducibility some feminists claim for a pregendered body on its head. That body is already indebted, Butler argues, to "the taboo that mobilizes the . . . beginnings of Western metaphysics,"[116] a taboo that invests the stuff of sex with prohibitions that instantiate compulsory heterosexuality in and as matter.

Butler's history, which extends from Plato through *Bodies That Matter*, invites a return to the wounded nuns who illustrate Boyle's claim for the relational contingency of the quality of poisonousness. The gender of Boyle's nuns, their tendency to metonymic association with the "Innocent Ingredients" that must be fused to become the glass that cuts them, seems to emblematize the disciplinary restriction of penetrability to persons whose sex is even redundantly presupposed. But I mean to suggest the opposite. The exemplarity of Boyle's pierced nuns is not restricted to the passive, even abject materiality borrowed from Plato's womb to police heterosexual difference from its foundation. Boyle employs nuns to house "the tender Membranes of the Stomach and Guts," but when he specifies the qualified person who determines the reciprocal toxicity of ground glass, he differentiates tender guts only from the "stronger Texture" of the innards of animals. Boyle's tender membranes, that is, are human membranes. His pierced nuns claim always relative penetrability as the attribute that renders any person susceptible to entrance by particles hyperliteralized as cutting fragments of glass. Such susceptibility involves a porous texture—of gut, of skin—that may admit minuscule bodies as well as a perceptual apparatus that may transpose particulate impressions into ideas. Boyle's nun, I mean to suggest, models eighteenth-century empiricism's avatar of the subject of sensory knowledge.

In the ground glass scenario, the candidate for materialized abjection somewhat akin to that proposed by Butler is the "much stronger Texture" incarnated by animal guts "Lin'd or Stuff'd with Gross and Slimy Matter." But if, like the existential antagonist of slime Jean-Paul Sartre, we affiliate "Slimy Matter" with the feminine, then we must acknowledge that slimy guts render animals *im*penetrable.[117] As the mechanical attribute texture, sliminess muffles glass's sharp points, offering as the contrary of empiricism's representative person animals whose dumb flesh fails to elicit qualities from entering matter. Unimpressionable texture, animal texture, is not, in Boyle's account, the human texture that mobilizes empiricism as a doctrine of relational qualities.

As we have seen of operationally indivisible chymical units, the corpuscle is not impenetrable. But even noble corpuscles' perviousness to a very piercing menstruum does not line up with the disciplinary imposition of passivity. All the way down to the level of matter's smallest operational unit, penetrability does not divest it of reactive powers, whether these are entailed by texture, relative stickiness, or tingeing potency. To transmute gold into silver, Boyle employs a piercing solvent that extricates tingeing parts from the bulk of the metal. But a long-standing alchemical schematics of transmutation figured as "penetrating the metals"[118] does not correlate with the Platonic binary that determines one of these positions as metaphysically weightier, passive, and unresponsive. Both the alchemical solvent (or alkahest, philosophical mercury, or philosopher's stone) and the metal react as substance and as structure. However piercing the former, and however penetrable the latter, their attributes are incarnated by corpuscles. An always relative attribute, reactivity instantiated by texture offers no ground for the imposition of binary metaphysical difference. Metal may be denominated gold after its entrance by tingeing parts, but this outcome heralds not the metal's takeover by ideal form but rather the alienability of an ontological predicate extricable from gold as a unit of matter. As the revision that leverages Boyle's refusal of peripatetic essence, qualified matter heralds not objects passively invested with ontology, but denominating attributes that matter cannot secure as such.

Because Boyle's account of sensation does not insulate perceptual ideas from the physicality of stimulating matter, corpuscular philosophy offers an alternative genealogy of the mind-body relation. According to Boyle's discussion of sensory understanding, ideas are effected by corpuscular contact. While they are indeed incorporeal, these ideas come into being as perceptions, not as transcendent Platonic entities divorced from the world of things. Rather than instituting ideality as a category antagonistic to matter—and thereby enshrining its metaphysical distinction—chemical science articulates a corpuscular account of ideas as well as body. With Irigaray, Butler takes Plato's *Timaeus* as the reigning philosophical and historical blueprint for Western metaphysical dualism. But Boyle's corpuscular doctrine of qualities formulates an alternative whose mainstream cultural impact is secured by Locke's *Essay Concerning Human Understanding*. In a corpuscular metaphysics, ideas are enmeshed with the texture that stimulates them, while mechanics and artisans enact *scientia* mindlessly.

A model empirical person, Boyle's nun sits at the cusp of micro-figure and macro-impression blown up to concretize the fatal relational effect portended by the emphatically Epicurean trait of shape. Her empirical experience is retheorized one level down by Boyle and his alchemical colleague Isaac Newton (1642–1727) as the less lethal but still invasive sensation of taste. Taste heralds corpuscular agency which, even as it acts on receptive flesh, exceeds the mechanical capacities of figure. Boyle first posits an experience akin to the nuns' encounter when he justifies why ingested crystals of niter (saltpeter or its salts) burn: "being dissolved and agitated by the spittle that usually moistens the tongue, their smallness may give them great access to the pores of that organ, and the sharpness of their sides and points may fit them to stab and cut, and perhaps fear [*sic*] the nervous and membranous parts of the organ of Tast."[119] Like glass fragments, these particles' mechanical affections "fit them to stab and cut," defining experimental receptivity to taste as the effect of "how deep the particles of these Crystals may pierce into the spungy organs."[120] While Boyle's "spungy" tongue reciprocally determines "how deep" niter will stab, cut, tear, and "pierce," Newton supersedes Boyle's materialist roster of core particulate attributes by assigning uneasily dissoluble corpuscular parts a differential tendency to stick or attract. In *De natura Acidorum* (1691), Newton denominates the category of acids in terms of this attribute: "for we call acid what attracts and is attracted strongly."[121] The trial of acid's qualitative identity may occur inside the chymist's own mouth: "these same particles when applied to a sense-organ tear away its parts . . . and inflict pain, and so are called acids, leaving behind, of course, the subtle earth to which they were adhering, because of the greater attraction to the moisture of the tongue."[122] Newton theorizes corpuscular stickiness that enters into relation with the chemical constituents of the experimenter's own organ of taste.

When attracting particles are placed on the chemist's tongue, the event of denomination becomes inescapable: they "inflict pain, and so are called acids." By ingesting matter that sticks to his own constituent parts, the experimenter transforms attraction into a sensible quality. As he reacts, the chemist apprehends particulate attraction as a painful taste or, rather, the pain that *is* taste. Newton reasserts the experiential vividness of stickiness as taste in Query 31 of his *Opticks* (1730): "Do not the sharp and pungent Tastes of Acids arise from the strong Attraction whereby the acid Particles rush upon and agitate the Particles of the Tongue?"[123] Penetrating "Particles" are smaller than Boyle's shards of glass, whose size, Boyle stipulates, surpasses that of dust. A microreaction motivated by Newton at subcorpuscular scale, acid's "sharp and pungent" sensation refines normative empirical impression not as wounded guts but as taste. But if acid *minima* do not "rush upon" receptive flesh as violently as do cutting fragments, penetrability remains the enabling condition of empirical understanding. Receptive or porous texture incarnates the impressionability of both gut and "spungy" tongue. Newton's supreme contribution to the scientific revolution—by positing the causal agency of attraction, he universalizes the laws of gravity—is sustained for empiricism by the variable affinities of corpuscles that react to engender taste.

Even the "ignorant Maid" who, we recall, jettisons the precious evidence of Boyle's alchemical assay after he is "call'd out of my Chamber" does not affirm the transhistorical discipline of matter that would preclude her from imposing her own stamp upon it. Instead, Boyle's maid mimics the alchemical gesture par excellence: she "swept this Paper, as a foul one, into the fire." The maid populates an episteme propelled by the capacity of bodies like "fire" to divest other bodies, even those Boyle prizes, of their denominating qualities. It may be impossible to move her, as well as the distracted woman who poisons the nuns' peas, from the outskirts of Boyle's science.[124] But the long chymical legacy she inhabits perpetuates neither isolated incorporeal substance nor unilateral philosophical reductionism. Instead, persons made of corpuscles enable the susceptibility of qualities to sense, an alternative relation of mind to body Locke adopts as the empirical origin of human understanding.

Semina and Mechanism: Hooke's *Micrographia*

With Hooke's *Micrographia*, this chapter shifts from attributes of singly insensible particles to a dilemma of empirical likeness. No stark discrepancy distinguishes the paired images accompanying Hooke's "Observ. XIX. *Of a* Plant *growing in the blighted or yellow specks of* Damask-rose-leaves, Bramble-leaves, *and some other kind of leaves*" and "Observ. XX. *Of* blue Mould, *and of the first Principles of Vegetation arising from* Putrefaction"[125] (Fig. 2). Indeed, the images afford no ocular cues to motivate these headings (the upper image is the mold; the lower, the plant). To stipulate their difference, Hooke writes, the reproduction of blue mold "require[s] no seminal property," while tiny plants proceed from "*equivocal* generation" necessitating some incident "seminal principles."[126] Under the microscope, however, the vegetables disclose no qualitative sign of their variant origins. The figural equivalence of the plates in Observations 19 and 20 portends empirical understanding that cannot clearly demarcate seminal from mechanical genesis. As I argue in Chapter 2, Locke's *Essay* extends similar difficulties to the qualitative identity of the category of man.

Hooke served as Boyle's assistant and engineer of the air-pump deployed in Boyle's *Spring of the Air* from 1655 until he became curator of experiments for the newly instituted Royal Society in 1662. His agreement with Boyle on the physical antecedents of plant growth is illuminated by Shackelton, who confirms recent historiographical reappraisals of the scientific revolution by citing "[t]he survival of Paracelsian *semina* . . . into the natural philosophy of members of the Royal Society."[127] The notion of *semina* is shaped by the Paracelsian innovator Joan Baptista Van Helmont (1580–1644), whose own crucial experiment, Boyle argues, shows the assimilation of aqueous corpuscles into the stuff of living bodies: "The Ingenious *Helmont* indeed mentions an Experiment . . . which he made by planting a Branch of Willow into a Pot full of Earth, and observing the increase of Weight he obtain'd after divers years, though he fed the Plant but with Rain water."[128] Because nothing other than "Rain water" augments the total "Weight" of the enlarged vegetable, pot, and soil, Helmont claims, water must be incorporated into the texture of the plant. For Epicurus as well as Helmont and Boyle, vegetable growth proves the existence of minuscule parts,[129] because measurable gains in the plant's size result from the insensible accretion of "the same Corpuscles, which, convening together after one manner, [to] compose that fluid, Inodorous, colourless, and insipid Body of Water [now] being contexted after other manners."[130]

But in Boyle's account of the changes water assumes upon ingestion by plants, qualitative metamorphoses undergirded by "the same Corpuscles" are not all ascribed to mechanically re-"contexted" parts. In a narrative enactment of water's progressive instantiation of vegetable difference, which anticipates prose on plant growth by his Helmontian colleague Kenelm Digby, Boyle proceeds:[131]

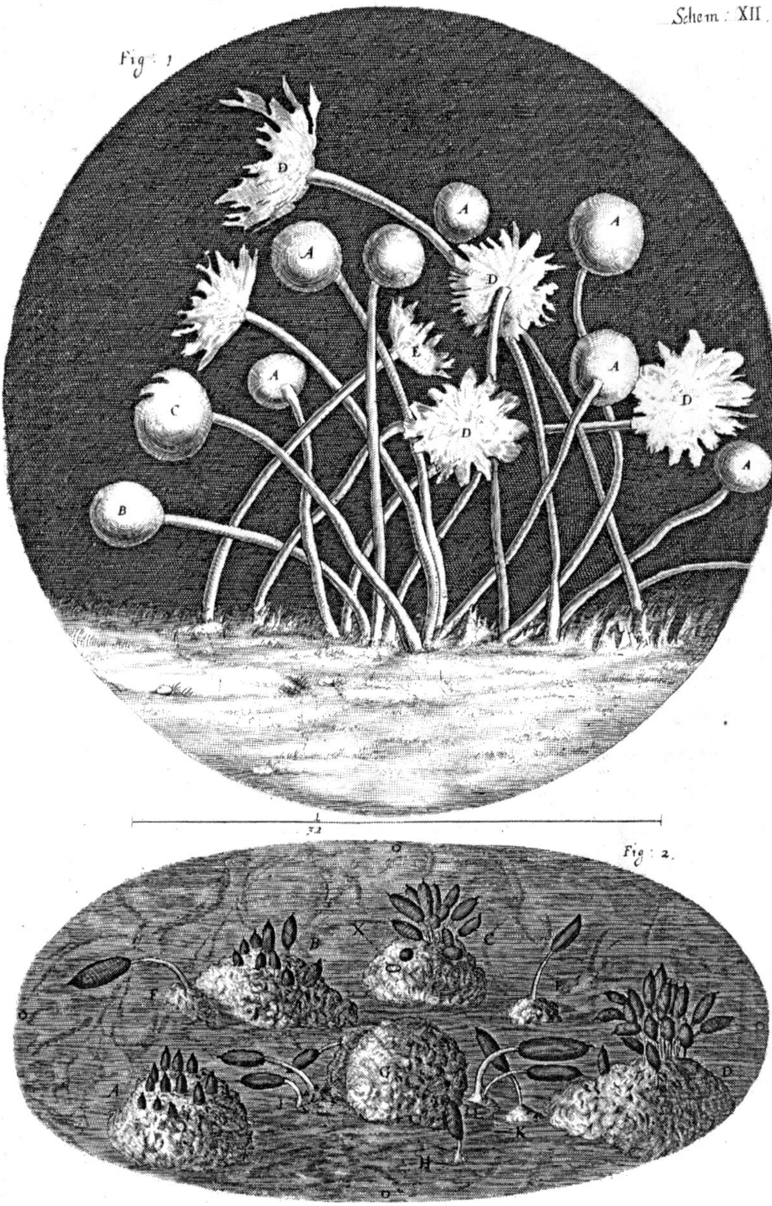

Figure 2. "Blue Mould," and "*a* Plant *growing in the blighted or yellow specks of* Damask-rose-leaves, Bramble-leaves, *and some other kind of leaves,*" Robert Hooke, *Micrographia* (1665), Scheme XII. Courtesy of the Lilly Library, Indiana University, Bloomington.

> Supposing then this Liquor, at its first entrance into the roots of the Vine, to be common Water . . . And first, this Liquor . . . is turn'd into the Wood, Bark, Pith, Leaves, &c. of the Vine; The same Liquor may be further dry'd, and fashon'd into Vine-buds, and these a while after are advanced unto sour Grapes, which express'd yield Verjuice . . . These soure Grapes . . . turne to well tasted Grapes; These if dry'd in the Sun and Distill'd, afford a fætid Oyle and a piercing *Empyreumatical* Spirit . . . If the juice of the Grapes be squeez'd out and put to Ferment, it becomes a sweet and turbid Liquor . . . the fermented Juice of Grapes is partly turned into liquid Dregs or Leeze, and partly into that crust or dry feculancy that is commonly called Tartar . . . The same Vinous Juice . . . Degenerates into that very sour Liquor called Vinegar . . . and oftentimes in this Vinegar you may observe part of the matter to be turned into an innumerable company of swimming Animals.[132]

From the side of human sense, qualities effected by corpuscular texture include sourness, tastiness, fetor, acidity, and alkalinity. But Boyle's stress on the serial requalification of the "*same*" water—which yields even the quasi-metaphysical split of "*Empyreumatical* Spirit" from "Dregs"—finally accommodates a more emphatic species of difference: "swimming Animals."

In a move that justifies water's final transmutation, Boyle posits causal agents that pass under cover of the vegetable that harbors them: "among the constituent parts of an Animal or Plant there may lurk some Seminal principles or rudiments, that is, small parcels of Matter of such a texture, that . . . they are not by sense . . . distinguishable from the rest of the compounded body."[133] As Newman notes, "Seminal" "rudiments" that hide inside bigger bodies resemble mercury or gold dissolved by corrosives in solution: "invisible *semina* of frogs or plants can lie latent in matter just as the atoms of a metal hide invisible in an acid."[134] Such hidden *semina* sustain Boyle's refusal of the materialist cosmogony authored by Epicurus because they enable biblical creation propelled by lurking parts: "I see no Necessity to conceive that the water mention'd in the Beginning of *Genesis*, as the Universal Matter, was simple and Elementary Water . . . we should Suppose it to have been an Agitated Congeries or Heap consisting of a great Variety of Seminal Principles and Rudiments."[135] Far from a pool of indifferent mechanical fragments, primordial water is "an Agitated Congeries" harboring the contingent eventuality of world creation. Likewise, the insensible "Variety" of any holistic thing is revealed when it rots, releasing latent *semina* from the strictures of imprisoning texture: "these Seminal principles or rudiments being set at liberty . . . perhaps by a lucky concourse of other circumstances may fall to act according to their own nature."[136] Like his aqueous "Heap," Boyle's "lucky concourse" articulates the intimacy of structure and substance as the interplay of materialist and seminal modes of causation. If luck refers to world creation inaugurated by random atomic collision, it intercedes in Boyle's account to enhance the fecundity of rudiments with seminal powers. Boyle draws from Helmont, for whom putrefaction and fermentation supply the motion to liberate seminal ends, when he observes: "Putrefaction opens many Bodies." Putrefying bodies are opened or, in alchemical idiom, "unlockt"[137] to activate seminal influence from within particulate things.

In Observation 20, Hooke denominates blue mold "*Microscopical* Mushroms" [*sic*], plants that lack "any such thing as seed in any part of them."[138] Hooke's chapter elucidates the microcausation that produces "a very pretty body shap'd and concreeted by Mechanical principles, without the least shew or probability of any other seminal *formatrix*."[139] But by extricating any lurking "seminal *formatrix*" from the mold-mushroom's engendering parts, Hooke introduces vegetable resemblance that cuts in another direction. The "very pretty" concretion of mushrooms resembles not seminally engendered plants but vegetating metals. Hooke identifies a key microlikeness to growing mold: the formative agency revealed, as we have seen with Boyle's scaly plates, by silver dissolved and then precipitated out of nitric acid: "we find that many pretty beards or *striæ* of the particles of Silver may be precipitated upon a piece of Brass put into a *solution* of Silver very much diluted with fair water, which look not unlike a kind of mould or hoar upon that piece of metal; and the hoar frost looks like a kind of mould." Hooke names "Figures of growing Salts, and the Silver Tree"[140] as reactions whose accumulating texture anticipates mold's. Of the Silver Tree or Tree of Diana, an experiment in which the dissolved metal precipitates and gathers, plantlike, on coils of copper placed in the solution,[141] Hooke writes: "it is a very pretty kind of Germination which is afforded us in the Silver Tree, the manner of making which . . . is well known to the Chymists, in which there is an Ebullition or Germination, very much like this of Mushroms."[142] As Principe remarks, "treelike (or dendritic) chemical 'growths'" like those evoked by Hooke "were familiar parlor tricks in the seventeenth century."[143] Hooke's tree visibly confuses chymical "Ebullition" and vegetable "Germination": both moving forces, they make mechanical mushrooms grow upward. A likeness sustained by "Figures" propels

Hooke's sequence of mutually formal and microcausal resemblances, from "pretty beards or *striæ* of the particles," which resemble "hoar frost," to frost that in turn "looks like a kind of mould."

Hooke's mold sits at a causal juncture that licenses passage from mechanical to seminal reproductive agency: "the figure and method of Generation in this concrete seeming to me, next after the Enquiry into the formation, figuration, and chrystalization of Salts . . . a *medium* through which he must necessarily pass, that would with any likelihood investigate the *forma informans* of Vegetables."[144] By surveying the progressively complex "formation" of things, Hooke defines a "*medium*" whose distinctiveness may be difficult to isolate: "the steps or foundations of our Enquiry, [are] *Fluidity, Orbiculation, Fixation, Angulization*, or *Crystallization Germination* or *Ebullition, Vegetation, Plantanimation, Animation, Sensation, Imagination*."[145] *Micrographia* skips from geometrical reductionism emblematized by such processes as "*Orbiculation*" to qualities whose emergence cannot be mechanically rationalized, "steps" whose arrival at "*Imagination*" does not rigorously insulate mind and starting body. At the crux of this passage, the mechanical mushroom instances "*Crystallization Germination* or *Ebullition*" that anticipates "*Vegetation*" already performed by silver corpuscles in solution. Because the silver tree enacts the schematics of plant growth in advance of the vegetable, Hooke's removal of *semina* from his mushroom sustains its backward likeness to forms engendered by metal.

Hooke places Observation 19, whose tiny plants *are* effected by seeds, before the mechanical mushroom. But plants growing from Damask rose leaves or other putrefying vegetables also occupy a chymical relay, because they occasion inanimate matter that does not disclose or mimic, but itself assumes, seminal powers. With Boyle, Hooke credits the apparition of creatures from rotting things to lurking seminal potency now "unfetter'd and left at liberty to move . . . [which] moves and acts after quite another manner than it did when a *coagent* in the more compounded *machine* of the more perfect Vegetable."[146] The Damask rose may be a "more compounded *machine*," but this machine is composed in part of seeds. To amplify the discrepancy between alien *semina* and the soon-to-be-liberated agency of environing texture, Hooke instances insect matter deposited on "musty *Casks*" from without:

> [T]he *putrifying* substances on which these Eggs, Seeds, or seminal principles are cast by the Insect, become, as it were, the *Matrices* or Wombs that conduce very much to their generation, and may perchance also to their variation and alteration . . . If therefore the

> *putrifying* body, on which any kind of seminal or vital principle chances to be cast become somewhat more then meerly a nursing and fostering helper in the generation and production of any kind of Animate body, the more neer it approaches the true nature of a Womb, the more power it will have on the by-blow it incloses.[147]

What Hooke claims as "the true nature of a Womb" departs from Butler's passive Platonic receptacle. As "putrifying substance" that acts "more then meerly a nursing and fostering helper," Hooke's "*Matrices*" ascribe formative powers to ambient matter—or, alternatively, divest seminal principles of what would seem to be their distinguishing difference from mechanical particles, their capacity to perpetrate an end. Hooke does not posit Platonic ideality imposed on a metaphysically antipathetic matrix, but seminal directives whose trajectory "more then . . . fostering" corpuscles divert. The recombination of extrinsic *semina* and a womb that induces "alteration" heralds inanimate texture that may continue to engender unexpected things: "notwithstanding its own production was as 'twere casual, yet it may germinate and produce seed, and by it propagate its own, that is, a new Species."[148]

Newman's suggestion that hidden *semina* act like tingeing parts echoes Hooke's depiction of putrefactive opening which authorizes further increments of change: "*Putrifactive* and warm steams should . . . produce a living Magot, and that, by degrees, be turn'd into an *Aurelia*, and that, by a longer and a proportion'd heat, be *transmuted* into a Fly."[149] This "*transmuted*" chrysalis recalls noble corpuscles whose tingeing potency qualifies lower metals. Like a piercing menstruum, "*Putrifactive* and warm steams" unlock the agency of *semina* that conjure golden "*Aurelia*" from less auspicious beings.[150] But just as ennobling corpuscles may be alienated or undone, seeds that transmute insects may be redirected by more than nourishing matter. Hooke's transmuted chrysalis recombines the chymical agency of gestation and the seminal reactivity of particles, because the womb on which *semina* are cast sustains the power to induce a different end.

Returning to mushrooms concreted in the absence of seeds, Hooke assigns blue mold an analogical zero degree: its generation "is purely Mechanical, and . . . as necessary upon the concurrence of those causes as that a Ship, when the Sails are hoist up, and the Rudder is set to such a position, should when the Wind blows, be mov'd in such a way."[151] Hooke's sailboat represents sheer physical necessity, mechanical movement resistant to the decorporealizing insinuation of teleology or will.[152] But because this necessity is enacted, for

Hooke, by the silver tree, blue mold does not incarnate the "purely Mechanical" causality that would be assured by Hooke's evacuated ship. The paired images in Scheme XII represent dendritic forms effected by particles as well as by seeds, by seeds that act like particles and particles that act like seeds. Rendered perceptible by Hooke's microscope, the mold and the tiny vegetable do not visibly transmit the anterior discrepancy of substance and structure. Indeed, Hooke's two plates cement the failure of corpuscles to manifest seminal as opposed to mechanical causation as qualitative difference. Locke's *Essay* will proceed to broach even more vexing dilemmas of sensory—that is, empirical—discrimination.

This chapter has shifted from singly insensible particles to bodies at the cusp of perception because *Micrographia* brings us to Locke. Hooke's vegetables are formed both extrinsically and intrinsically, by seeds as well as by agents, like the cask, that shape them with equally profound determining power. These plants emerge neither from necessity nor innate ends, neither solely from texture nor solely from seeds. What world do they project for empiricism? In this world, human perception does not disclose how corpuscles engender denominating attributes of things, including human things. In this world, matter can bear ends, but those ends can be diverted by other ambient matter. And in this world, knowledge is produced by texture that may become something else in relation.

CHAPTER 2

John Locke and Matter's Power

> The Knowledge of Bodies we must get by our Senses, warily employed in taking notice of their Qualities, and Operations on one another.
>
> —John Locke, *An Enquiry Concerning Human Understanding* (December 1689)

In the history of science and philosophy, the influence of Boyle's corpuscular chemistry on John Locke's *An Essay Concerning Human Understanding* is a critical given.[1] This is less the case in the field of literary studies. The following paragraphs cannot fully recapitulate Locke's tutelage, chymical collaboration, and intellectual ventures with Boyle, whose residence at Oxford from 1654 until his displacement to London in 1668 overlapped with Locke's thirty-year tenure at Christ Church, which began in 1652 and was first interrupted when Locke moved to London in 1667 as advisor and physician to Lord Ashley, later Earl of Shaftesbury. Besides jointly participating in the experimental philosophical activities of a group of men based at Wadham College (hosted by John Wilkins and including Christopher Wren, Richard Lower, Thomas Willis, and Robert Hooke), Boyle and Locke collaborated on Boyle's experimental histories of blood and air. As Peter R. Anstey, Kenneth Dewhurst, and Lawrence M. Principe establish, Boyle and Locke may have undertaken alchemical (that is, chymical) projects together as early as May 1660, a joint enterprise extending through the end of Boyle's life.[2] Locke reviewed the Latin edition of Boyle's *Specific Medicines* (1686) in the *Bibliothèque universelle et historique*; he wrote the publisher's advertisement for Boyle's *General History of the Air* (1692) after Boyle's death in 1691. The scientific relations between Boyle and Locke, friends and Fellows of the Royal Society, show the deep appeal, for Locke, of Boyle's experimental and philosophical repudiation of scholastic substance.[3]

Because Locke's library catalog, roughly half his books, and many of his notebooks survive, Locke's bibliographic connection to Boyle is also well known. Locke's catalog, which represents his library as it was finally assembled when he stayed at Otes with Sir Francis and Damaris Masham from 1691 until he died in 1704, records sixty-two works by Boyle, the largest number by any single author. Anstey attests that Locke scrutinized many of them: "It is clear from Locke's medical notebooks that he was an avid reader of Boyle's natural philosophical works and that in the early to mid-1660s he read each publication."[4] John W. Yolton declares of Locke's intellectual debt to Boyle: "Reading the theoretical part of Boyle's *Origine* [*of Forms and Qualities* (1666–67)] alongside Locke's *Essay* . . . the conclusion is inescapable that Locke had his copy of this work by Boyle open beside him when writing out . . . portions of the *Essay*."[5] In what follows, I do not speculate on the logistical particulars of Locke's bibliographic recourse to Boyle; along these lines, John Harrison and Peter Laslett, editors of the library catalog, warn: "There is nothing to show what titles were in Locke's room in Exeter House [London] in the summer of 1671, when he made the very first sketches for his *Essay*."[6] All the same, it is difficult not to sympathize with Yolton's insistence on the proximity of Boyle's corpuscular philosophy to Locke's *Essay*, because Locke adopts the anti-essentialism of Boyle's *Sceptical Chymist* (1661) and *Forms and Qualities* in terminologically precise as well as theoretically ambitious ways. This chapter claims Locke's notes, experimental ambience and colleagues at Oxford, and ongoing collaboration with Boyle as evidence adequate to support his practical, philosophical, and sometimes concretely textual debt to Boyle's corpuscular chemistry.[7] But to assume the force of Boyle's influence on Locke is not to suggest that Locke's appropriation of corpuscular philosophy is unvexed.

The *Essay*'s impact on the eighteenth-century novel is a less tendentious proposition. Ian Watt's *Rise of the Novel* predicates the novel's generic specificity on the epistemic shift crystallized by empiricism: " 'realism' had come to denote a belief in the individual apprehension of reality through the senses."[8] Watt designates the novel's representative stance "formal realism" ("formal," he stipulates, because it claims a narrative vantage not affiliated with "any special literary doctrine or purpose"): "the novel is a full and authentic report of human experience, and is therefore under an obligation to satisfy its reader with such details of the story as the individuality of the actors concerned, the particulars of the times and places of their actions, details which are presented through a . . . largely referential use of language."[9] Recent engagements with Watt contest the humanizing self-evidence of entities like "individuality," "actors," and "actions."[10] Michel Foucault has inspired challenges to the apparent neutrality of formal realism, leading some readers instead to claim it as an instrument of discursive power.[11] After Fredric Jameson, Marxist critics excavate from realism repressed and irresolvable social contradiction.[12] This literary-historical work continues to reconceive, rather than dispute, the profundity of the novel's involvement with Locke. The current chapter likewise affirms that eighteenth-century British prose fiction elaborates Locke's sensational epistemology as plot, character, and form. But the Locke to whom I refer is, specifically, a Boylean Locke.

Locke's *Essay* marks the single most influential textual bridge between corpuscular science and the eighteenth-century novel. An experimentalist and philosopher, Locke transposes Boyle's anti-elemental chemistry into a denial of essential substance that encompasses metals and organized bodies. The *Essay* projects deep mutual repercussions for epistemology and ontology: How do empirically sensed qualities add up to something's identity? Does the source of qualitatively adjudicated identity reside inside things at all? This chapter tracks Locke's treatment of corpuscular matter from two sides. From the side of the particle, Locke renames Boyle's mechanical affections—autonomous, core physicality like the particle's shape and size—*primary qualities*. For Locke, the physicality of primary matter guarantees the truth of perceived reality. But made vivid by words, insensible corpuscles become vulnerable to the charge of fiction. Corpuscles propel Locke's monitory account of fiction while, in the major dialectical turn this chapter charts, also subverting the fixity of species of things. From the side of those perceptible things, Locke's anti-essentialist deployment of texture extends into the event of empirical sensation. This is not static witnessing of passive, isolated, stably qualified objects, but encounters that approximate identity in the very act of perceiving. The Locke who emerges from this chapter does not license an empirical scenario in which mute objects submit to mastery by a transparent gaze.

Locke names sensed attributes like sweetness *secondary qualities*. As the aspiration that extends through the rest of my book, this chapter aims to reorient our critical appreciation of Locke's secondary qualities and their literary afterlife. Far from transparently coded representational input, secondary qualities implicate sensed ideas in the matter that effects them. These are not illusory ideas alienated from real things but sensory impressions Locke inserts back into bodies—not as such, but as corpuscular matter's power to stimulate empirical knowledge. Because this power is empirically sensed in time and through relations, it erodes the fixity of both species and words. As a result, Locke bequeaths the eighteenth-century novel not objects that contain essences but identity approximated from the outside. This is not the mimetic transparency of realist perception and realist reference credited to science and imported back to an always more precariously fictional novel. Rather, sensory understanding in novels as well as science entails qualities that do not sustain a divide between empirically known objects and terminally inaccessible—or, imaginary—causes. Ambivalent as Locke may be about this crossover, his *Essay* projects secondary qualities that migrate between ideas and matter as the source and sensational outcome of corpuscular texture's power to produce feeling in persons.

Corpuscular Locke: Matter, Power, Figure

The paradox defining Locke's adoption of what he calls "the corpuscularian Hypothesis" (4.3.§16) is laid out by the educational reformer John Amos Comenius's illustrated vernacular-Latin primer *Orbis Sensualium Pictus* (1658). A textbook which aspires to disseminate "universal knowledge," *Orbis Sensualium Pictus* puts in practice Comenius's claim that "the understanding possesses nothing that it has not first derived from the senses."[13] Knowledge that is universal or, as Comenius writes in reference to a postlapsarian state of linguistic fragmentation, "disgraced with no foul Casme," must recur to "the open field of things themselves" because only "sensuall observations"[14] promise to repair the chasms perpetuated by abstract, jargonized, or sophistic discourse. By providing definitions in Latin and English, with an image to accompany both, Comenius's picture book offers young learners, its translator attests, "nothing but rudiments . . . the grounds of the whole world, and the whole language, and of all of our understanding about things."[15] Comenius's "little *Encyclopeadia* of things subject to the senses"[16] concretizes the democratic reach of "sensuall" knowledge championed by his ally Samuel Hartlib and the reformist grammarian Hezekiah Woodward, whose *Light to Grammar* (1641) promotes childhood pedagogy that is "sensuall, materiall, practicall" and, as the fundamental guarantor of its universality, "experimentall."[17] The "experimentall" immediacy of perception grounds the universalizing educational ambitions of the Hartlib circle before the English Civil War and during the Interregnum.[18] Sensual learning marks a vital antecedent to Locke's *Essay*, whose own appeal to the sensory origins of knowledge Locke affirms by invoking what "we experiment daily in our selves" (2.23.§15).

Yet one pairing of words and image in Comenius's *Orbis Sensualium Pictus* claims an uneasy relation to the experimental imminence of language's sensual rudiments. Under the Latin-English header "Specularia" / "Looking Glasses," Comenius includes the visual aids "Looking-Glasses" (*Specula*), "Spectacles" (*Perspicilla*) and "a perspective glass" (*Telescopia*). He proceeds to the fourth item, accompanied like the others with a number indicating the location of its "material" counterpart in the illustration on the facing page: "A flea appeareth in a multiplying-glass, [4.] like a little Hog" / "In *Microscopio*, Pulex apparet ut Porcellus."[19] The print underscores the exceptional problem posed by this scenario (Fig. 3), since the number four, which hovers at the center of the frame composed by the room's cutaway walls, seems to refer to nothing at all. While within the room the viewer espies a mirror, a pair of spectacles, and a telescope, the illustration includes no microscope, no "flea," and no "Hog." Such an absence points to what would seem an obvious violation of Comenius's sensual order of language acquisition: if the flea is too small to register among the other objects in this frame, how can its prelinguistic immediacy be grasped by a child learner? The compensatory likeness of the flea to a hog would render the smaller entity familiar; but because the room's interior is scaled to the size of the macroscopic objects—including his own face—the surrogate viewer inside perceives, the flea cannot assume the dimensions of "things subject to the senses."[20]

This problem is specific to Comenius's deployment of analogy. If the student has already encountered a hog, then the insect borrows imaginative force from that empirically known animal. But since the room's objects anchor it in the quotidian world, the illustration instead manifests the failure of analogy to transpose microscopic objects into the "experimentall" scene modeled by the man before the looking-glass. The illustration literalizes the flea's exclusion from the world of sensual familiarity occupied by the hog, because these two objects can never possess sensory self-evidence within the same experiential frame. The sensory likeness assumed by the flea precludes its appearance within a world of perceptual knowledge whose scale is set by one's face in the mirror.

The insect whose analogical vividness amplifies its empirical failure to appear captures the paradox defining Locke's relation to corpuscular philosophy. The crux of this paradox is anticipated by Boyle's experimental proof that

Figure 3. "*Specularia* / Looking-Glasses," John Amos Comenius, *Orbis Sensualium Pictus*, trans. Charles Hoole (1685). Photo courtesy of the Newberry Library, Chicago. Wing ZP 645.M46.

bodies are composed of insensibly tiny parts, a proof, Chapter 1 shows, in which a substance like silver dissolves, disappears, and is precipitated back out of a solution. Because silver vanishes but is subsequently retrieved in its metallic state, the *reductio* experiment conjures the existence of durable corpuscles from the temporary event of their ocular disappearance. Boyle's proof shows the experimental salience of imperceptibility: as the very revelation of its existence, particulate matter is crossed with its own vanishing. For Locke as for Comenius or Woodward, however, conscious perception is the bedrock of a new epistemology whose universalizing reach draws from the incontestable immediacy of "sensuall observations" accessible even to a young child. The attendant conundrum can be starkly recapitulated. Locke defines sensory ideas as the building blocks of human understanding at the same time that he adopts a "corpuscularian Hypothesis" whose proof of particles pivots on their disappearance from sense. The epistemological standard on which Locke, with Comenius, Woodward, and Hartlib, bases the universalizing renovation of language—whether in the name of a Puritan commonwealth or, as Locke enjoins, of postrevolutionary "Peace" (3.9.§21)—is the unalloyed imminence of sensory perception. Boyle's corpuscles can never meet this standard.

As a result of this paradox, I will argue, Locke sometimes decries the analogical embodiment of insensible matter as an incentive to "Fancy" (2.23.§13) or "fiction" (2.31.§2). But this is no wholesale indictment of analogy as, according to Jonathan Lamb, "Arthurian romance."[21] Locke's standard of reality is not mathematical reductionism (a position Lamb assigns Locke's alchemical colleague Newton), but primary matter whose proneness to figuration Locke sometimes endorses. The *Essay*'s ambivalence must be more carefully specified in relation to chymical science. The *reductio* experiment proves that bodies like silver are made of particles by recovering the dissolved metal from a corrosive menstruum—not by recourse to analogy. Yet the primary attributes size, shape, or motion, which persist in singly indiscernible corpuscles, nonetheless owe their lucidity to Boyle's critical analogical move: shape, size, and motion are intelligible as *mechanical* affections. A productive reality inaccessible to human sense, it is, more precisely, the corpuscular endowment shape or *figure* that emblematizes Locke's equivocal reliance on analogy. To suggest that figure signals both the *Essay*'s most insistent limit and its most subversive impetus, I first turn to Locke's negative characterization of primary corpuscular attributes no person can perceive.

By adopting corpuscular philosophy as the *Essay*'s working theory of matter, Locke also agrees with Boyle's doctrine of qualities: for Locke as for

Boyle, it is "the Bulk, Figure, Texture, and Motion of . . . insensible parts" that "produce various Sensations in us" (2.8.§10). While mechanical affections, with the addition of amassed particles' "Texture," delimit the ineluctable physicality of the object, perceptible qualities like heat or coldness exist as ideas whose reliance, for example, on the relative warmth or coolness of a perceiving hand shows their relational determination by the human body and mind. Sensed qualities like cold are produced by texture's relation to receptive persons, not the presence of elemental essence or what Locke calls "Substance" (2.23.§1 and passim). Underscoring the contingency of perceived attributes like heat or whiteness, Locke distills Boyle's selective deployment of materialist physicality to assert the freestanding reality of the mechanical affections Locke renames "*primary Qualities*" (2.8.§9, emphasis Locke's):[22] "let not the Eyes see Light, or Colours, nor the Ear hear Sounds; let the Palate not Taste, nor the Nose Smell, and all Colours, Tastes, Odours, and Sounds . . . vanish and cease, and are reduced to their Causes, *i.e.* Bulk, Figure, and Motion of Parts" (2.8.§17).

Despite the exact correspondence of Boyle's and Locke's mechanical affections or primary qualities, Locke's *Essay* does not consistently endorse the explanatory reach of one of these insensible "Causes": "Figure." Figure delimits the charged interface between Boyle's corpuscular philosophy and Locke's empiricism. While Boyle's defense of the core physicality of texture might seem, on the face of it, antipathetic to empirical understanding, Locke's *Essay* hedges against the capacity of figure to render particles *too* familiar. Analogically vivid attributes like shape encroach upon the prohibition that defines the entities Locke calls simple ideas, the primordial, uncorrupted perceptions that serve as "all the Materials of Knowledge or Thought we have or can have" (2.22.§9): simple ideas resist translation into words, or "The *Names of simple* Ideas *are not capable of any definitions*" (3.4.§4; emphasis Locke's). To illustrate the extralinguistic quiddity of simple sensation, Locke takes the case of light:

> Those who tell us, that *Light* is a great number of little Globules, striking briskly on the bottom of the Eye, speak more intelligibly than the Schools: but yet these Words never so well understood, would make the *Idea*, the Word *Light* stands for, no more known to a Man that understands it not before, than if one should tell him, that *Light* was nothing but a Company of little Tennis-balls, which Fairies all day long struck with Rackets against some Men's Fore-heads, whilst they passed by others. For granting this explication

> of the thing to be true; yet the *Idea* of the cause of *Light*, if we had it never so exact, would no more give us the *Idea* of *Light* it self, as it is such a particular perception in us, than the *Idea* of the Figure and Motion of a sharp piece of Steel, would give us the *Idea* of that Pain, which it is able to cause in us . . . And therefore should *Des Cartes*'s Globules strike never so long on the *retina* of a Man, who was blind by a *Guttâ Serenâ*, he would thereby never have any *Idea* of *Light*, or any thing approaching to it. (3.4.§10)

This condensed reenactment of particulate "explication" (Locke conflates Boyle's corpuscle and René Descartes's "globule") does not dispute its claim, in an echo of *Sceptical Chymist*'s anti-scholastic mandate, to "speak more intelligibly than the Schools." Locke elsewhere defends the superior explanatory salience of "Rays that rebound from any visible Area of any Body . . . enough to move the *Retina* sufficiently to cause a sensation in the Mind."[23] Rather, Locke denies the capacity of such explication to communicate feeling. The "particular perception" of light—which is, in Locke's terminology, equivalent to its simple "*Idea*"—cannot be approximated by even the most "exact" rendition of that sensation's particulate antecedents. Locke's familiarizing animation of "little Globules" foregrounds the absolute failure of analogy to engender the experience of light.

Locke's insistence on the gap between explication and idea compels him to spoof the analogical vividness of particulate microcausation. "Fairies" who employ "Rackets" to bat "a Company of little Tennis-balls" against the sensoria of blind men parody the mechanical intelligibility of body assumed to hold all the way down to its infinitesimal parts. Locke marshals this comically populated microscenario to assert its irrelevance to a life-sized viewer's sensation. Because they are inconvertible into the medium of sense, fairies brandishing rackets may as well "be true" as any other particulate schematics of the cause of the idea of light. Locke thus satirizes both the indefinite analogical reach of physical attributes minimized as still-legible figure and the epistemological stance that drives his own *Essay*: explanatory renditions of particles are irrelevant to units of human understanding which must be experienced to be known. Regardless of how intelligible or how vivid they may be, explanations of causes cannot elicit the particular perception that is Locke's simple idea of light. Locke's ludic tennis game thus anticipates appraisals of the *Essay* that affirm its embrace of the limitations summoned in Patrick Romanell's synopsis: "the only wise course to take under the circumstances is to make the best of a cognitively limited situation by concentrating and improving on what we *can* know *empirically* and is *useful* to us."[24] By collapsing billiard-ball atomism incarnated by "little Tennis-balls" and a ludicrous enactment of mechanism still inaccessible to human sense, Locke's fairies deny the epistemological use-value of causal discourse that cannot itself be felt.

Yet the fairy passage invites a more positive appraisal of Locke's relation to particulate causes. The *Essay* appropriates the corpuscular hypothesis to serve a crucial prophylactic end, because the divide between explication and simple ideas is secured by the anterior distinction of particulate physicality from sensible qualities that transpire only in us. That is, imperceptible texture insulates even gripping renditions of "Figure and Motion" from the sensations such corpuscles produce in persons. For Locke, the lack of mimetic consistency between particles and perceived ideas assures that corpuscular explanation cannot be converted into feeling. Indeed, as Anstey remarks, "the prospects of actually getting epistemic access to the corpuscular realm were extremely remote." But Locke's fairies gesture toward a prophylactic deployment of the break between ideas and particles not acknowledged by Anstey's conclusion: "Locke does not regard the corpuscular hypothesis or the mechanical philosophy as having any special heuristic value."[25] The problem for Locke is not blocked "epistemic access" but rather proneness to analogical figuration that makes corpuscles too much like everyday bodies. Thus analogized, primary particles transgress Locke's founding defense of simple ideas known only in the medium of sensation.

One indicator that analogical embodiment may be difficult for the *Essay* to contain is Locke's own recourse to figure: "And therefore should *Des Cartes*'s Globules strike never so long on the *retina* of a Man, who was blind by a *Guttâ Serenâ*, he would thereby never have any *Idea* of *Light*, or any thing approaching to it." The reader who imagines this submicroscopic scenario makes analogically vivid her own failure of sensory apprehension. Entities like "*Des Cartes*'s Globules" (for Locke as for Boyle, a red flag for speculative systematizing) represent discourse that can never stimulate persons who have not already experienced the sensory ideas for which words stand. More strongly, the "Man, who was blind" marks Locke's analog for all human percipients, whose imperviousness to linguistic impression renders them unmoved by words that elucidate the corpuscular antecedents of feeling. As testimony to the creeping instrumentality of figure, Locke exploits analogy to denounce the empirical redundancy of analogy.

What is figure? Strictly speaking, it is "the Termination of Extension" (2.13.§5) in two or three dimensions. However, as the potential for primary matter to be evocatively limned, figure renders corpuscular texture amenable to mechanical reduction, on the one hand, and vivifying illustration, on the other. In the first mode, a simple idea like color may be exhibited as something else. Locke proposes: "Had we Senses acute enough to discern the minute particles of Bodies . . . that which is now the yellow Colour of Gold, would then disappear, and instead of it we should see an admirable Texture of parts of a certain Size and Figure" (2.23.§11). Perceived by extrahuman "Senses," color is transformed into "Texture," "Size and Figure"—the freestanding physicality of the particle's core attributes. For Boyle, texture may be intuited from sensation, as when he ascribes niter's burning taste to particles that stab or tear. With Locke, Robert Hooke elaborates the corpuscular etiology of sensed qualities as also a sensory experience, but an alienating one. While Locke suggests that superhuman persons would "see" texture instead of gold, Hooke's *Micrographia* (1665) includes images like "Schem. VI" (Fig. 4), which accompanies his "search after the true cause of Colours."[26] The image represents a "Vessel holding a ting'd liquor," in which the letter *I* identifies a clear fluid while *H*, *G*, *F*, and *E* identify "determinate" "particles" of the "*tinging* substance" which would elicit in real life the impression of evenly diffused color. In this instance, Locke's argument for the break between mechanical cause and sensory idea is easily grasped: while Hooke's Figure 7 is indeed perceptible, it is rendered in black and white. Color's source—"*refraction*," or the power of the suspended particles to "more easily or more difficultly transmit the Rays of light"[27]—admits a visual schematics of corpuscular attributes that do not themselves stimulate color. Mimetically detached from the sensible idea whose cause it delineates, the visibility of Figure 7 is a strictly heuristic feature.[28] Like gold extrasensorially apprehended as texture, the very legibility of Hooke's blown-up particles enforces the break between mechanical causes and simple ideas.

Figure's second mode, analogical vivification, lacks the reductionist momentum of mechanical causation. (Although, indeed, Hooke's "*tinging* substance" could possess some extramechanical source of colorific influence.) In this register, Locke claims human percipience as matter's lowest limit, as when, writing of time and space, he proposes that "the least Portions of either of them, whereof we have clear and distinct *Ideas*, may perhaps be fittest . . . as the simple *Ideas* of that kind." According to this sensationalist standard of smallness, the "least" unit of space is "*a sensible Point*" (2.15.§9, emphasis

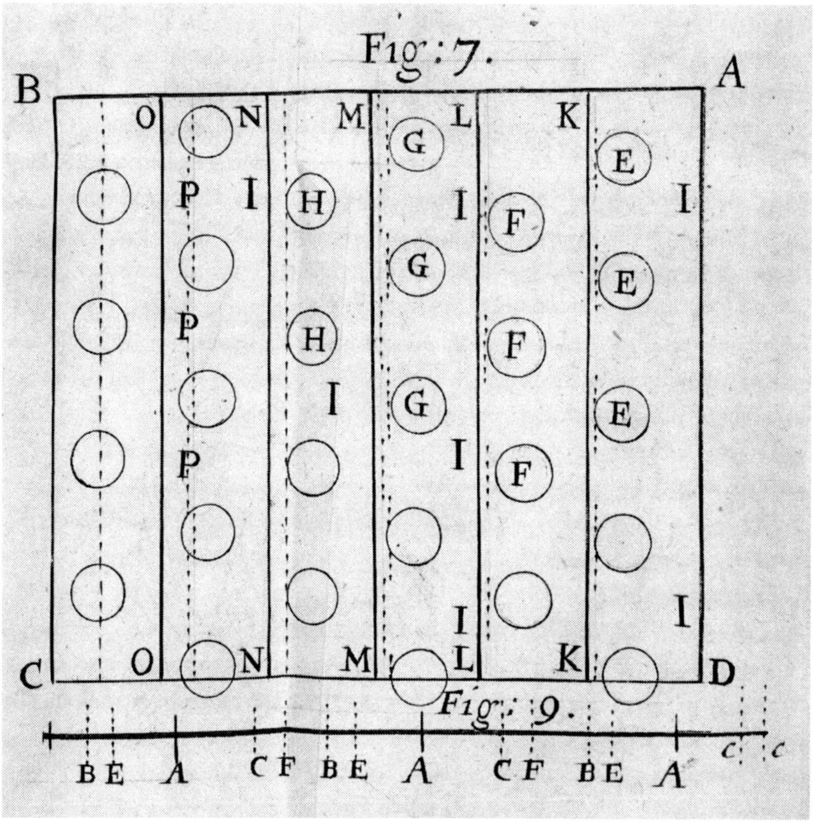

Figure 4. "Vessel holding a ting'd liquor," Robert Hooke, *Micrographia* (1665), Scheme VI (detail). Courtesy of the Lilly Library, Indiana University, Bloomington, Indiana.

Locke's). The calibration of simple ideas to "the least Particle of Matter . . . we can discern" (2.15.§9) arrests human understanding at its tiniest unit of "*sensible*" discernment. Like the flea in Comenius's *Orbis Sensualium Pictus* (were the insect visible), Locke's "*Point*" marks the smallest apprehensible entity in any human-scaled environment. The analogical continuity of bodies arrested at perception's least unit—from hogs to fleas to sensible points—affords a liaison between exceedingly small and larger objects the logic of whose rendition of the former is evident when Hooke writes of seeds of thyme: "The Grain affords a very pretty Object for the *Microscope*, namely, a Dish of Lemmons plac'd in a very little room."[29] The homely objecthood of Hooke's "Dish of

Lemmons" underscores the non-reductionist impetus of this likeness; by contrast with Hooke's tinged fluid, microscopic amplification of the seed yields Hooke's affirmation of its sensory likeness to bigger things.[30]

Jonathan Swift's *Gulliver's Travels* (1726) inverts Comenius's analogy when Gulliver, diminished to one-twelfth human scale, witnesses atop giant beggars "the Lice crawling on their Clothes. I could see distinctly the Limbs of these Vermin with my naked Eye, much better than those of an *European* Louse through a Microscope, and their Snouts with which they rooted like Swine."[31] Swift refers to a dramatic episode in *Micrographia*, whose "Obsv. LIV. *Of a Louse*" confronts the reader with an oversized image of the insect. While Hooke's illustration is magnified, Gulliver "see[s] distinctly . . . with my naked Eye": these "Lice" glean their likeness to "Swine" because the viewer shrinks. Swift's trick illuminates one liability of the familiarizing attribution of "Limbs" and "Snouts" to insensible insects. Whereas corpuscular blueprints like Hooke's tinged fluid alienate tiny bodies from perceived effects, the mimetic embodiment of infinitesimal entities all the way down may come at the expense of explaining anything.

Lice that "rooted like swine" foreground the constitutive exception to corpuscular philosophy's anti-mimetic refusal to entrench sensed qualities in objects. This is the fact that mechanical affections are themselves analogically derived from sensible bodies. Of Locke's primary qualities, the texture Locke's superhuman observer of gold would encounter, Lisa Downing observes: "[Their] sensory criterion highlights the fact that corpuscularian concepts—size, shape, motion, solidity, situation, etc.—are clear, empirical concepts, drawn straight from our sensory experience of bodies."[32] As Downing suggests, while sensible qualities like color do not inhere as such in corpuscles, "ordinary sensory experience"[33] nonetheless funds Boyle's and Locke's insistence on the inextirpable physicality of the particle's core features.[34] The materialist impetus of Boyle's mechanical affections and Locke's primary qualities must be analogically sustained. Rooting lice amplify matter's consistency: instead of an alien world where gold turns into texture, miniature Gulliver's environment looks, feels, and smells just the same as before—only bigger. In Brobdingnag, Swift converts the corpuscle back into the same body that, perhaps, it already was.

Figure's two modes, mechanical reduction and analogy, are thus not entirely opposed. In the crucial instance of primary qualities, mechanical affections import their quotidian intelligibility from everyday bodies. Such borrowing propels the corpuscular explanations delineated by Boyle's French contemporary Nicolas Lémery, who offers refinements of particulate shape to become chemistry's foremost innovator of the agency of figure.[35] Lémery's influential textbook *Cours de chimie* (1675) limns the medicinal properties of "l'or fulminant" (potable gold), a boiling solution which, ingested by mouth, induces perspiration:

> I think that it is [more] truthful to say that the vapors lift with them some parts of gold with which they are intimately mixed, and by this we understand better how such a small quantity of spirits is capable of provoking sweat; because supposing that it lifts through the pores one grain of gold and two grains of spirits, these spirits being, so to speak, armed with bigger parts of gold, will be much more capable of forcing the resistances that oppose their passage, than if they were alone; in the same manner that a large piece of wood being carried by the current of a river, will strike with much more violence against the hull of a boat, and will rock it more than a wave would by itself, no matter how rapid.[36]

With gold particles that resemble wave-born flotsam, figure displaces medical causation that derives healing potency from metallic essence. By replacing therapeutic substance with the analogical familiarity of boats and rapids, Lémery's "l'or fulminant" shows that the causal elaboration of figure can have it both ways. Lémery imports explanatory self-evidence from the quotidian experience of body while nonetheless claiming a mechanical rationale for medical effects. "L'or fulminant" does not abdicate the intelligibility of mechanical explication, but it grants particles the sensory immediacy parodied by Swift's rooting lice.

Before returning to Locke's equivocal treatment of corpuscular figure, we must examine his revision of Boyle, because the *Essay* elaborates a terminological and philosophical departure from Boyle's doctrine of qualities. Most crucially, Locke does not rigorously affirm Boyle's distinction between mechanical affections or texture, which compose a thing's insensible but real physicality, and sensible qualities, which texture effects as ideas "in us." While Boyle divides mechanical affections in things from sensible qualities in persons, the *Essay* differentiates primary qualities—the same mechanical attributes, plus texture, invoked by Boyle—from "*secondary Qualities*" that *also* reside within the object: "Such *Qualities*, which in truth are nothing in the Objects themselves, but Powers to produce various Sensations in us by their *primary Qualities*" (2.8.§10). For Locke, the distinction between "*primary*" and "*secondary*" recurs to the *same* particles, only differently considered: primary qualities are mechanical affections and texture, while secondary qualities reflect the "Powers" of those particles to move human sense.

By this logic, Lockean ideas seem to be locatable in persons. But throughout the *Essay*, ideas and secondary qualities migrate back and forth between perceivers and objects, as Locke warns of "*Ideas*, [which] if I speak of sometimes, as in the things themselves, I would be understood to mean those Qualities in the Objects which produce them in us" (2.8.§8). In the *Essay*, sensed "*Ideas*" can inhabit "things themselves," but what Locke stipulates the former word means is matter's retroactively specified power to trigger sensation "in us." Locke's secondary qualities do not often refer to sensations like taste, but rather the *power* of texture to produce them: because light and heat, for example, are nothing "but powers to excite such *Ideas* in us, I must, in that sense, be understood, when I speak of secondary *Qualities*, as being in Things; or of their *Ideas*, as being in the Objects, that excite them in us" (2.31.§2). Remarkably, Locke justifies the reversal of his terms: "secondary *Qualities*" may be "in Things" or, he writes, the "*Ideas*" things stimulate in persons may be "in the Objects."

Secondary qualities confuse "powers to excite" and perceived ideas: the latter exist "in us," but they also signify the productive agency of the physical object. When referred to as ideas, however, secondary qualities reflect the polemical thrust of Boyle's selective deployment of classical materialism. From the vantage of sense, matter's power to move persons is, Locke affirms, "nothing in the Objects" other than corpuscular texture. Secondary qualities exist in matter by virtue of human ideas, while they exist in humans by virtue of physical texture that does not harbor ideas as such. In response to enduring critical controversy, I affirm that Locke's secondary qualities are effectively in things and in persons at once.[37] This is also to argue for the non-Platonic genealogy of the mind-body relation propelled by chemical science and routed through Locke's empiricism. Neither strictly idea nor strictly body, secondary qualities open insensible matter to human figuration of its power to produce empirical understanding.

Secondary qualities do more than define texture as the power to stimulate sense. Locke also essays an unprecedented refinement on the side of perception. Unlike Boyle, Locke posits a category that can be called primary ideas, which *do* faithfully reflect an object's mechanical attributes. (As Chapter 1 intimates, such ideas may be heralded by macroscopic shards of glass that illustrate the relational agency of corpuscular figure.) While "Bulk, Figure, Texture, and Motion of parts" (2.8.§14) cannot be singly apprehended, for

Locke these coordinates can nonetheless be transposed into some perceptible scale: "the *Ideas of primary Qualities* of Bodies, *are Resemblances* of them, and their Patterns do really exist in the Bodies themselves; but the *Ideas, produced* in us *by* these *Secondary Qualities, have no resemblance* of them at all" (2.8.§15). The distinction Locke claims for primary qualities, unlike secondary powers, is the mimetic "*resemblance*" of body and idea. Locke's primary-secondary distinction can, again, be only retroactively applied, because "these *Secondary Qualities*" refer to the same particles that provoke "*Ideas of primary Qualities.*" For Locke, primary-secondary difference classifies ideas retroactively and extrinsically to distinguish modalities of the same materiality that produces them: primary ideas are "*Resemblances* of . . . Patterns" as opposed to secondary ideas that bear "*no resemblance*" to texture's power. The most Cartesian turn in the *Essay*, primary ideas exploit the privileged lucidity of geometrical "Patterns" to claim resemblance not championed by Boyle, for whom human sense is estranged from particulate units.[38] For Boyle, the relation between texture and sense is fundamentally productive, not mimetic. Even in the case of things like ground glass, for Boyle what humans perceive *are* ideas, even though ideas are produced by bodies.

For Boyle, the perdurable identity of metal that can be recovered after it dissolves locates corpuscles at the confluence of ocular disappearance and experimental sense. Because it can *and* cannot be perceived, silver precipitated back from an acid solution proves that the metal is made of insensibly tiny parts. With Locke's primary ideas, the relay between particles and perceptions occurs with the resemblance of sensory input to patterns. Yet primary ideas fail to deliver empirical access to a mathematical or Cartesian realization of body, because primary ideas drop out of the *Essay* almost entirely. Locke may try to supplement analogy with ideas that transmit geometrical truth, but no patterns surface in the *Essay* to obviate the instrumentality of figure.

No Connexion: Figure and Fiction

When, in a chapter entitled "*Of the Degrees of Assent,*" Locke examines probability, he turns to "*Things . . . falling not under the reach of our Senses.*" In a more sanguine appraisal of particulate agency than his tennis-playing fairies allow, he affirms the likeness of submicroscopic causation to macroscopic "Truths" already vindicated by "Observation" (4.16.§12):

> *Analogy* in these matters is the only help we have, and 'tis from that alone we draw all our grounds of Probability. Thus observing that the bare rubbing of two Bodies violently one upon another, produces heat, and very often fire it self, we have reason to think, that what we call Heat and Fire, consists in a violent agitation of the imperceptible minute parts of the burning matter: Observing likewise that the different refractions of pellucid Bodies produce in our Eyes the different appearances of several Colours; and also that the different ranging and laying the superficial parts of several Bodies, as of Velvet, watered Silk, *etc.* does the like, we think it probable that the Colour and shining of Bodies, is in them nothing but the different Arangement [*sic*] and Refraction of their minute and insensible parts. (4.16.§12)

Locke articulates "Probability" as the prospective consistency of causes—like "violent agitation"—incarnated as "burning matter" and the ideational correlates they engender in persons. He flags the relational contingency of the latter as "what we call Heat and Fire" and color "produce[d] in our Eyes," while primary qualities expressed as secondary powers refer to "imperceptible . . . matter" and "insensible parts." Here the *Essay*'s endorsement of the analogical reach of "Bodies" seems unequivocal: sticks, prisms, or fabric "produce" heat, colors, or "shining" to model a causal link extrapolated all the way back to "imperceptible minute parts" whose generative texture cannot be perceived.

To account for color, Locke draws from Boyle's *Experiments and Considerations Touching Colours* (1664), a text he perused while studying with the author at Oxford. *Colours* defends the so-called modificationist argument that color is effected, as Locke writes, by "superficial parts" that disturb incident white light. To elaborate this hypothesis, Boyle asks that his reader

> conceive the Physical superficies of a Body, where . . . its Colour does as it were reside, to be cut Traversly by a Mathematical plain . . . and then as some parts of the Physical superficies will be Protuberant, or swell above this last plain, so others may be depress'd beneath it, as (to explane my self by a gross Comparison) in divers places of the Surface of the Earth, there are not only Neighbouring Hills, Trees, &c. that are rais'd above the Horizontal Level of the Valley, but Rivers, Wells, Pits, and other Cavities that are depress'd beneath it; and that such Protuberant and

Concave parts of a Surface may remit the Light so differingly, as much to vary a Colour.[39]

Even more emphatically than Lémery, Boyle resorts to extracorpuscular experience. To demonstrate how "Protuberant and Concave" superficies effect color, Boyle's scenery almost displaces the intuition of particulate texture, threatening instead, like Brobdingnag, to exchange corpuscular schematics for a habitable blowup of the microworld. Another "gross Comparison," Boyle's commonplace analog for Locke's second example, velvet, appeals to an even more everyday likeness:

> [I]f Differing parts of the same piece of Black Velvet be stroak'd Opposite ways, the piece of Velvet will appear of two Distinct kinds of Blackness . . . of which Disparity the Reason seems to be, that in the Less obscure part of the Velvet, the Little Silken Piles whereof 'tis made up, being Inclin'd, there is a Greater part of each of them Obverted to the Eye, whereas in the other part the Piles of Silk being more Erected, there are far Fewer Beams Reflected Outwards from the Lateral parts of each Pile . . . I have near Great Towns often taken notice, that a Cart-load of Carrots pack'd up, appear'd of a much Darker Colour when Look'd upon, where the Points of the Carrots were Obverted to the Eye, than where the Sides of them were so.[40]

The unexpected congruity of a "Cart-load of Carrots" and "Little Silken Piles" renders obtrusive the sensory immediacy lent velvet by vegetables. But by contrast with Cartesian fairies, carrots instantiate a zero degree of humble mundaneness. One return on *Colours*'s prefatory promise "to deliver things rather *Historical* than *Dogmatical*,"[41] carrots locate history's experiential base not in Swiftian satire, which touts travel narrative as Gulliver's encounter with the distortions unleashed by analogy, but rather in a motivation of color as anybody's understanding of "Erected" or "Obverted" carrots. By deploying strenuously humdrum sensation as "the delivery of matters of fact,"[42] Boyle downplays the intelligibility of geometric schema sustained by obverted "Points" or "Sides."

As Locke's own illustrations of fire and velvet show, the *Essay* does not consistently deny the capacity of "gross Comparison" to bridge sensible qualities like heat and corpuscular causes like "bare rubbing." But the *Essay* rejects other renditions of microcorpuscular causality that seem just as empirically routine as lateral stacks of carrots. One striking example is ice. While insisting that inanimate objects, or "corporeal Agents," act and are acted upon by "nothing else but Modifications of Motion," Locke bemoans the imprecision of verbs that fail to specify the precise kind of movement that produces any given effect: "And therefore *many words, which seem to express some Action*, signify nothing of the Action, or *Modus Operandi* at all, *but* barely *the effect*, with some circumstances of the Subject [object] wrought on, or Cause operating . . . And when a Country-man says, the Cold freezes Water, though the word Freezing seems to import some *Action*, yet truly it signifies nothing, but the effect, *viz.* that Water, that was before fluid, is become hard and consistent, without containing any *Idea* of the Action whereby it is done" (2.22.§11). Yet of all the verbs Locke could have chosen, it is difficult to find one that more deeply embeds the relation of sensible "effect" and corpuscular "Action" than "freeze." While what Locke demarcates as "barely *the effect*"—the conversion of water into ice—marks the dominant definition of this verb, "freeze" also possesses transitive meanings, dating from the fourteenth century, which include "to congeal," "to paralyze," and "to stiffen, harden."[43] A contemporary activation of the latter semantic range is found in John Milton's *Comus* (1634), for example, which evokes the power of feminine chastity to induce stasis: "What was that snaky-headed *Gorgon* shield / That wise *Minerva* wore, unconquer'd Virgin, / Wherewith she freez'd her foes to congeal'd stone."[44] Milton's enjambment of local cognates for freezing contravenes Locke's insistence that the verb communicates "nothing of the Action" whereby liquids might become fixed. Even the most obviously metaphorical resonance of "freez'd," to be paralyzed, justifies how fluidity, a sensible effect Boyle and Locke attribute to insensibly fast corpuscles that displace and replace each other, would cease.

As Boyle argues when he proposes that "*Fixity* may be *Mechanically* produced," one "means of fixing, or lessening the Volatility of, bodies, is by preserving that rest among the parts . . . And this may be done by preventing or checking that Heat, or other motion, which external Agents strive to introduce into the parts of the proposed body."[45] A signifier of water's conversion into ice as well as a homologue for congelation or paralysis, freezing figures the rationale for its own sensible effect: "rest among the parts." As the verb that would represent grammar's failure to isolate the "*Modus Operandi*" whereby particles produce empirical sensation, Locke chooses a word which—at least, since fourteenth-century English—conflates the primary cause "rest" and the secondary quality frozenness. This marks a resonantly infelicitous choice (not least because the *Essay*, as we have seen, ascribes heat to the "violent agitation" of particles). Locke reaffirms his ignorance of the mechanics of freezing when, writing of "Particles of *Water*," he denies human cognizance of the agency that makes its units "cohere" into ice: "He that could find the Bonds, that tie these heaps of loose little Bodies together so firmly; he that could make known the Cement, that makes them stick so fast one to another, would discover a great, and yet unknown Secret" (2.23.§26). By invoking the placeholder "Cement" as a cover for the failure of both analogical and mechanical explanation—since agents like cement make things "stick so fast" only by obfuscating the subcausality that unites their smaller parts—Locke rejects the solution proposed by another text he owned, Boyle's *Certain Physiological Essays* (1669), whose "History of Fluidity and Firmnesse" broaches the same analogical dilemma to suggest: "if the Corpuscles that make up the finest Glue imaginable are not kept together by a Cement, we may be allow'd to ascribe their Adhesion to the immediate Contact and Rest of the component parts, (which is a cause intelligible, and at least probable)."[46] In fact, Locke revises the *Essay* to affirm the legitimacy of Newtonian attraction.[47] The persistent problem is not the failure of the verb "freeze" to offer an adequate refinement of motion, but the analogical reach of human ideas of body all the way to primary matter.

Writing of Locke's uneven affirmation of the validity of corpuscular causes, Margaret Wilson observes: "Locke's talk of secondary qualities flowing from primary ones cannot be freed from the implication that a body's sensible appearances flow from its real essence [mechanical affections and texture]. But this implication is evidently contradicted by his claims about the arbitrary relation of sensation to the operations of matter."[48] Locke's treatment of freezing permits a differently focused statement of this contradiction. Locke denies the probability of Boyle's "immediate Contact and Rest of the component parts" even though its mechanical intelligibility is connoted by the very verb he selects to illustrate grammar's insulation of corpuscular causes from sensational effects. While Boyle invokes particulate rest to debunk some mystified intracorpuscular "Cement" (although, as we have seen in Chapter 1, he and Newton admit subcorpuscular stickiness as a primary particulate attribute), Locke insists that the cause of water's transformation into ice remains an "unknown Secret." To enforce his point that microcausation is "incomprehensible" (2.23.§26), Locke denies figure to the verb "freeze" only to reintroduce it in the guise of the aggravated analogical impotence of cement. His insistence that freezing refers to effects but not causes—an insistence that represses the extant semantic range of the word—ratifies the *Essay*'s most extreme denial of the plausibility of any link between primary and secondary qualities: "our Understandings can discover no conceivable connexion between any *secondary Quality*, and any modification whatsoever of any of the *primary* ones" (4.6.§7).[49] And yet Locke relies on analogy to bolster the causal logic empirically evinced by fabrics, sticks, and texture. His ambivalent treatment of the "connexion" between primary and secondary qualities signals unease with a peculiarly stubborn analogical tautology: sensed ideas like velvet's blackness lend empirical vividness, circularly, to the corpuscular mechanics of their own imperceptible causes.

Locke's best-known expression of ambivalence is found in a scenario that complements the fairy passage. Whereas fairies parody the sensory inutility of microexplanation modeled by Cartesian tennis balls, in the following episode Locke projects the variant irrelevance that would result if corpuscles shed the assistance of analogy:

> Nay, if that most instructive of our Senses, Seeing, were in any Man 1000, or 100000 times more acute than it is now by the best Microscope . . . so he would come nearer the Discovery of the Texture and Motion of the minute Parts of corporeal things; and in many of them, probably get *Ideas* of their internal Constitutions: But then he would be in a quite different World from other People: Nothing would appear the same to him, and others: The visible *Ideas* of every thing would be different . . . And if by the help of such Microscopical Eyes, (if I may so call them,) a Man could penetrate farther than ordinary into the secret Composition, and radical Texture of Bodies, he would not make any great advantage by the change . . . He that was sharp-sighted enough to see the Configuration of the minute Particles of the Spring of a Clock . . . would no doubt discover something very admirable: But if Eyes so framed, could not view at once the Hand, and the Characters of the Hour-plate, and thereby at a distance see what a-Clock it was, their Owner could not be much benefited by that acuteness; which, whilst it discovered the secret contrivance of the Parts of the Machin, made him lose its use. (2.23.§12)

Locke's "Man" enters a world of primary qualities that seems not to resemble Brobdingnag. There, Swift spoofs the sameness of the bodies surrounding

Gulliver: their secondary qualities do not dissolve into texture but instead become, in an echo of Boyle's gross comparison, much, much grosser. Unlike Gulliver, Locke's acutely sighted man inhabits an environment that offers totally alien sensory input. Occupying a "quite different World," this man's "visible *Ideas* . . . would be different" from those of "others." But the insoluble rupture between "the radical Texture of Bodies" and "*Ideas*" of objects does not carry through Locke's scenario. In the end, the gravest irrelevance Locke claims for "Microscopical Eyes" is not the anti-analogical break propelled by incomprehensible texture whose ideas cannot be reconciled to normal understanding, but the failure of his microscopically sighted man to read the "Hand" and "Characters" of a clock. The ensuing redundancy stems not from nonsensical sensory input but rather from the man's failure to tell "what a-Clock it was." Indeed, Locke affirms that the man is not empirically at a loss: he "discover[s] the secret contrivance of the Parts of the Machin." Because mechanical "Structure and Impulse" renders texture still intelligible, Locke stages epistemological irrelevance that fails to divorce micromatter from the "visible *Ideas*" to which ordinary men are accustomed. The problem is not the radical inscrutability of primary qualities that throw the viewer into a wholly foreign perceptual world but the difficulty facing any viewer who tries to occupy two levels of ocular resolution at once. Yet whether trained on clocks or corpuscles, both levels of focus disclose familiar things. Locke's micro-eyes refute analogy, it turns out, by empirically ratifying it.

Locke animates a problem of focus, not a radical break in the intelligibility of things. The perceptual problem may thus be resolved, in a move Locke proceeds to make, with the "extravagant conjecture" (2.23.§13) of superhuman beings whose eyes traverse an entire range of resolution at once. The microscopically eyed man's impasse is corrected not by familiarizing matter but by rendering texture and characters simultaneously legible. To address longstanding debate over the phenomenal intelligibility of Locke's primary matter, should humans acquire the sensory capacity to access it, I conclude that Locke's primary qualities are no less empirically intelligible, in principle, than the face of a clock. His primary particles *are* possessed of the mechanical endowments that enable the microscopically eyed man to see texture as a machine.[50] In the same scene that would disprove it, Locke's acutely seeing man vindicates the analogical move that authorizes the intelligibility of micromatter in early modern corpuscular science.

Locke's stipulated isolation of sensation from microcausal discourse—"simple *Ideas* cannot be described, nor their Names defined" (2.20.§1)—is contravened by the assimilation of particulate texture to the model of perceptual intelligibility provided by clocks: reading. Reading heralds the denial at the core of the *Essay*, the denial of sensory impact to the medium of language itself. Locke reaffirms the solely sensory derivation of simple ideas or "the Materials of all our Knowledge" (2.2.§2): "The simple *Ideas* we have are such, as experience teaches them us; but if beyond that, we endeavour, by Words, to make them clearer in the Mind, we shall succeed no better, than if we went about to clear up the Darkness of a blind Man's mind, by talking; and to discourse into him the *Ideas* of Light and Colours" (2.4.§6). "Words" cannot transmit simple sensation; the guarantor of their sensible nullity is again a "blind Man[]" who registers the sound of "talking" but can never experience speech as "*Ideas*." In this compressed iteration of the tennis-ball scenario, "talking" amounts to futile bombardment of the unseeing man's impervious sensorium, while the "failure to discourse [sight] into him" assures Locke's denial of autonomous phenomenal impact to words. Locke's divorce of sensory ideas from language restricts the enthusiasm, most topically after the Civil War, of a "*Man inspired by GOD*," because "this Revelation, if it be of new simple *Ideas*, cannot be conveyed to another, either by Words, or by any other signs" (4.18.§3). Whether "new simple *Ideas*" transmit fanaticism or other factional feeling, Locke's denial of their capacity to be shared proceeds from his insulation of corpuscular explanation from the indefinable quiddity of secondary effects.[51] In this prophylactic sense, Locke's *Essay* is deeply Boylean. Words that figure the corpuscular causes of sensible qualities cannot discourse the experience of those ideas into persons.

But as I have argued, the very passages in which Locke restricts body's analogical likeness implicate the *Essay* in figural "conjecture." Such conjecture cuts two ways, as the reciprocal intelligibility of numerical characters and the particles perceived by the microscopically eyed man suggests. Despite the model afforded by Locke's clock, which levels seeing and reading, the *Essay* offers no dedicated treatment of the latter practice. Locke most tellingly frames reading as a particular mode of perceptual susceptibility when he restricts the intake of simple ideas to one sensory orifice at a time. The sound of a word, which is yoked to its referent not naturally but arbitrarily, cannot stimulate simple ideas effected by organs other than the ear:

> [T]he signification of Sounds, is not natural, but only imposed and arbitrary. And no definition of *Light*, or *Redness*, is more fitted, or able to produce either of these *Ideas* in us, than the sound *Light*, or

> *Red*, by it self. For to hope to produce an *Idea* of Light, or Colour, by a Sound, however formed, is to expect that Sounds should be visible, or Colours audible . . . Which is all one as to say, that we might Taste, Smell, and See by the Ears: a sort of Philosophy worthy only of *Sanco Panca*, who had the Faculty to see *Dulcinea* by Hearsay. And therefore he that has not before received into his Mind, by the proper Inlet, the simple *Idea* which any Word stands for, can never come to know the signification of that Word . . . A studious blind Man, who had mightily beat his Head about visible Objects, and made use of the explication of his Books and Friends, to understand those names of Light, and Colours, which often came in his way; bragg'd one day, That he now understood what *Scarlet* signified. Upon which his Friend demanding, what *Scarlet* was? the blind Man answered, It was like the Sound of a Trumpet. (3.4.§11)

To enforce the break between texture and its sensible effects, Locke superimposes upon primary-secondary difference a much more hostile articulation of the linkage between word and idea: the relation of signified to signifier. The "imposed and arbitrary" relation of words and referents negatively mimics the liaison between ideas and stimulating texture. What isolates words from sensory ideas that language cannot "produce" is, for Locke, the impossibility of synesthesia. Locke iterates primary-secondary difference as a breach between signifier and signified reinforced by the antipathy of sound (spoken words) and sight (ocular experience). Because no amount of verbiage renders "audible" the experience of color, Locke enforces the breach between primary and secondary qualities as the irreparable discrepancy of sound and sight.

It is striking that Locke's two avatars of synesthetic susceptibility receive their impressions from "Books." While it would be "Hearsay," not reading, that makes "*Sanco Panca*" "see," both he and Locke's studious "Blind Man"—who, unlike his famous counterpart the Molyneux man, refuses to restrict himself to ideas stimulated only by "the proper Inlet"—are incited to sensory confusion by discourse that promises to communicate feeling. (In this, the illiterate Sancho Panza substitutes for Don Quixote.) The bibliographic spur to the two men's synesthetic projections illuminates one motive for Locke's dictum: "*simple* Ideas *are not fictions* of our Fancies, but the natural and regular productions of Things without us, really operating upon us" (4.4.§4). By straddling sound and sight, words and ideas, fiction and the "regular productions of Things without us, really operating upon us," Sancho Panza and the studious man collapse synesthesia and the purveyance of "*fictions*." As perpetrated by these consumers of words, fiction transgresses the primary-secondary divide. Discursive "explication" of sensation's primary causes blurs into these men's experience of secondary qualities. For Locke, fiction is the improper sensation of incommunicable materiality.

The "fictitious" implicates the link between cause and effect because analogy facilitates what is, for Locke, the incorrigible error of the blind man's claim to feel scarlet: "[*Scarlet*] was *like* the Sound of a Trumpet." Analogy transgresses the sensory discipline exemplified by Locke's Molyneux man, who, born blind and abruptly restored to sight, cannot visually identify a three-dimensional globe and a cube based on his prior experience touching them. Martha Brandt Bolton points out that Locke's negative answer to the question of whether new visual ideas can, without tactile mediation, be reconciled to earlier haptic input curtails the newly sighted man's "perception of *figure*." As Bolton writes, "Locke's commitment to a radically atomistic theory of sense experience"[52] precludes the transfer of sensory knowledge from the medium of touch to that of sight even though, Bolton suggests, "[g]iven that both ideas resemble the same cause, it would be difficult for Locke to deny that their contents have significant overlap." She concludes: "[Locke] may well have thought that a man just made to see would not receive any ideas of the figures of bodies he looked at."[53] It is worth noting, with Bolton, the ensuing oddity of the attribute of shape or figure: although Locke includes it among the primary attributes of particles, figure does not proffer sensory input that would engender Locke's primary or resembling idea. Despite the *Essay*'s initial call for a category of ideas that mirror geometrical features of objects, Locke excludes shape from immediate visual understanding. As Bolton concludes, figure emerges as neither a primary nor simple idea because its ocular apprehension must be facilitated by touch: "visual ideas of figure may require further operations of mind."[54] Because its correlation with touch cannot be established by sight alone, figure enforces the impossibility of synesthesia. By the same logic, figure annuls the promise of mimetically apprehended particulate attributes afforded by Locke's primary ideas.

The Molyneux man respects the radical specificity of ideas received from each anatomical inlet, even when they replicate the shape of objects he has already felt. The scarlet-hearing man improperly collapses sight and sound. But as the opposed proclivities that structure the agenda of Locke's *Essay*, their disparity founders over the attribute of figure. The auditor who likens scarlet and trumpets may illustrate the preposterousness of analogy, but the Molyneux proof precludes immediate ocular understanding of what Locke claims as his clearest simple idea, pattern. By animating Locke's refusal of analogy, his Molyneux man denies primary matter the geometrical lucidity promised by figures like cubes, the same lucidity Locke grants the corpuscular engine perceived by his microscopically sighted observer.

For Locke, the collapse of color and sound consummates the anatomical transgression provoked by analogy. (The tethering of ideas to anatomical inlets implicates pornography as a denial of Locke's prohibition of synesthetic knowledge.)[55] But as we have seen with the likeness of superficial parts and velvet piles, synesthesia sustains corpuscular explanation by ascribing endowments of body to imperceptible parts. In Locke's *Essay*, analogy marks an insoluble error at the same time that it limns the insensible endowments that entail the capacity of secondary qualities to be empirically known. We now turn from the sensory impact of fiction that usurps matter's productive power to sensed attributes of bigger things.

Extrinsical Essence: Figure, Color, Denomination

Because words alone make Don Quixote feel, he violates Locke's sensory derivation of the origin of ideas: "And so each Sensation answering the Power, that operates on any of our Senses, the *Idea* so produced, is a real *Idea*, (and not a fiction of the Mind, which has no power to produce any simple *Idea*;)" (2.31.§2). This formulation distills Locke's prophylactic deployment of corpuscular science: "a real *Idea*"—a simple idea that gives the mind "the ultimate Materials of all its Compositions" (2.12.§2)—must be "produced" by insensible texture in concert with "any of our Senses." When Locke cites "the Power" of this object, he refers to secondary qualities whose core physicality guarantees that the ideas they stimulate are not "fiction."

Yet Locke's fiction is paradoxically delimited by Boyle's science. An experimenter who shared alchemical recipes with Boyle, Locke would have known processes like sublimation, distillation, calcination, and the *reductio* proof, which conjure reactants back from imperceptibility to show they are composed of minute particles. It is not corpuscles per se but the conferral of sensationally known endowments to insensible parts which, for Locke, poses the threat of synesthesia, the communication of empirical ideas received at the proper anatomical inlet to things—like words—that cannot engender sense at all. But Locke's denial of analogy cuts into the reality of the primary matter that guarantees the truth of his "real *Idea*." The Molyneux man plays out an anti-analogical prohibition so stringent that it precludes the intelligibility of pattern: despite his haptic knowledge of cubes and spheres, Locke refuses him immediate ocular comprehension of geometrical shape. In the *Essay*, figure is an endowment liable to the charge of synesthesia at the same time that, like all of Locke's primary qualities, figure underwrites the truth of simple ideas. Analogy propels the impropriety of fiction, but analogy is not so easy to extricate from the primary physical endowments whose "power to produce any simple *Idea*" assures the integrity of the Lockean real.

In this section, I turn from Locke's ambivalent treatment of corpuscles to human knowledge of macroscopic things. If ideas received at the correct inlet obviate the fictionalizing danger of analogy, does perception stably demarcate empirical understanding? Going further, Locke will ask whether sensible qualities provide the basis for ontology or, as he calls it, essence. As we will see, the ontology that accompanies empirical knowledge does not, contrary to Shapin and Schaffer or Barad, insulate modest spectators from passive, stably qualified objects. On the contrary, empirical perception fails to uphold a transparently realist ontology. This failure does not jeopardize the collectively felt consistency of simple ideas. Rather, it implicates the clusters of impressions that comprise the entities Locke designates complex ideas, which admit variation not at the level of human sense but at the level of human experience. Secondary qualities produce empirical knowledge at the same time that they preclude the ascription of static essence to all complexly encountered things.

The complex idea that most preoccupies Locke's *Essay* is, as he writes, man. Comenius's *Orbis Sensualium Pictus* helps anticipate how Locke queries the category man to precipitate insoluble taxonomic effects. After the entry "*Homo.* / Man.," *Orbis Sensualium Pictus* offers a sequence of terms which proceed in the following order: "*Septem Ætates Hominis.* / The seven Ages of Man."; "*Membra hominis Externa.* / The Outward parts of a Man." / "*Caput & Manus.* / The Head and the Hand." / "The Flesh and the Bowels. / *Caro & Viscero.*"; "The Chanels [*sic*] & Bones. / *Canales & ossa.*"; "The outward and Inward Senses. / *Sensus externi & interni*"; and "The Soul of man. / *Anima hominis.*"[56] This sequence of definitions and images offers the child reader a heuristic breakdown that moves roughly from the outside in. Beginning with biblical creation and man's fall from paradise ("Man") and the progression from infancy to old age ("seven Ages of Man"), the plates proceed from external appendages to skin, bones, veins, the five sense organs, and finally the brain ("outward and Inward Senses"), where we find the "Common-Sense [7.] /

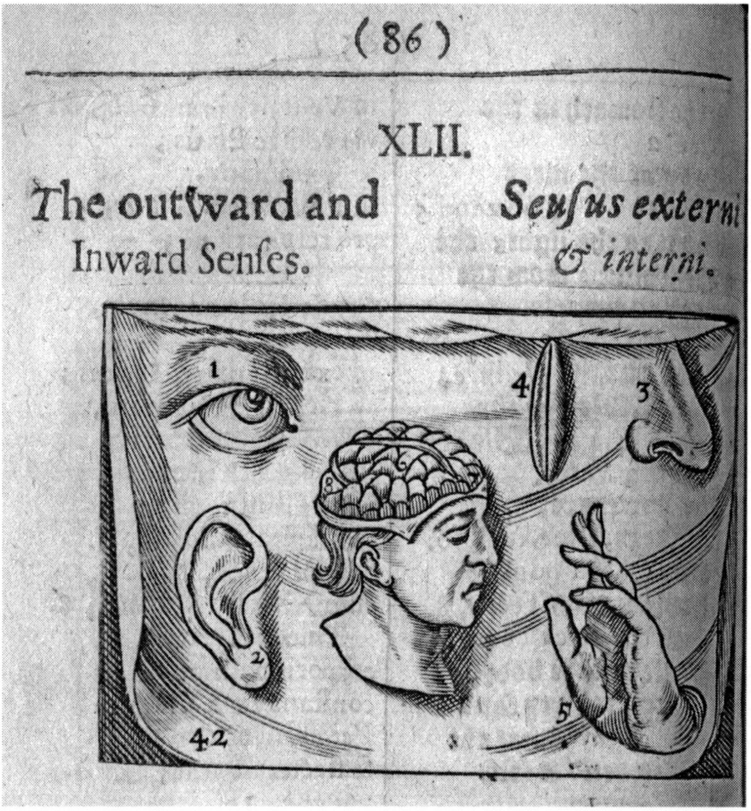

Figure 5. "The outward and Inward Senses / *Seusus* [sic] *externi & interni*," John Amos Comenius, *Orbis Sensualium Pictus*, trans. Charles Hoole (1685). Photo courtesy of the Newberry Library, Chicago. Wing ZP 645.M46.

under the fore-part of the head," the "Phantasie [6.] / under the crown of the head," and the "Memory [8.] / under the hinder part of the head" (Fig. 5).[57] Because the numbers six through eight sit atop a decapitated head, whose flayed skin exposes the outer surface of the brain, even Comenius's most interiorized image of the human person correlates mental function and topography. Although the operations of "Common-Sense," "Phantasie," and "Memory" are located "under" the head's front, center, and rear, Comenius's numerical indicators illuminate the persistently conspicuous breakdown of the operations just beneath this representative skull. The sites of recombination, imagination, and

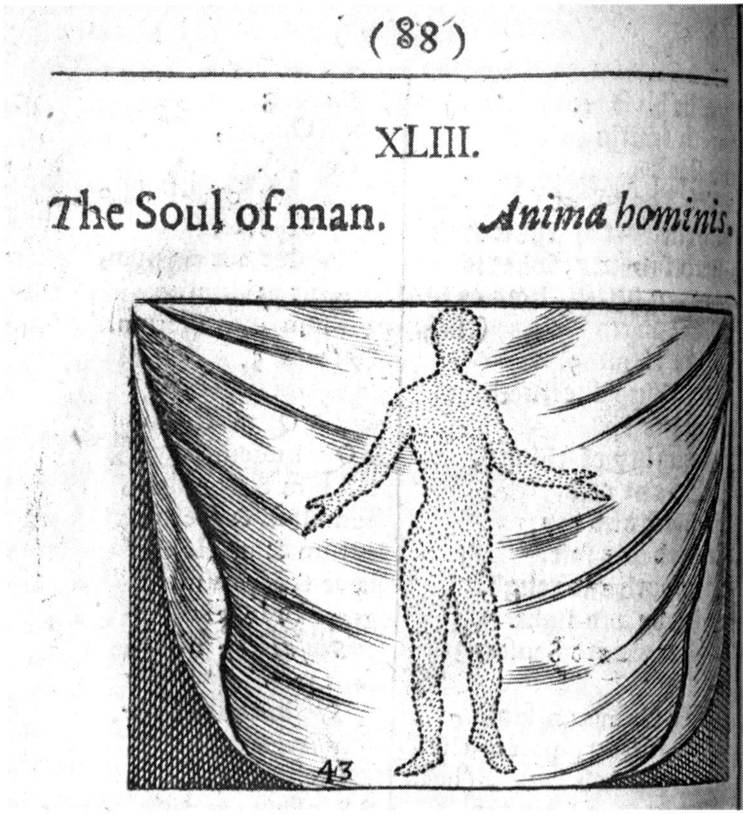

Figure 6. "The Soul of man / *Anima hominis*," John Amos Comenius, *Orbis Sensualium Pictus*, trans. Charles Hoole (1685). Photo courtesy of the Newberry Library, Chicago. Wing ZP 645.M46.

retention of ideas, which are both concrete and superficial, render the epistemological standards of sensual learning one with the organization that realizes them. Sensual language acquisition is performed by a brain whose own most abstruse functions are visible. (The imminent perceptibility of Comenius's perceiving brain may affirm Locke's denial, found in "Anatomia" [1668], of epistemological utility to the practice of dissection.)[58]

The topographical order of empirical understanding marks the penultimate rendition of man in Comenius's series. The visual logic of the final image, "The Soul of man," is both consistent and surprising (Fig. 6). Contained within the hatched outline of a standing human body are horizontal and, at about clavicle level, vertical rows of short dashes which compose a haze of elongated dots. Several challenges are confronted by this image: first, of course, is the object's immateriality, possibly affirmed by Comenius's field of dots, which fills in morphological contours while refusing further substantive or anatomical detail. Yet the most striking thing about the compromise whereby the soul acquires an elusive sensory counterpart is its violation of the metaphysical stricture, forcefully iterated by Descartes, that spirit is not extended. For in the eyes of Comenius's child reader, the defining attribute of soul appears to be its shape. The perceptibility of the perceiving brain is paralleled, in Comenius's scheme, by the (anti)perceptibility of spirit whose coincidence with a physical body becomes sensually manifest as figure. As a contradiction triggered by sensual pedagogy, spirit acquires one classically materialist attribute of matter.

The endowment gained by Comenius's soul is figure. That figure operates not simply to offset this object's invisibility, but also performs crucial taxonomic service, is evident in the word-image pairing *Orbis Sensualium Pictus* places directly after the soul: "*Deformes. & Monstrosi* / Deformed and Monstrous People."[59] The text accompanying this illustration proceeds (Fig. 7):

> Monstrous / and deformed people, / are those which differ / in the body from the / ordinary shape; as are, / the huge Gyant, 1. / the little Dwarf, 2. / One with two Bodies, 3. / One with two heads, 4. / & such like Monsters. / Among these are / reckoned, / The jolt headed, 5. / The great Nosed, 6. / The Blubber-lipped, 7. / The Blub-cheeked, 8. / The Google-eyed, 9. / The Wry-necked, 10. / The Great-throated, 11. / The Crump-backed, 12. / The Crump-footed, 13. / the steeple-crowned, 15 / add to these the Bald-pated. 14.[60]

What makes the transition from the soul to "Monstrous People" startling, while it illuminates the compromise entailed in the sensual apprehensibility of the category of man, is the fact that Comenius's definition of monstrous people makes no mention of spirit at all. "Deformed and Monstrous People" are extrinsically rendered, delineated by morphological departures "from the ordinary shape" manifest as anomalies like "the huge," "the little," and "One with two." Yet all three bodies remain, as the plate's title indicates, discernible as "*People.*" They therefore push the child learner in the direction of metaphysical queries: what are the spiritual pretensions of "Monstrous People?" As the criterion for monstrosity, does "shape" preclude these people's

Figure 7. "*Deformes. & Monstrosi* / Deformed and Monstrous People," John Amos Comenius, *Orbis Sensualium Pictus*, trans. Charles Hoole (1685). Photo courtesy of the Newberry Library, Chicago. Wing ZP 645.M46.

possession of souls? Must a human soul possess the contours delineated in Comenius's preceding plate? While it may be futile to venture the transhistorical urgency of such queries, we can imagine Comenius's child audience registering his implicit delineation of a morphology of spirit. More to the point, Locke asks the same questions, precisely to indict the metaphysical breach propounded by the humanizing criterion of shape: if "Crump-backed" signals an adequate indicator of monstrousness, then figure would define extension as the normalizing attribute of spirit. Locke deplores the sensory apprehension of monstrosity evidenced by morphological deviation as an historical incentive to infanticide: "I do not know how they can be excused from Murther, who kill monstrous Births, (as we call them,) because of an unordinary Shape, without knowing whether they have a Rational Soul, or no; which can be no more discerned in a well-formed, than ill-shaped Infant, as soon as born" (3.11.§20).

The figural adjudication of the humanity of "ill-shaped Infant[s]" marks one of two sustained instances of the abdication of essence demanded by sense-based epistemology. (The *Essay*'s other object is gold.) Locke radicalizes Boyle's skeptical chemistry not only to exacerbate the break between corpuscular texture and the simple ideas it effects in persons, but also to extend Boyle's refutation of the mystified substance of elements to animate bodies. Infanticidal parents who "make bold, as everywhere they do, to destroy ill-formed and misshaped productions" (4.4.§16) evidence both the extrinsic nature of humanity's determination and the impossibility of a surer test. These are exemplified by a story Locke tells to refute the essentializing agent he designates substance:

> *When the Abbot of St. Martin . . . was born, he had so little of the Figure of a Man, that it bespake him rather a Monster. 'Twas for some time under Deliberation, whether he should be baptized or no . . .* This Child we see was very near being excluded out of the *Species* of *Man*, barely by his Shape. He escaped very narrowly as he was, and 'tis certain a Figure a little more odly [*sic*] turn'd had cast him, and he had been executed as a thing not to be allowed to pass for a Man. And yet there can be no Reason given, why if the Lineaments of his Face had been a little alter'd, a rational Soul could not have been lodg'd in him; why a Visage somewhat longer, or a Nose flatter, or a wider Mouth could not have consisted, as well as the rest of his ill Figure, with such a Soul, such Parts, as made him, disfigured as he was, capable to be a Dignitary in the Church. (3.6.§26)

The rhetorical stress in this passage falls on the chanciness of the infant abbot's "escape[]." Because "Shape" dictates either a body's belonging to "the *Species* of *Man*" or its classification as "a thing," the abbot's "ill Figure" almost prevents him from "be[ing] allowed to pass." Yet the abbot proves the Comenian—and empirical—standard of monstrosity wrong over the course of his adult life. Since his possession of a soul is demonstrated by his rise to the office of "a Dignitary in the Church," the abbot's morphology might retroactively be "alter'd" even further: his subsequently vindicated humanity reconciles species belonging and a "longer" face, "flatter" "Nose," and "wider Mouth." But if it transpired while he was a "Child," this metamorphosis would violate the empirical boundaries of human spirit.

Locke tracks a drama of qualitative identification that takes place over experimental time. Grammatically passive, the infant abbot suffers "near[ly] being excluded" and "be[ing] executed," while "he escaped very narrowly" because others judge him "allowed" to be human. If he does anything over the narrative arc of his tale, he manifests attributes which prove him "capable" of harboring a soul. This is not a plot of disguise but an empirical determination of prospective humanity that mimics the tests of elemental substance narrated by Boyle's *Sceptical Chymist*. The latter assays devolve into a scene of naming, for the scholastic or spagyrical tendency too easily to confer elemental substance is encapsulated by Boyle's complaint: "Qualities sleight enough may serve to Denominate a Chymical Principle [element]."[61] Operators who "Denominate" despite insufficiently rigorous analysis of the object foreground the doubled role of naming and qualitative identity in Boyle's chymistry: "Portions of . . . matter seem to differ from One Another, but in certain Qualities or Accidents, fewer or more; upon whose Account the Corporeal Substance they belong to receives its Denomination."[62] *Sceptical Chymist* indicts vulgar chymists for analytic sloppiness that also perpetuates denominational confusion: "there is no just cause why I should rather give the body propos'd the Name of this or that Element or Principle, because it has a resemblance to it in some obvious Quality, rather then deny it that name upon the account of divers other Qualities."[63]

Boyle's lament that "obvious" qualities are enough to make elements is echoed by Locke's condemnation of shape as a humanizing determinant of man. But Boyle pursues a more subversive agenda, because his perceived qualities do not proceed from elemental ingredients at all, but rather from corpuscular texture. Boyle revises chymical reference to confer names as a result of laboratory analysis, not scholastic essence: "Bodies . . . we know are Denominated and Distinguish'd not so much by any Imaginary Substantial Form, as by the aggregate of these Qualities."[64] With the displacement of neo-Aristotelian "Substantial Form" by "the aggregate of these Qualities"—and, as the physical cause of the latter, texture—comes Boyle's suggestion that texture may, under certain conditions, *change*: "the Qualities or Accidents, upon whose account Chymists are wont to call a portion of Matter by the name of Mercury or some other of their Principles, are not such but that 'tis possible as Great . . . may be produc'd by such changes of Texture . . . as the Fire may make in the small Parts of a Body."[65] Boyle does not simply argue that chymists employ lax criteria to designate elements. Rather, the changes wreaked on "Texture" by temperature, atmosphere, solvents, or mixture bely the very existence of elemental essence. Vulgar chymists who posit elements falsely project a regime of linguistic fixity, when instead "one and the same parcel of the Universal Matter may by Various Alterations and Contextures be brought to Deserve the Name, somtimes [*sic*] of a Sulphureous, and sometimes of a Terrene, or Aqueous Body."[66] Made of corpuscles whose texture may be recomposed, the "same parcel of . . . Matter" is no abiding referent of any elemental "Name." Over time—like the time of the abbot's metamorphosis from a man into a Comenian monster—this parcel of matter may elicit qualities that solicit its denomination as an entirely different kind of body.

Locke's deep appropriation of Boyle's anti-elemental chemistry is evident in the *Essay*'s definition of the categories "real essence" and "nominal essence." Real essence corresponds to insensible but physical primary qualities, were those knowable, while nominal essence comprehends human understanding of a thing gleaned from its secondary qualities.[67] Locke's two so-called essences claim a peculiarly buttressed mutual dependence:

> Supposing the nominal Essence of *Gold*, to be Body of such a peculiar Colour and Weight, with Malleability and Fusibility, the real Essence is that Constitution of the parts of Matter, on which those Qualities, and their Union, depend . . . Here are *Essences* and *Properties*, but all upon supposition of a Sort, or general abstract *Idea*, which is considered as immutable: but there is no individual parcel of Matter, to which any of these Qualities are so annexed, as to be *essential* to it, or inseparable from it. That which is *essential*, belongs to it as a Condition, whereby it is of this or that Sort: But take away the consideration of its being ranked under the name of some abstract *Idea*, and then there is nothing necessary to it,

> nothing inseparable from it. Indeed, as to the *real Essences* of Substances, we only suppose their Being, without precisely knowing what they are. (3.6.§6)

For Locke as for Boyle, no "parcel of Matter" instantiates "immutable" essence. Neither substance nor permanently fixed texture, essence is a referential "Condition" deployed to taxonomize bodies as "of this or that Sort." Were there no extrinsic mandate to classify, no sensible quality would be "necessary" to any given thing, for any quality's engendering texture can change even in the absence of mixture. The impetus for Locke's denial of essence to any "individual parcel of Matter" is the mutability of corpuscular texture Boyle affirms in *Sceptical Chymist*: "how sleight a variation of Textures without addition of new ingredients may procure a parcel of matter divers names, and make it be Lookt upon as Different Things."[68] For Locke as for his alchemical collaborator Boyle, "*Gold*" is not made of particles that immutably fix the metal's "annexed" attributes. The fungibility of chymical texture propels Locke's denial that sensed identity—whether of gold or man—refers to innate substance.

Locke posits "*real Essence*," the "Constitution of the parts of Matter" on which the secondary qualities or powers of any set of perceived properties "depend," as opposed to "*nominal Essence*," the ensuing set of humanly sensed attributes. But each of Locke's essences undoes the other. While nominal essence refers to categories like man, monster, air, water, earth, and fire, real essence uproots their defining qualities from any given thing. Real essence is real in the sense that it refers to matter, but it is not an essence in the critical sense that its fixity is only linguistically enforced. An extrinsically imposed condition, essence recurs to the denominational "Sort" to which any object's secondary qualities might intermittently correspond. While texture effects the perceived "Colour," "Weight," "Malleability," and "Fusibility" of gold, Locke, like Boyle, denies that these qualities claim an essential home in any parcel of matter. By extending the reach of anti-elemental science to men, Locke's *Essay* subverts ontology even more profoundly than Boyle's *Sceptical Chymist*: "the *Species of Things to us, are nothing but the ranking them under distinct Names, according to the complex* Ideas *in us*; and not according to precise, distinct, real *Essences* in them" (3.6.§8, emphasis Locke's).

Locke's reliance on *Sceptical Chymist* as the warrant for the *Essay*'s anti-essentialist understanding of species is evident in his repetition of the clause "*in us*." In the *Essay*, not only secondary qualities but taxonomic categories are relationally grounded "in us." But while Boyle's skeptical critique applies to "these differing substances that are call'd Elements,"[69] Locke also grants such things as monsters and men an ontology instituted only by "*distinct Names.*" Boyle shields seminal bodies from texture whose immutability cannot be secured, whereas Locke asserts the failure of names to demarcate the "distinct . . . sort" of humankind: "I would gladly know what are those precise Lineaments, which . . . are, or are not capable of a rational Soul to be joined to them. What sort of outside is the certain sign that there is, or is not such an Inhabitant within? For till that be done, we talk at random of *Man*" (4.4.§16). As the abbot shows, Locke does not mean that "precise Lineaments" can actually be specified. Any empirical cutoff that delimits the outer figure of human essence—as Locke's metamorphosis of a "well-shaped Changeling" into "a *Monster*" (4.4.§16) makes clear—is arbitrary. For Locke, "Shape . . . the leading Quality, that seems more to determine that Species" (3.11.§20), is no less random than the crude and reproducible signs exhibited by chemical pretenders to elemental essence. Humanity's leading quality, figure, exposes the fatal arbitrariness of the determination whether the "*Foetus* . . . were, or were not to be nourished and baptized" (3.3.§14). Empiricism offers humanity no sure empirical criterion. The problem is not simply the randomness of the qualitative standard, but the failure of texture—even as this comprises persons—stably to uphold it.

Locke exploits skeptical chemistry both when he cites Boyle's fungible parcel of matter and when he underscores the indeterminacy of the sensible qualities that define the identity of perceptible objects: "He that shall but observe, what a great variety of alterations any one of the baser Metals is apt to receive, from the different application only of Fire; and how much a greater number of Changes any of them will receive in the Hands of a Chymist . . . will not think it strange, if I count the Properties of any sort of Bodies not easy to be collected, and completely known" (3.9.§13). For Locke, a "Chymist" like Boyle demonstrates that the qualities of "baser Metals" may be changed, as *Sceptical Chymist* asserts, by the "application only of Fire." New qualities elicited not by mixture but by one parcel of matter's revised texture portend the indeterminate "number" of "Properties" that must "be collected" exhaustively to comprehend anything's nominal essence. Identity stipulated from the outside in thus propels an individualizing turn, because it depends on both experiment and "Experience" (3.9.§13): every empirical knower "with Reason thinks he has the same right to put into his complex *Idea* . . . those Qualities, which upon Trial he has found united" (3.9.§13). Locke repeats: "every one has a right to put into his complex *Idea*, those Qualities he has found to be united together" (3.9.§13). Locke's iteration of this claim illuminates its corollary: entities may rank as species if the observer perceives qualities that match a norm comprised of the assembled attributes Locke calls a "complex *Idea*"; but norms vary according to what each observer happens to have "found." Different persons, that is, may elicit different qualities from an object. Essence is not only extrinsic in the sense that it must be adjudicated from the outside, but also in the sense that it depends on the observer's ability to conjure qualities from objects. The "Hands of a Chymist" stand in for the analytic capacities of all persons, more or less adept, who approximate the essence of things in the very act of perceiving them.

According to Locke, every perceiver possesses the "right" to assemble her own complex ideas of species. Insofar as nobody's "Collection" (2.31.§13) of qualities exhausts the determination of a thing's essence, the *Essay* gestures toward the collective compilation of natural histories as empiricism's outer epistemological horizon. Indeed, the word "right" recalls the universalizing reach of the puritan pedagogical project that restores foundational linguistic authority to the experience of unsophisticated persons. At the same time, "right" resonates with Boyle's inversely personified pretenders to essence, those aspirants to elemental status that invite chemical denomination. While qualitative change may "procure a parcel of matter divers names," the same parcel may also "be brought to Deserve the Name" of an element: it may become "worthy to be call'd by the Name of a Principle."[70] As Boyle writes, one parcel may "be lookt upon as more worthy to be called an Elementary Principle"[71] or "seem to have a better title . . . to the name,"[72] while "*Spagyrists*" award others "specious Titles"[73] and another only "passes for a Chymical Principle."[74] Motivated adjectivally, Boyle's texture may be deserving, worthy, and meritorious, the empirical appeal of its qualities gauged against the integrity of the "Name" to which it pretends. But the circularity of this process cannot be overlooked. Titles bolster elemental essence by dissimulating the changeability of texture. Far from static indicators of anterior essence, elemental denomination corroborates the worthiness of matter only after chemists and other perceivers solicit congeries of qualities from it.

The contingency of elemental monikers spurs Boyle to characterize aspiring referents as deserving, worthy, or meritorious. Indeed, the prospective entitlement of Boyle's parcel of matter offers a trope for other objects that solicit essential denomination. We have already encountered Locke's baby abbot, whose infantile passivity renders him, like Boyle's chymical ingredients, something whose title must be tried. Locke broaches another case that extends the operation of naming to include other possibly human beings: "a Child having framed the *Idea* of a *Man*, it is probable, that his *Idea* is just like that Picture, which the Painter makes of the visible Appearances joyned together; and such a Complication of *Ideas* together in his Understanding, makes up the single complex *Idea* which he calls *Man*, whereof White or Flesh-colour in *England* being one, the Child can demonstrate to you, that *a Negro is not a Man*, because White-colour was one of the constant simple *Ideas* of the complex *Idea* he calls *Man*" (4.7.§16). Indeed, Comenius's illustrations of men and monsters are all fair-skinned. His child reader's unwittingly exclusive understanding of the category of man proves, for Locke, that the truth of the ensuing proposition is "only verbal" (4.7.§16). The child deploys denominating criteria whose leading quality is not shape but color to exemplify the inescapably topical influences that define "Complication[s] of *Ideas*" derived from "visible Appearances." Because the child's understanding is stocked with simple ideas to which he is exposed "in *England*," his notion of "Flesh-colour" retains a "constant" association with the simple idea "White." "Flesh-colour" is an ambiguous entity: it is an arbitrary signifier, insofar as its referential support varies with geography; but because it is susceptible to replacement by a word like "white," it may be wrongly confused with a simple sensory idea. Flesh-color falsely claims the naturalness of a simple idea to produce as humanity's criterion the condensation "White-colour." For if nonwhiteness would, according to the letter of this reflexive act of linguistic compression, disqualify only human *flesh*, it operates—like the ill figure of Locke's abbot—extrinsically to nullify the humanity of the whole person. Locke thereby exposes the universalizing pretensions of sense-based pedagogy as always imminently counter-universalizing ones.

It is possible to identify a chymical genealogy of human rights discourse propelled not by humanizing spirit but rather by the indeterminacy of the collection of qualities requisite to garner any given object a title. That is, the right of persons to bundle their perceptions into complex ideas of essences is reflected by the right of any given parcel of matter to elicit qualities that might earn it elemental denomination. The circularity of identity fixed neither by particulate texture nor by names (which refer to differently collated sets of qualities) propels the arbitrariness of the title "man." An eighteenth-century novel like Daniel Defoe's *Roxana* (1724) illustrates the extendable reach of titles that recur to both semantic convention and newly recombined qualities. Roxana, Defoe's protagonist, aspires to be a "*Man-Woman*,"[75] a neologism that weds the economic autonomy of Locke's masculine individual to Roxana's near indefinite deferral of the corporeal and domestic labor of childbearing. Roxana's ambition is supported by her power to evince the endowments that sustain her hybrid denomination: Roxana is, concretely, a parcel of matter whose qualitative experience by others—obsessively relayed as the perceptual impact of her performance—enables or jeopardizes her linguistic pretense. In no sense, Defoe's novel makes clear, does *Roxana* adumbrate a spiritual or affective claim to essence, because her autonomy excludes Roxana from female humanness stipulated as sentimental motherhood.[76] Many readers equate Roxana's monstrosity with her decision to leave her five children in the care of their aunt. But the sentimental derivation of her humanity comes with significant preestablished conditions—namely, that Roxana not solicit a denomination vindicated, in part, by her capacity to survive egregious failures of patriarchal protection. Roxana marks a chemical aspirant to the title man-woman evidenced not by the disclosure of depths that elicit sympathy but by her manifestation of secondary powers.

In closest methodological accord with Monique Wittig's contention that "there is a plasticity of the real to language" (a contention Wittig credits, in part, to Locke),[77] I mean to suggest that Roxana's self-naming implicates secondary powers: Defoe represents a parcel of matter that, placed in relation, evinces the properties of a man-woman. The inextricability of something's texture from its claim to denomination may be suggested by Boyle's and Locke's use of the verb "pass." For Boyle, some matter merely "passes for a Chymical principle," while Locke's abbot almost fails "to pass for a Man." This congruence underscores the external adjudication both species solicit. Eighteenth-century novels, I will now suggest, animate the corpuscular Locke with characters known in the event of their appeal to the perception of others, characters whose approximation of identity might sustain their claim to some different name.

Knowing and Naming: Haywood

The *Essay*'s distinction between real and nominal essence offers stable ground to neither side. Nominal essence sums an open-ended set of experienced qualities, while real essence refers to changeable texture. For the latter, Locke often deploys the cognates "internal Constitution" (3.6.§9), "internal frame and Constitution" (3.6.§36), and "true internal Constitution" (3.6.§9). But the *Essay* correlates internality and truth to deny this formal promise: because species boundaries are approximated from the vantage of empirical sense, "if we suppose it to be done by their real internal Constitutions, and that Things existing are distinguished by Nature into Species, by real Essences, according as we distinguish them into *Species* by Names, we shall be liable to great Mistakes" (3.6.§13). Locke cites "internal Constitutions" whose imperceptibility renders human taxonomy faulty; and because of the arbitrariness of differently collated complex ideas, he revokes the correspondence of real essences to humanly limned species. Species demarcated "by Names" do not line up with internal differences instituted "by Nature," because the names do not rationalize essence but rather authorize "Mistakes." Rather than stressing that sensory attributes determine how percipients "distinguish . . . *Species*," Locke throws the essentializing agency of texture wholly into doubt. He cites the "odly-shaped Foetus" (3.6.§27), the non-elemental textures of "parcels of yellow Matter coming from *Guiny* and *Peru*, under one name [gold]" (3.6.§32), and "the Issue of a Cat and a Rat" (3.6.§23) to conclude that "*the boundaries of the Species, whereby Men sort them, are made by Men*" (3.6.§37, emphasis Locke's). The fictionalizing creep of analogy emblematized by figure extends to species delineated by man-made "*boundaries*."

Locke revokes the confluence of essence and interiority: " 'tis not this real Essence that distinguishes them into *Species*; 'tis *Men*, who, taking occasion from the Qualities they find united in them . . . *range them into Sorts, in order to their naming*" (3.6.§36, emphasis Locke's). Does some unknowable real essence—which will never line up with the boundaries of humanly instituted "*Sorts*"—engender the ideas that stimulate humans to assign names? Or, given Locke's insistence on the variability of complex ideas of species, is the imposition of real essence on corpuscular structure the greatest mistake? Interiority operates in the *Essay* as both the site of the real and a placeholder for ontological substance. The formal fallout of Locke's ambivalent corpuscularianism remains: Does innerness yield texture whose secondary powers might be analogically conceived, or does innerness perpetuate the specter of essence in a new, post-scholastic guise? In Locke's *Essay*, it does both.

Locke derives qualitative identity not from texture locked inside the boundaries of things but from secondary powers activated in relation to "extrinsical . . . Bodies" (4.6.§11). But objects do not stimulate persons to perceive in one-to-one relation, because they are qualified by an indefinite network of more or less proximate things:

> [W]e in vain search for that Constitution within the Body of a Fly, or an Elephant, upon which depend those Qualities and Powers we

> observe in them . . . This is certain, Things, however absolute and entire they seem in themselves, are but Retainers to other parts of Nature, for that which they are most taken notice of by us. Their observable Qualities, Actions, and Powers, are owing to something without them . . . and we must not confine our thoughts within the surface of any body, but look a great deal farther, to comprehend perfectly those Qualities that are in it. (4.6.§11)

Locke cites such "extrinsical" but life-sustaining "other parts of Nature" as the "Air" and the "duly tempered motion of Particles coming from" the sun (4.6.§11). By asserting the capacity of extrinsic things to elicit "observable" properties from flies, elephants, persons, or gold, Locke turns the agency of texture inside out. No longer situated "within the Body," the corpuscular sources of "Qualities and Powers" are diffused throughout "Nature." The thrust of Locke's argument, which belies Comenius's sequence of neatly demarcated images, is that the figural boundaries of things as humans perceive them are delusory. Because qualities are elicited from bodies by relations to other bodies—including, of course, the human body—the cause of no thing's sensible attributes sits "within [its] surface." The atmospheric theorist Boyle makes the same claim: like Locke, Boyle orients texture towards the relational pull of the whole world. Boyle's "History of Particular Qualities" (1670) offers a local instance of the relational nature of qualities, which proceeds at some length as features emerge serially from one parcel of matter:

> [T]he same body . . . may by vertue of its Shape, and other mechanicall Affections . . . have such differing respects to different Sensories, and to the Pores &c. of divers other Bodies, as to display severall very differing Qualities. The Example I speake of, is afforded me by the destillation of Putrefied Vrine . . . the spirits of it swimming in a Phlegmatick Vehicle have a pungent Saltnes upon the Tongue, and a very strong, and to most persons, an offensive smell in the nostrells; and when they are freed from the water, they are wont to appear white to the Eye; and to very tender parts . . . they feele exceeding sharpe, and seeme to burne almost like a Caustick . . . the same saline Particles invisibly flying up to the Eyes prick them, and make them water; and invading the nose often cause that great Commotion in the head . . . that wee call sneezing. The same Corpuscles if they are much smelt to by a

> woman in Hystericall Fits, doe very often suddainely releive her . . . The same Corpuscles being taken into humane Bodies have the Qualities that in other medicines we call Diaphoretick, and Diuretick; The same particles being put upon filings of Brasse produce a fine Blew, whereas upon the Blew or Purple juices of many plants they presently produce a Greene; being put to work upon Copper whether crude, or calcin'd, they doe readily dissolve it, as Corrosive menstruums are wont to doe other Metalls, and yet the same Corpuscles being blended in a due proportion with the acid Salts of such menstruums, have the vertue to destroy their Corrosiveness . . . I might here add . . . how the same Particles applyd to severall other Bodies to which they have differing Relations, have such distinct operations on them, as may intitle these saline spirits to other Qualities.[78]

Boyle makes, although he cannot exhaust, the point: "the same Particles" evince different qualities in relation. Placed in proximity to "different Sensories" and "other bodies," the "same Corpuscles" manifest qualities that seem almost to transform one object into a sequence of "distinct" things. The narrative momentum of this passage derives from the incongruity of the successive qualities justified by the bodies and organs in whose vicinity the thing resides: "the Tongue," "the nostrells," "the Eye," the "very tender parts," "the nose," "a woman in Hystericall Fits," the intestines of "humane Bodies," "filings of Brasse," the "juices of many plants," "Copper," and "acid Salts." As Locke proposes when he locates the source of a fly's qualities outside its borders, the qualities elicited by this series of organs and objects—"pungent Saltnes," "offensive smell," whiteness, causticness, tears, sneezes, relief of hysteria, diuretic potency, blueness, greenness, acidity, and alkalinity—do not proceed solely from inside "Putrefied Vrine." The anti-elemental thrust of Boyle's proof, insofar as Boyle refutes the chymical mystification of ingredients like salt, follows from his displacement of the source of these qualities from inside to outside of the object. As Boyle motivates their successive changes, qualities do not flow from innate essence but from the array of discrepant bodies and parts that galvanize an equally discrepant series of endowments from the same particles. The corpuscles comprising Boyle's putrefied urine may possess their own "Shape, and other mechanical Affections," but rather than secreting away the source of this object's qualities, its texture occasions taste, color, smell, and acidity, among others, only in concert with the external world.

In the register of form, Locke adopts Boyle's doctrine of qualities to deny the primacy of innerness as a determinant of either ontology or empirical knowledge. In eighteenth-century novels, Locke's move promotes the susceptibility of character to external denomination. My chapter concludes with two texts whose relational treatment of character's empirical determination is evident and even polemical: Eliza Haywood's *Fantomina* (1725) and *Love in Excess* (1719). These fictions plot the extrinsic and contingent adjudication of character, both as a signifier of real essence and an index of the attributes one parcel of matter exhibits. They derive their polemical bite from Haywood's isolation of the external source of women's qualities: in Haywood's plots, women evince denominating attributes in relation to the sensoria of men who wish to enjoy them. *Fantomina*'s unidentified "Young Lady of distinguished Birth, Beauty, Wit, and Spirit"[79] passes for the rake Beauplaisir as Fantomina, a prostitute; Celia, a credulous servant; Mrs. Bloomer, a lusty widow; and Incognita, a masked seductress. Because Haywood rationalizes the contextual variants that enable the same "Young Lady" to display this series of endowments for one perceiver, *Fantomina* places its reader in the position of the experimentalist who tracks Boyle's relational production of qualities from putrefied urine. Through Boyle's stand-in Beauplaisir, Haywood deploys the conceit of multiple women who are really one to mount a mock exposé of the rake who pledges fidelity to some of these personas while pursuing trysts with others. His faithlessness does not require that Beauplaisir be perceptually impaired: to imagine infidelity somehow perpetuated by the betrayed object, Haywood's reader sustains the premise that Beauplaisir vows loyalty to one person *and* the sequence of secondary qualities that elicit his serial attachments.

Haywood does not seal off an interior from which this object's incriminating charms emanate. It is not the real essence of Haywood's protagonist from which her contradictory series of attributes can be said to proceed, but rather the environments, objects, and organs in relation to which the parcel of matter known as a "Young Lady" is placed. Haywood's nameless young lady *is* the analogical—that is, fictional—figuration of secondary powers that are, as Locke stipulates, in perceiving minds and stimulating things at once.

Like Locke's empirical knower, Beauplaisir has the right to assemble whatever perceived attributes he can gather into his complex idea of an object. He may, however, lack some of the analytic tenacity of Boyle's skeptical chymist. That Beauplaisir's lack of denominational rigor could be a structural effect of patriarchal power is suggested by Haywood's *Love in Excess*. While *Love* does not represent as radical a personification of secondary powers as *Fantomina*, the novel's first volume assigns female characters names (Amena, Alovisa, Anaret, and Ansellina) that barely sustain the basic semiotic function of differentiation. These names exacerbate the failure of depth to distinguish women anterior to their placement among external—masculine—bodies. Indeed, *Love*'s women are differentiated in relation to only one rakish percipient, Count D'elmont. Haywood further suspends the individualizing function of names when the coquette Melantha, upon overhearing the impassioned Count's plan to break into the bedroom of his virtuous ward Melliora, substitutes herself in place of the Count's intended victim. In the dark, the Count enters the bed and ravishes Melantha, as Haywood relates: "Tho' the Count had been but a very little time in the arms of his supposed Melliora, yet he had made so good use of it . . . that all his fears were at an end, he now thought himself the most fortunate of all mankind."[80] As this narrator specifies, the consummation of the Count's desire does not depend upon the inner essence of the person with whom he is "fortunate." In this scene, the woman he enjoys and the woman he "thought" he enjoys are qualitatively indifferent, since the former inhabits the position sufficient to merit the latter title. From the Count's vantage in the darkened bedroom, "his supposed Melliora" *is* the body that makes him fortunate. At this juncture, what makes Melantha Melliora—or, what makes Melliora—is not innate essence, or interiority at all, but a more or less strictly external relation.

Haywood prepares her reader for the Count's stupefaction when, as he lies in bed with the supposed Melliora, another one appears in the doorway: "What surprise was ever equal to his, at this view—He stood like one transfixed with thunder, he knew not what to think, or rather could not think at all, confounded with a seeming impossibility. He beheld the person, whom he thought had laid in his arms, whom he had enjoyed, whose bulk and proportion he still saw in the bed . . . come from a distant chamber . . . He looked confusedly about, sometimes on Melliora, sometimes toward the bed . . . 'am I awake,' said he, 'or is every thing I see and hear illusion?' "[81] Confronted by Melliora, the Count proves more eager to adopt the Cartesian hypothesis that "every thing I see and hear [is] illusion" than to surmise that the woman "he had enjoyed" is really somebody else. Having elicited the relevant qualities from the person under the covers, the Count *did* effectively enjoy Melliora: for the man who has held her in his arms, Melliora's appearance outside the bed *is* a Cartesian "impossibility." But in Haywood's plot, this impasse does not ratify the ghostly ideality of Descartes's skeptical mobilization of doubt. Rather, Haywood aggravates the divide between Melliora's primary and secondary qualities. The former are geometrical or mechanical attributes, the "bulk and proportion he still saw in the bed," whose color, smell, smoothness, and surface detail are masked by the covers. The concealed body whose dimensions the Count "saw" recalls Locke's primary idea, the *Essay*'s untenable commitment to a Cartesian schematics of primary matter as mathematical pattern. For Haywood, the collapse of primary idea and primary texture takes shape as another generic pattern, the bulk and proportion of the rake's used object.

In the doorway, the Count witnesses the detachment of Melliora's secondary qualities from the person to which they should be attached. He thus takes his visual idea of Melliora as illusion, the spectral break of her denominating features from the body to which, he still believes, they should be tethered. His "surprise" at this rupture between perceived ideas and engendering texture anticipates Beauplaisir's avowal, when *Fantomina*'s anonymous young lady faces him in her own persona, that she "was a Person whom he had never, more than at a Distance, admired." Beauplaisir's "wild Astonishments"[82] proceed from the same source as the "transfixed" Count's stupefaction. While Melliora's identifying attributes migrate away from the body to which they should be bound, Beauplaisir has never suspected the young lady's power to solicit discrepant names. For these men and the rigorously indeterminate number of women they have enjoyed, astonishing dénouements fall within the empirical compass of the novel because interiority fails to contravene the extrinsic, and analytically easy, criteria that denominate feminine objects as a function of their proneness to be enjoyed. The congeries of qualities delimiting this complex idea cannot sustain the referential coherence of female names.

A novel whose empirical norms are dictated by rakes propels women's mutually real and linguistic indeterminacy. In this novel, secondary qualities migrate to an unprecedented degree, vindicating Locke's placement of them at once in objects and in the percipients who register matter's powers. Indeed, where is Celia, if not "in" masculine enjoyment that only belatedly insinuates this secondary power into Haywood's young lady? Haywood's feminism resides not in the facticity of an immutable body, but in secondary powers whose perplexing effect on men—and readers—derails the referential consistency of empirical denomination. When Beauplaisir confronts a young lady who is none of the women he enjoyed, or D'elmont confronts Melliora at his door and in his bed at once, Haywood plays with the understanding of patriarchy's model percipient. As known by lovers and guardians alike, female objects subvert denominational stability sustained only from the extra-empirical vantage of Haywood's narrator.

Haywood affirms Locke's repudiation of innate essence to represent women as they are known by men who enjoy them. From this agreement proceeds the rigor with which she articulates female character as a function of masculine discernment. Haywood's fiction instances the plasticity of the real, in Wittig's words, to denomination, because women like Melliora and Melantha *are* the same at the moment when the Count most vitally knows them. Whereas men's empirical understanding precludes the unity of female character in Haywood's plots, Roxana musters her secondary powers to pass as a man-woman. Defoe and Haywood plot denomination that transpires as sensible qualities are elicited by others under patriarchy, over time, and in relation. My next chapter turns again to character's form, to chemical medicine and to persons who cannot see corpuscles that lurk inside themselves.

CHAPTER 3

Morbific Matter and Character's Form

> To see a man eat and drink liberally, to be jocund, frolick, seemingly enjoying a Jubilee, to exercise Venery, and several Disports . . . and yet to carry about him the very Picture of Death within, is a Plague of the Plague.
>
> —George Thomson, *Loimotomia: or the Pest Anatomized* (1666)

The physician George Thomson (c. 1619–1676) wrote *Galeno-pale: or, a Chymical Trial of the Galenists* (1665) before London's Great Plague of 1665–66 entered its most lethal phase. *Galeno-pale* denounces the humoral medical therapies employed by the Royal College of Physicians. By offering an alternative rationale for illness and cure, Thomson limns a prototype of the plague patient that, I argue, models character for the eighteenth-century novel. Thomson's rejection of Galen's arsenal of therapeutic methods, which include bleeding, evacuating, and purging, centers on the fact that they deal with entities that are perceptible, even though, Thomson counters, "very many Diseases have in them . . . something of a more spiritual and invisible Nature, then to be carried off by gross Purgatives." In an appeal to the merely literate as well as the learned reader, he compares the agency of disease to a range of inscrutable natural forces: "Hath not the *Apoplexie* (that destroyes a man in the twinckling of an eye) something in it like the poison of a *Basilisk*? Is there not in the *Palsie* something of the stupefying Nature of the Fish *Torpedo*? Have not some Plagues destroyed men suddenly like some *Mephitical* or pernicious Damps, arising in the subterranean and deep caverns of the Earth, which happens to those that dig in mines?"[1] The "*Basilisk*," "Fish *Torpedo*," and "*Mephitical* . . . Damps" portend fatal powers that act insensibly at a distance.

The things Thomson names are natural, but their toxic operation is "spiritual and invisible." A basilisk "that destroyes a man in the twinckling of an eye" may sound fanciful, outside the purview of the anti-occult remit claimed by positivist historians of the new science. However, the plague's "spiritual"—that is, subtle—and "invisible" workings propel the revisionary mandate of chemical medicine.

By presuming a person comprised, like other things, of minuscule parts, corpuscular science inversely delineates human bodies pervaded by tiny holes. This chapter's first two sections survey anti-Galenical pamphlets by doctors of chemical medicine or iatrochemistry, which coincide with the emergence of the Great Plague to promote a therapeutic warrant occasioned by imperceptibly porous patients. Because such therapy entails the admission and extrusion of insensible parts, it delimits a person whose contents are empirically inaccessible. My chapter's last section suggests that persons who admit imperceptible things determine the form of character in Daniel Defoe's fictional history *A Journal of the Plague Year* (1722). Far from a bounded individual, Defoe's model for character is chemical medicine's pervious person. This person admits not the progressive refinement of interiority into legible self-representation or sentiment but particulate content that cannot be known, even from the inside.

In Chapter 2, I argue that for Locke, human understanding of corpuscles entails the figuration of matter's power, all the way down to the primary endowment of shape. This chapter turns from empirical conceptions of corpuscular texture to the person who empirically knows—and, in the event of the plague, knows too late. Minuscule particles, I argue, hollow out empirical persons because they define the body's interior as a site of sensory understanding. This is not the presumed accessibility to himself of Locke's so-called rational individual but self-knowledge thwarted by minuscule contents that percipients can never apprehend.

Plague's Pervious Person

Declaring his fascination with small objects, Boyle writes early in his scientific career: "I must confess my wonder . . . is more exercis'd in the coyness of the sensitive Plant, and the Magnetical Properties of a small and abject Loadstone, then the bulk of the tallest Oakes, or . . . vast Rocks."[2] Over twenty years later, Boyle's *Experiments and Considerations about the Porosity of Bodies*

(1684) extends this predilection to assert the natural historical value of not only small things, but small nothings: "And I scarce doubt, but if such little things had not escaped the sight of our Illustrious *Verulam*, he would have afforded a good Porology (if I may so call it) a place, (and perhaps not the lowest neither,) among his *Desiderata*."[3] To repair this lacuna in the "Outline of a Natural and Experimental History"[4] detailed by Frances Bacon's *New Organon* (1620), Boyle vindicates the contribution of "Porology" to a corpuscular schematics of preventative and therapeutic medicine. In an argument ballasted by William Harvey's proof of the blood's circulation in *Exercitatio anatomica de motu cordis et sanguinis in animalibus* (1628), as well as Boyle's contact with Oxford physiologists from 1656 until his departure for London in 1668, *Porosity* claims the necessity of tiny pervasive holes to the sustenance of organic beings: "the Nutritive liquor [must] be distributed through the whole Body of the part that is to be nourished and augmented. And to this distribution 'tis requisite that the Body abound with Pores into which the congruous particles of the Juice may be intimatly admitted, & penetrating even into the innermost recesses."[5] According to the corpuscular rationale for growth, fluid-borne "particles" circulate in the bloodstream to enter "congruous" micro-openings. Nutritive particles that "penetrat[e] even into the innermost recesses" project physical personhood whose sustaining attribute is not impenetrability but perviousness, not impermeability but nourishing susceptibility to intimate admittance.

Gauged in terms of his early affiliation with the Hartlib circle as well as his fidelity to Baconian "*Desiderata*" that solicit the wisdom of artisans and mechanics, Boyle's homespun analogies reflect the empirical and democratic aspirations of puritan reformers' sense-based pedagogy.[6] To evoke the outer surface of a body that nonetheless "abound[s] with Pores," Boyle proffers a mundane comparison: "And perhaps it may be allowable to conceive, both the Skin that covers the Limbs, and the Membranes that invest the internal parts of the Body, to be like worsted stockings, Wast-Coats, *&c.* Which in their ordinary state have a kind of continuity, but upon occasion can have their Pores every way enlarged and stretched, in this or that manner, as the Agents that work on them determine them to be."[7] As a likeness intended to resolve the incongruity of perviousness and enclosure, the analogy of "Skin" or "Membranes" and "worsted stockings" transforms the apparent borders of the human person—and, within the person, the apparent borders of anatomical "parts"—into a "kind of continuity." Because worsted garments appear intact until they are "enlarged and stretched," Boyle offers a mechanical justification for the penetrability of skins and membranes illustrated by his reader's quotidian experience. The presumed ubiquity of the analog reflects empirical understanding restricted to British climes: whereas Locke's exemplary child equates white and flesh-color, Boyle's adult reader likens human skin to wool stockings. While Boyle's avatar of empirical impressionability remains the poisoned nun discussed in Chapter 1, knit stockings represent human texture whose porousness is sustained by dint of its continuity with domestic practice.

Knitting, whose revelation of pores anticipates skin's power to exude and admit, entails the curative strategies of chemical medicine as opposed to the Galenic methods institutionalized by the London College of Physicians. Like socks, apparently impervious flesh is shot through with insensible holes: for Boyle as for iatrochemists like George Starkey (1628–1666) and Thomson, bodily surfaces that let in particles justify both illness and health. Unlike Galenists, chemical practitioners posit particulate ingress and egress—in the event of plague, the traffic of "morbifick matter"[8]—that elude macroscopic therapies like vomits. As Boyle affirms, pestilence exacerbates the perviousness of chemical medicine's patient: "the Skin of a Living man may be easily penetrated by external steams whose approach the Eye does not perceive, and whose operations, though not inconsiderable, may therefore be unsuspected."[9] Because "the Eye does not perceive" the body's penetration by "steams," the person specified by Boyle's porology demands further refinement: this person is *insensibly* pervious. Chemical medicine, whose ascendency coincided with the Great Plague, propels a formal articulation of personhood that limns not only patients during the pestilence but also, as Defoe's *Journal* will show, characters in the novel.

With their attack on the Galenical doctors, the chemical practitioners tried and failed to win a royal charter for a rival Society of Chemical Physicians.[10] These institutional and experimental ambitions were tested, at the same time they were diffused in print, by the plague's efflorescence in London from the week of April 25 to May 2, 1665 (the first in London's weekly Bills of Mortality to herald an increase in plague deaths) until, symbolically at least, the return of Charles II to the city on February 1, 1666.[11] The aggregate number of plague deaths in this period, roughly 100,000, felled twenty percent of greater London's population.[12] These ravages succeeded what Charles Webster evokes as "the outburst of English medical writings and translations which appeared at the end of the Civil War . . . a torrent of vernacular [medical] literature"[13] including English translations of pharmaceutical secrets as well as,

Webster states, the translation from Latin or composition in English of a "disproportionately" large number of medical texts "of the Paracelso-Helmontian bias."[14] The chemical and medical innovators Theophrastus von Hohenheim or Paracelsus (1493–1541) and Joan Baptista Van Helmont (1579–1644) promoted a radically anti-Galenic conception of disease and treatment, adumbrated for lay readers to repudiate truth claims based on classical as opposed to experimental knowledge. The emergence during the English Civil War, P. M. Rattansi writes, of "an anti-College attitude [that] was given open and bold expression in print, in works directed at a lay audience,"[15] conjoined democratizing demands for medical authority with insistence on the agency of a disease suffered by imperceptibly porous patients. Plague concretely compelled the former—as most orthodox physicians, following their wealthy patrons, fled London—while galvanizing the therapeutic, empirical, and narrative elaboration of the form of persons into whom pestilence insensibly enters.

In what follows, I trace the formal contours of Thomson's and Starkey's chemical remedies. Thomson's friend and co-pamphleteer, the Harvard-educated Starkey was an expert practitioner of Helmontian chemistry as well as, William R. Newman and Lawrence M. Principe have shown, Boyle's tutor in the early 1650s.[16] Thomson, a proponent of Helmontian etiology, fanned the debate with the manifestos *Galeno-pale*, *Loimotomia*, and *Loimologia*, which mobilize iatrochemists' defining challenge to classical expertise: "He is unworthy the Title of a Doctor . . . that labours not with his own hands to attain such Remedies sufficient to save his Patients Life."[17] More particularly, Starkey's *Natures Explication and Helmont's Vindication* (1657) claims chymical cookery to ground the author's pretense to medical competence: "And as the fire taught *Helmont* to understand *Paracelsus*, so it hath also taught me to understand them both, and by it must every one that would understand Nature truely, and not notionally."[18] Truth assured by the chemist's "hands" bests notional theory, because while "Goosequil Doctors," Starkey asserts, exploit "smooth and fallacious language" to defend humoral medicine, his own understanding is proved by pyrotechnical labor. Knowledge claims must be presaged by experimental toil, as Starkey states of Helmontian "Tractates, (being troublesom enough to me to digest into that method they are in, but more troublesom, nay toylsom and chargeable to me to get that experience out of which I then wrote, and do here write)."[19] Showing his agreement with the overall impetus of revolution-era epistemic reform, Thomson charges that College physicians spin "a curious fine texture of empty words";[20] the fraudulence of classical practice is masked by "the babble or confused noise of their

Galenical garrulity and multiloquacity,"[21] perpetrated by "smooth Operators, Polyglossi, multiloquous Doctors."[22] While experimental understanding is authorized by "Actions . . . able to bear the test of the Fire best,"[23] classical conceits reflect, Starkey writes, "lazie Ignorance."[24] Thomson likewise indicts as "*Lazie Recipes*" the entire panoply of Galenical interventions, which comprehend "Bleeding, Issues, Vesicatories, Cupping-glasses; uncorrected Purgatives . . . languid and ineffectual Cordials, Juleps, Apozems, Electuaries of the Shops, and Diet-drinks."[25]

For Thomson, the doctrinal orthodoxy of College physicians typifies "a base slavish Spirit unworthy any man, but especially a Physician."[26] Such "*implicite Obedience*"[27] to Galenical dogma epitomizes the cowardice of pretenders who dispense medicines without, Thomson writes, "that true experimental knowledge which he may (setting aside this timorous childish apprehension of danger and hurting himself) acquire, by frequent repetitions of ingesting that into his own ventricle, which may instruct him far better then anothers."[28] The "experimental" derivation of expertise confirms the chemist's ability, Thomson writes, "to make it appear by Optical Practice."[29] To underscore the fortitude of practitioners who acquire knowledge "by frequent repetitions of ingesting," Thomson asserts his eagerness to enter the laboratory with the College physicians: "Upon these conditions ye shall finde us ready to instruct you, and teach you how to handle a Retort and Cucurbite."[30] Only the first-person experience of handling and taking things in—experience claimed by all "who profess our selves Philosophers by the Fire"[31]—guarantees mutually medical and narrative truth.

Yet as events proved, chemical medicine's ultimate trial was not the laboratory contest projected by Thomson but the arrival of the plague in early summer 1665. Starkey affirms the providential origin of this historical contest: "this hand of God . . . will prove a signal note of distinction, between Physicians elected, and sent forth by God, and those mercenary Hirelings, who either run unsent, or were created by the Schools."[32] Thomson stakes his life as security for the empirical derivation of "Essential Physical Truths (which if I maintain not demonstratively let me perish)."[33] Plague descends on London, "from which (to their infinite shame) the Ablest of the Galenists cowardly and unworthily run away,"[34] to assist the proof of iatrochemical labor whose cures Thomson shows himself willing to disseminate by fearlessly publishing his home address: "If any desire to be accommodated with these or more noble Chymical preparations in this sad time of Contagion, let him repair to the Place of my Abode without Algate, nigh the Blew Boare Inn."[35]

In 1666, Starkey's confidence that plague would ratify his Helmontian prowess was curtailed by his death from the pestilence, possibly caught while assisting Thomson. Thomson, however, survives to recount knowledge he does not ingest by mouth, but which enters through the pores of his practicing hand. Upon the death of one patient's servant, "a youth about 15 yeares of age," Thomson is granted "liberty to open this defunct body."[36] Having conducted the dissection, he "presently found some little sensible alteration tending to a stiffness and numness in my hand, which had been soaking and dabling in the Bowels and Entrals"[37] of the corpse. Thomson washes his hand and holds it over "a dish of burning Brimstone," but it is too late: he has admitted the illness without feeling it, because "This material virulent substance endued with exceeding Lepomery, or subtility of parts, hath free passage through the pores of mans Body, which are patent both for admission and emission of minute particles."[38] At this point, Thomson reverses course to evoke the insensible surrender of his "Archeus"—a life-spirit, theorized by Helmont, that mediates between the entering force of disease and the vulnerable person of the patient:

> [T]hose slie, insinuating, venemous Atoms, excited by the heat of the body, opening the pores of my skin, had quickly free ingress; the Archeus, the Porter of my hand, that should have better guarded it, forthwith tergiversating, and taking its flight, being extreamly terrified at the Alarum of so fierce and potent an Enemy, and afterward in an abject manner conducted it to the principal place of the Souls residence, the Stomack . . . and now do I carry about me the very Pest, closely spreading like a Gangrene, diffusing its malignity into all my members, covered over for some short space as it were in the ashes of silence, while I in the mean time visited, visit others visited, administring that help to them, which I (then more perplexed at my Neighbours Calamity, then sollicitous for my own) suffer my self to want, relating with joy what an Inquest I had made into that Subject which had made a Conquest of me.[39]

Thomson's first-person "I" articulates the novelty of his exposure to plague as a formal development. Since his outer surface poses no defense against "insinuating" particles, he animates a pronoun that no longer assures semantic or physical closure. Indeed, Thomson's present-tense assertion of his own porousness—"now do I carry about me the very Pest"—requires some temporal adjustment, because the crux of his story transpires during the interval when his infection is not known. From the retrospective vantage that corrects for the "short space" of an empirical lacuna sustained before "slie . . . Atoms" exert their powers, Thomson inhabits the form of a narrator whose own contents cannot be sensed.

While his sickness is unperceived, Thomson "in the mean time visited, visit[s] others visited." This pun propels the formal and semantic comedy of an obliviously "visited"—that is, infected—doctor who visits plague on his doubly visited patients. Likewise, Thomson's "Inquest I had made into that Subject which had made a Conquest of me" plays on a corpse that really exercises "Conquest" over the anatomist: if Thomson opens the dead body, its "slie, insinuating, venemous Atoms" proceed to enter him. The form of the first-person plague narrator licenses Thomson's collapse of the passive and active valences of the verb *make*, for while he has "made" an inquest of the evidence he gleans from the body, its insinuating particles have "made" inroads into him. Thomson's suspended self-awareness proceeds from the insensible perviousness of a person ignorant of what he contains. His openness to "Malignity that lay lurking within me"[40] summons a narrative "me" defined by corpuscles that cannot be felt even after they enter.

Later in *Loimotomia*, Thomson stresses the dramatic irony engendered by his symptoms' belated appearance to himself: "The space that this venom lay cryptick within me, closely and silently working in this subterranean Microcosm before it acted publickly was Eight hours, and then after a very sound sleep from Eleven till Two in the Morning, a grievous oppression at the stomack, with deep and difficult sighs, seized upon me, which doth still confirm me in this judgment, That the Pest never falls to acting a Tragical Scene openly, till such time it hath taken up its lodging place in the Stomack."[41] Thomson calls himself a "subterranean Microcosm," echoing terminology popularized by Paracelsus and Helmont, for whom individual persons recapitulate in miniature the cosmic chemical order of the universe.[42] Defined by the new medicine, Audrey Davis writes, as "a container of chemical solutions,"[43] Thomson is a vessel who contains another, even smaller microcosm. "Silently working" particles eventually appear "publickly" to an audience of one: his internal "oppression" becomes Thomson's own object of empirical understanding, disclosed by a stomach that stages this "Tragical Scene" for the patient himself. In Thomson's *Loimotomia*, the plague narrator internalizes a new scene of experimental knowledge.

Thomson delimits innerness as the result of oppression that commences "when those destructive Atoms entered into my hand."[44] Matter insensibly enters "my hand, which was for some time demerged and souzed in the poysonous liquamen of a pestilent Carkass,"[45] to render the plague victim's interior an object of his own empirical apprehension. His depth is thus summoned not as psychology but as the formal repercussion of sensory understanding turned inward during plague-time. In the end, Thomson recovers to justify the mechanism of his survival: "it was sent packing . . . through the pores of the Skin."[46] Before addressing character in *Journal of the Plague Year*, I trace the late seventeenth-century shift from Helmontian iatrochemistry to empirical medicine. Although doctors like Locke and Thomas Sydenham divest patients of actors like the Archeus, they continue to work on persons from whom morbific corpuscles must be "sent packing."

Seeds of Disease and Corpuscular Cures

Physician of Charles I, appointed physician of Charles II, and College of Physicians and Royal Society member Walter Charleton (1619–1707) expresses his adherence to Helmont in his annotated translation *A Ternary of Paradoxes* (1650).[47] Embracing an etiology of action at a distance whereby "almost all created Entities, have a certain *adumbration of sense*, or obscure sensibility, they largely declare as well by *Sympathy*, as *Antipathy*," Charleton argues for the therapeutic agency of sapphires: "A *Saphire* enobled with a deep cœrule tincture, if it be applied to, and a small time rub'd upon a *Carbuncle*, whereby the *Plague* pathognomically discovers it self, and after a while be removed, the absent Jewel then ceaseth not Magnetically to allect and extract all the pestilential virulency."[48] As Walter Pagel explains of Helmont's anti-dualist metaphysics, the capacity of the sapphire "Magnetically to allect . . . pestilential virulency" is both spiritual and material, the "product and expression of that sense and sympathy which dwell in each object of the created world."[49] Therapeutic action "by ray and look," Pagel recounts, is natural but not akin to the patient's "digestion"[50] of a curative agent. Instead, the sapphire is rubbed in a circle around its cognomen, the plague "*Carbuncle*," and "removed" to draw toxic contents from the patient's sore, as Charleton relates: "the virulency exhales, magnetically attracted from the infected body as it were through a trunck, or conduit-pipe."[51] Helmont and Charleton endorse healing power that transpires "magnetically" from a distance: the gem works to "associate and unite its own *influential ray* to the pestilential *vapor*, and so captive it, that afterwards being withdrawn, it may forcibly command it from the heart."[52]

The sympathetic confluence of "*ray*" and venomous vapor holds the latter "captive" while it exits the infected person: as an intimately binding modality of particulate attraction, magnetism exceeds mechanism to map a warrant for cure whose form is exemplified by carbuncles that pull virulency back through the patient's pores.

Following the decline of the overtly Helmontian worldview associated with Charleton and Thomson, Boyle continues to advance the displacement of humoral by corpuscular medicine. As Antonio Clericuzio suggests, the Helmontian doctors' anti-Galenism and Boyle's mechanical philosophy accommodate corpuscles and an "ontological" understanding of disease that attributes illness to seminal virulence rather than humoral imbalance.[53] Charleton affirms the directive enacted by the exit of toxins through a "conduit-pipe": because it works "by an *excentrical* attraction," Charleton writes, "so much the more excellent and efficacious a *Saphire* is esteemed, by how much . . . it has suckt out the venome of the pestilence."[54] A sapphire is "efficacious" because "it has suckt": Helmontian therapy operates eccentrically, from without, to attract ontologically discrete "venome" from deep inside the patient. But this person's flesh exerts countervailing force: "he draws in, through the invisible pores of the skin, the *pestilential Atomes* . . . [He] greedily sucks in the pestiferous aer, and invites *death* into the inmost closet of *life*."[55] Like Thomson, whose hand is infected after being "soused" for hours in a carcass, Charleton's plague victim unwittingly but "greedily sucks in" ambient pestilence. Gems and porous patients who suck at cross purposes define a formal structure of personhood that excavates the victim's unfelt "inmost closet."

Thomson promotes "the Protrusion of the venemous Atoms from the Center to the Circumference"[56] as his curative method. With Charleton, he cites the "much commended . . . Emerald and Saphire," stones which, "as (*Helmont* advises) do magnetically extract the virulence and malignity in the body."[57] In this capacity, the Oxford physiologist, theorist of fever, and brain anatomist Thomas Willis (1621–1675) criticizes Galenic bleeding and purging "in time of Plague" because "the Veins being emptied either way, will readily suck in whatever poysonous Atoms lurk in the outward Pores of the Body."[58] According to Willis's application of vacuist causality, veins "emptied" by improvident Galenists are primed to "suck in" invisible pestilential particles at the threshold of the pervious patient. To explain atmospheric infectiousness in plague-time, Thomson proposes that infinitesimal constituents of "Concretes" disaggregated by "an inward Ferment, or torn into invisible parts by violent torture of the fire, are all greedily suckt up into the spongy Spaces . . .

of the Air."[59] Like Charleton's victim, who mechanically "invites" plague's entrance, air sucks in morbific parts to fill its rarefied spaces. Helmontian medicine asserts the seminal influence of "venemous Atoms" that enter and exit persons as a result of mechanical forces like suction, exhalation, and, as we will see, impression.

Thomson attributes illness to a "pestiferous seed" whose identity and expression in the patient proceed from "matter that never fails . . . to explicate it self in this or that manner, and at last to come to this or that determinate, individual thing it was at first destinated [to be], so certain as a Hen-egg, Turky, or Goose-egg."[60] The seminal directive enacted by Thomson's "seed" or "Hen-egg" justifies plague as an ontologically distinct entity with the power to "explicate it self" by realizing a "determinate" set of symptoms in persons. While Thomson, following Helmont, insists that "Diseases in mans body have as certain a Seed, as any Mineral, Vegetal or Animal,"[61] he refers their potency to the mechanical as well as seminal attributes claimed by "infinite small particles disagreeing to our Nature, in their Figure, Crasis, Power, Texture and Proportion, making uncessantly a strong impulse upon the vital Aura."[62] Some of these properties, like "Figure" or "Proportion," resemble Boyle's mechanical attributes, inalienable features of particles whose mechanical intelligibility is funded by the corpuscle's founding likeness to empirical bodies. Other endowments Thomson assigns his small particles, however, like "Crasis" (mixture or composition) and "Power," reflect the capacities of seeds of disease. Plague particles recombine seminal and mechanical agency, welding Helmontian science to the far from unilaterally reductionist endowments entailed by corpuscular physiology.

Planting, Thomson's analogy for an ontological modality of infection, involves the "injection or superaspersion of some matter agreeable and consentaneous to"[63] the patient, who receives the extrinsic seed. The stipulation that the receiving person be "consentaneous to" pestilential matter returns us to Helmont's Archeus, defined by Starkey as "that Vulcan in man, which doth stir up and feed that heat of life."[64] The Archeus, writes Thomson, becomes complicit with plague by facilitating the victim's even deeper receptivity to entering disease: "[The Archeus] joins with the Occasional, Antecedent matter, to dissolve the goodly structure of this Microcosm; for the antecedent matter could do us no harm, were not the Continent united to it, so as to become capable to receive the true figure, character or impression of this determinate disease."[65] Once "united," the Archeus assists a "determinate disease" by perpetuating its ontological integrity in a markedly mechanical fashion: "Neither is there wanting in Diseases an Architectonical Spirit, which is the Faber and Vulcan that hammers out and forges every kind of malady, observing an exact Copy . . . this Stamp, Figure, Idea or Character doth direct the Archeus."[66] Stamped by the "Figure, Idea or Character" of the plague seed or "Spermatick power,"[67] the Archeus serves to realize disease's blueprint or "Architectonical Spirit." Delimited as a forge as well as a "Nest" where pestilence is "first brooded, hatched, fed, and at length perfected,"[68] the Archeus replicates disease in persons as both printing and generation. Thomson's mutually corpuscular and seminal vision of disease enlists both "Spirit" and mechanism to promulgate the ontological consistency of its "exact Copy." (It is notable that Thomson's invading architectonic spirit and stamped Archeus share the formal etiology of contemporary viruses which co-opt the host's DNA.)

From the vantage of reductionist approximations of the new science, the plague particle's architectonic power appears in flagrant violation of the modernizing criterion of mechanism. But Newman and Principe offer a corrective that situates finer distinctions of mechanism from extra-mechanism within the field of chemistry: "Van Helmont's argument [for the Archeus and *semina*] may at first sound like a blanket appeal to vitalism in opposition to mechanism. Yet it can be read more compellingly as an attempt to distinguish superficial physical change from intimate chymical interactions that result in change of substantial identity. This distinction underlies much of Van Helmont's chymistry, which distinguishes the mere mechanical grinding and spatial translation of corpuscles from the 'deep connection' and 'marriage' of substances brought about by seminal interaction."[69] As we have seen with the confluence of sapphire and plague sore, Helmont motivates sympathetic coalition to assure the profundity of chemical mingling. A more or less consentaneous agent, Helmont's Archeus also resembles a printer or metalsmith, a worker historically "analogous," Jole Shackelford writes, "to the human cog in the social machinery."[70] When Thomson declares that the doctor must "make it his principal pains and study to be throughly acquainted with the Practical use of those Medicines he possesses so exactly and accurately, as a Shoo-maker with his *Last*, or an Artificer with his *Tools*,"[71] he implicates "Medicines" that act like "*Tools*" as well as a mechanized kinetics of "Practical use" perpetuated by habitual knowledge. The plague particle that promotes a "destinated" end likewise confounds mechanical and extra-mechanical powers to express, as Newman and Principe write, "substantial identity" propagated by seeds of disease.

Writing of Boyle and Sydenham, who adopt Helmont's notion of morbific particles, Kenneth Keele remarks that they "came to the very frontier of a germ theory of disease. But one cannot say they crossed it."[72] Vivian Nutton offers a more finely tuned historiographical appraisal of the social and epistemic context in which the postulation of disease-causing creatures provoked "the charge of inventing specific beings, a new order of creation."[73] Nutton designates seeds like Starkey's and Thomson's an "agent *X*"[74] which sustains the specificity of infectious ends that are seminal but not animate. Of the ensuing shift to an emergent set of symptoms, Starkey suggests: "a disease is most of all the fury of the indignation of the Archeus, which finding a preterusual character impressed on its place of habitation, straight rages, and acts in its fury beyond all rule and measure."[75] To address symptoms proceeding from "rage," Starkey avows that "the highest and noblest way of medicine, is by pacifying the Archeus,"[76] a strategy whose restatement as popular remedy Willis affirms: "Therefore Wine and Confidence are a good Preservative against the Plague."[77] "Wine and Confidence" produce an unconsentaneous Archeus, rendering the receiver's inner cavity less easily impressed; wine also thins the blood to prevent the plague's direst effect, "preternatural Coagulation."[78] Thomson hypothesizes that pestilence, "a stop more or less put to the Circulation of the blood," proceeds from a "Gorgonian Venom that depredates, sacks, and confounds the Vital Spirits."[79] Engendered by seeds rather than creatures, "Gorgonian Venom" paralyzes still particulate "Spirits" mechanically to arrest the blood's circulation.

The plague's capacity to operate as *semina* and mechanism does not preclude the nascent autonomy of either logic. Starkey touts the medicinal properties of "alcalies" because, "being indued with an abstersive and resolutive vertue, they . . . resolve all preternaturall excrements and coagulations in all the Vessels. That they take away all filthy residence, which is in any of the veins, and that they do resolve all (though never so obstinate) obstructions . . . That their spirit is so penetrative, and efficacious, that withersoever it will not reach, nothing else will. And in a word, that as Sope cleanseth linnen, so they cleanse the whole body, and cut off, and cleanse away the material cause of all diseases."[80] Starkey's alkali works in the manner of an alchemical menstruum that opens metals to extricate their tingeing parts: like a perfected metal, the patient's "whole body" is penetrated and "cleanse[d]." More urgently, the resemblance of alkaline medicines to "Sope" reinforces the likeness of flesh and knit stockings proposed by Boyle's porology. Thomson assumes that human texture admits "so penetrative" medicine in his most schematic rejection of Galenical method: " 'tis the inward Medicament that must principally cure most outward Diseases, if they be ever radically and to a purpose healed."[81] As we have seen, the minuscule pores that license Thomson's "inward Medicament" propel the formal refinement of innerness itself.

Boyle's appraisal of Galenical orthodoxy lacks the outrage of Thomson's and his former tutor Starkey's—not a practicing physician, Boyle stayed at Oxford during the plague. But whatever Boyle's defense of chemical medicine may lose as testimony to his biographical virtue, it gains in the vigor of his defense of the perviousness of his patient. In *Of the Reconcileableness of Specifick Medicines to the Corpuscular Philosophy* (1685), Boyle demonstrates his experimental expertise while showing the sanative powers of imperceptibly tiny volumes of medicine: to cure a friend's recalcitrant son of the fever caused by worms or "verminous matter,"[82] Boyle mixes a minute amount of quicksilver into the child's small beer, which the boy ingests without realizing he is being treated at all. Restored to health by a beverage "impregnate[d] . . . with Mercurial particles"[83] he cannot taste, the patient proves that cures as well as illness operate by means of insensible corpuscles. This story encapsulates the thrust of Boyle's critique of Galenic therapy: "Medicines act not by naked Qualities but by small Particles."[84] His denial of causal salience to "naked Qualities" renders Boyle's defense of specific medicines, which correct particular ailments rather than humoral disequilibria, a corollary of his doctrine of qualities. Just as Boyle debunks denominations hazarded by scholastics and vulgar chemists who fail to analyze would-be elements into smaller parts, he refutes physicians whose diagnostic acumen is restricted to "Manifest Qualities, as heat, cold, tenuity of Parts, faculty of making large Evacuations by Vomit, Siege, *&c.*"[85]

On the redundancy of manifest qualities to corpuscular causes, Boyle echoes Thomson's *Galeno-pale*, which decries the method of cure by contrary qualities: "What an erroneous *Definition* have they made of a Feaver . . . as if there were no more to be done but to take an *Indication* from the preternatural *Heat*, and so to cool in the same degree . . . making that essential to a Feaver which is but a meer product."[86] Thomson reclassifies fever as a "meer product" to define sensible symptoms as secondary qualities: the chemical doctor understands illness by positing particulate causation that must be symptomatically accessed. Symptoms are produced by a more radical chemical reality, as when oil of vitriol (sulfuric acid) works therapeutically "not by cooling, as they [Galenists] grossly imagine, but by comforting the natural ferment of the Stomack, and by correcting that Alkalizate, Exotick, and putredinous matter therein."[87] Galenical doctors—or, as Thomson dubs them, "Mysochymists"[88]—deny the power of reactions that operate at the level of particulate microstructure. Such mysochymists "grossly imagine" that coolness obviates fever, Thomson repeats, "not considering those qualities of Heat and Cold are but Products and Consequents of the Disease."[89] Galenical physicians conceive of illness "grossly" because they deny that sensible qualities—that is, sickness itself—proceed from reactions and powers that cannot be perceived.

Thomson invokes the porosity of the "grand Emunctory of the whole Body the Skin" to iterate the formal warrant of chemical therapy: "our whole Body is porous and transmeable, especially our Skin: for did we not freely breathe out those hurtful Atomes, that are engendred within, and likewise insinuate into us from without, we should perpetually be obnoxious to *Feavers*, and other horrid Diseases, that would quickly destroy the World; for never did any one recover of a Feaver, but by *Transpiration* either *sensible* or *insensible*."[90] Like Starkey's alkaline medicines, Thomson's "Transpiration" prepares the way for medicine from which the Archeus drops out, leaving "hurtful Atomes" that impinge directly on the patient's texture. Boyle's mercury-imbibing boy, who "perceiving no change of Colour or Tast in the Drink, swallow'd it greedily,"[91] illustrates a corpuscular cure whose agreement with Thomson resides in treatment that is not, to use the word repeated by Boyle, "grosse."[92] The boy who obliviously ingests small beer illustrates, Boyle argues, "cure or relief [which] the Patient finds, is usually attain'd without violence, and without tormenting or much disordering him."[93] Indeed, in 1651 Boyle relates his own cure by a "Great Chymist," possibly Starkey, "without the wanted Martyrdome of Physicke."[94] Chemical remedies heal without torment because their operation eludes empirical sense; instead Boyle promotes expedients like "invisible Corpuscles [that] may pass from *Amulets*" or, more generally, "invisible Bodies [that], by passing thorough grosser ones . . . may produce lasting alterations in their Textures."[95] Like metals, humans are comprised of "Textures" whose inverse power to be penetrated is figured, in the event of plague, by knitting as well as magnetic exsuction.

When salubrious particles act on a person's texture, Boyle insists, they stimulate corrections the patient cannot experimentally know:

> And indeed the Physiologie, wherewith Physitians as well as others are wont to be imbu'd in the Schools, has done many of them no small Disservice by, accustoming them to grosse apprehensions of Natures wayes of working . . . [N]ot a few ev'n Learned Doctors

> will never expect, that any great matter should be performed in Diseases, by such Remedies as are neither obvious to the sence, nor Evacuate any grosse, or at least sensible matter. Whereas, very great alterations may be wrought in a Body . . . without the Ingresse or Egresse of any visible matter, by the intestine commotion of the parts of the same body acting upon one another, and thereby acquiring a differing Motion, Location (if I may so speak), or Figure, which . . . may alter the motion and Texture . . . and thereby produce great changes in the Body.[96]

Just as Thomson's Galenists "grossly imagine," Boyle's classical physicians entertain "grosse apprehensions." For Boyle, "at least sensible matter" is already too gross to explicate "Natures wayes of working," because sensation can never mimetically reflect the physical reality of corpuscular causes. It is, rather, the mechanical affections "Motion, Location . . . Figure, [and] Texture" that realize the efficaciousness of chemical medicine. Like qualities exhibited by any corpuscular body, symptoms are caused or relieved not by the "Ingresse or Egresse" of mass volumes, but by the "unperciev'd recesse of a few subtile Parts."[97] As with fluidity or any other sensible endowment, sickness is qualitative change that recurs to insensible texture. In the manner of vulgar chymists, Boyle's "Learned Doctors" attribute symptoms to "grosse, or at least sensible" agents. Instead of "large Evacuations by Vomit, Siege, *&c.*"—evacuations by emetic, bleeding, and purge that were historically "grosse" indeed[98]—Boyle champions corpuscular intervention sustained by an insensibly pervious person. The practice that rectifies texture cannot be felt.

Andrew Wear attributes the decline of Helmontian medical practice in England after the Great Plague to several factors: cultural attachment to bleeding, purges, and vomits; a chemical regime that was, despite its claim to intelligibility, "too mystical"; and the likelihood that chemical therapies promoted as "patient-friendly" may really have been "unpleasant and feared."[99] But the formal warrant of Helmontian therapy persists into corpuscular medicine along with its porous person. Indeed, while the Archeus recedes from corpuscular mechanisms of illness, Boyle affirms the "Medicinal Qualities" of "Expirations even from transparent Gems."[100] As an example of the "great Efficacy of Effluviums" evinced by "Mineral Exhalations," Boyle instances a gentleman whose prodigious nosebleeds are cured by "a Blood-stone, about the bigness of a Pigeons Egg."[101] The same stone "stop'd a Hæmorrogy in a neighbouring Gentlewoman, whom the violence of the Distemper kept from knowing that any thing had been applyed to her."[102] Boyle does not credit the obscure sensibility to which Helmont and Charleton attribute the curative powers of sapphires. But because the gentleman and lady are imperceptibly cured, the effluvial sympathy emblazoned in the gem's cognomen advertises the genealogy of Boyle's rupture with Galen. Like Helmont, Boyle compares the therapeutic agency of mineral "Effluxions" to the magnetic action of "vigorous Load-stones."[103] The bloodstone emits "agile and subtle parts"[104] that insensibly penetrate the gentleman; the gentlewoman is healed without feeling the stone's mineral exhalations enter her. Sydenham and Locke insist on the empirical inscrutability of corpuscular causes, but they continue to posit patients whose penetrable depths are amplified by the plague.

Dangerous Conversing: Defoe's *Journal*

Writing of the plague, the physician Thomas Sydenham (1624–1689) declares that "I dont pretend to define it exquisitly."[105] This disavowal communicates the gist of Sydenham's departure from corpuscular claims to causal insight: refusing the "vain pomp of nice Speculations" as "perhaps . . . no more useful to a Physitian, in the Cure of Diseases, than Skill in Musick to a Carpenter in *Building*,"[106] Sydenham endorses a strictly empirical discipline of medical knowledge that relegates inquiry into such causes as the atmospheric origins of plague among the "things about which the doating and arrogant crowd of Philosophers trifle."[107] With his colleague Locke, who sometimes accompanied Sydenham on his rounds after Locke arrived in London in 1667, Sydenham seems to affirm the inutility of speculations that refuse perceptual sanction.[108] Their jointly attributed manuscript *De Arte Medica* (1669) excludes from the mandate of experimental medicine "penetrat[ion] into the hidden causes of things,"[109] inciting Patrick Romanell to claim this text as a precursor to Locke's canonical formulation of empirical philosophy: "the major lesson of *An Essay Concerning Human Understanding* first appeared in clear and distinct outlines in the 'De Arte Medica' of 1669."[110]

Yet as doctors, Locke and Sydenham share the *Essay*'s ambivalent recourse to corpuscular figuration whose exile from the scene of empirical practice is difficult to enforce. In an echo of the *Essay*, Locke writes to the mathematician William Molyneux: "I wonder, that after the Pattern Dr. Sydenham has set them of a better Way, Men should return again to that Romance Way of Physick."[111] (Harold J. Cook reports: "When [Sydenham] was asked what someone should read before embarking on a practice, he is said to have replied *Don Quixote*, a book that would protect anyone from seeking illusions of the mind.")[112] Locke's equation of "Romance" and words that try to stimulate empirically inimitable feeling marks, as Romanell states, a reprisal of the *Essay*; but as with Locke's account of secondary powers, Sydenham does not unequivocally deny the relevance of particles to the etiology of plague. Decrying hidden causes, Sydenham writes, as "a fiction of the Brain,"[113] he defends strategies like moderate bleeding, purging, and treatment by manifest qualities.[114] But because plague is propagated by invisible means, Sydenham cannot avoid elaborating morbific matter's powers. By attributing plague to an "*Inflammation* of the most spirituous Particles of the Bloud,"[115] Sydenham postulates causes he cannot sense; he likewise effects a cure by entrapping effluvia he cannot see: "sometimes I put part of the sheet over the Face, to collect the vapors: for . . . the rays of the Morbific matter extend themselves to the circumference of the Body."[116] Sydenham's "rays" affirm the same formal mandate limned by Charleton and Boyle. Indeed, in an echo of Thomson's basilisk, Sydenham characterizes plague's insensibility by evoking its unfelt articulation of the victim's depths: "by reason of the great tenuity of its substance, 'tis fitted to pass through the innermost recesses of the Body like Lightning."[117]

By comparing plague to "Lightning," Sydenham extends the reach of insensible powers into the scene of empirical medicine. Although disease, he argues, is cured "not by the knowledge of the Causes, but by a convenient Method, approv'd by Experience,"[118] the plague's subtlety implicates the "innermost recesses" delineated by Starkey, Thomson, and Boyle. Sydenham insists that "eliminating the febrile matter through the pores of the skin . . . is . . . more troublesom and tedious" than bleeding and purging, but he evokes healing unassisted by medical art to echo Boyle's worsted socks: nature "gently expells [matter] through the habit of the Body."[119] With the empirical doctor Sydenham's vision of his patient's insensibly porous "habit," I turn to the form of character in Defoe's *Journal*.

* * *

H.F., the otherwise unnamed narrator of Defoe's *Journal of the Plague Year*, recounts a scene that heralds the flight of nearly 40 percent of London's population at the height of the fatality: "People of the better sort, and Horsemen attending them, and all hurrying away . . . was a very terrible and melancholy

Thing to see, and as it was a Sight which I cou'd not but look on from Morning to Night; for indeed there was nothing else of Moment to be seen, it filled me with very serious Thoughts of the Misery that was coming upon the City, and the unhappy Condition of those that would be left in it."[120] When H.F. writes that the flight of the upper classes from London in June 1665 "filled me with very serious Thoughts," the trajectory of these "Thoughts" shows Defoe's affinity with the discourse of empiricism. Locke's *Essay* offers an influential depiction of empirical understanding that proceeds from the outside in: "The Senses at first let in particular *Ideas*, and furnish the yet empty Cabinet."[121] H.F.'s references to "a Thing to see" and "a Sight" credit the origins of knowledge inaugurated when external things are "let in" by the senses, affirming the receptivity of a first-person "me" that can be "furnish[ed]" or, as H.F. writes, filled. As a warrant for his narration of the Great Plague, H.F. deploys prose that aims to trigger the response historically induced by his sight: "were it possible to represent those Times exactly to those that did not see them, and give the Reader due Ideas of the Horror that every where presented it self, it must make just Impression upon their Minds, and fill them with Surprize" (17–18). A simulation of H.F.'s sight offered to "those that did not see," *Journal* would repeat in these readers the same process Locke traces to justify how human understanding occurs: "For since there appear not to be any *Ideas* in the Mind, before the Senses have conveyed any in, I conceive that *Ideas* in the Understanding, are coeval with *Sensation*; which is such an Impression or Motion, made in some part of the Body, as produces some Perception in the Understanding."[122] Locke's keyword for external stimulation and its ensuing mental idea, "Impression" vindicates the origins of H.F.'s "very serious Thoughts" as well as his ambition to fill the minds of the *Journal*'s readers.

As Sydenham affirms, the plague also fills porous persons. But because plague particles insensibly mingle with air, they cannot, in Locke's words, "produce[] some Perception in the Understanding." Hidden within a transparent medium, morbific matter does not trigger empirical ideas. In "An Experimental Discourse of Some Unheeded Causes of the Insalubrity and Salubrity of the Air" (1685), Boyle argues that air is rendered deadly by "the impregnation it receives from Subterraneal *Effluvia*"; as Boyle hypothesizes, "Noxious Subterraneal Fumes . . . suddenly, and without warning belch'd up," enter the air unheeded to cause "the dreadful *London* Plague that broke out in the year 1665."[123] "*Subterraneal Reeks*" wield pestiferous powers, Boyle suggests, because they communicate toxins consumed with even less awareness than breath: "some parts of the substance of the Air (for I speak not of its Heat, Coldness, or other such Qualities) do not onely affect humane bodies . . . as they are taken in by Respiration, but as they outwardly touch the Skin: and the Skin being . . . full of Pores, and those perhaps of different sizes and figures, those Corpuscles that get in at them may have their operation, even upon the most inward parts of the body."[124] The imperceptibility of Boyle's corpuscle correlates with its capacity to induce form. "Parts" of air that do not trigger sensible ideas like "Heat, Coldness, or other such Qualities" portend plague's unfelt entry into persons, because the capacity to evade sense dictates how far tiny parts can "get in." Corpuscles minuscule enough to reach "the most inward parts of the body" define inwardness as a site of empirically inaccessible knowledge. As Thomson remarks, the plague "treads so softly within us, that it cannot be heard to walk."[125]

Defoe's *Journal* derives character's form from insensible effluvia that harbor what H.F., following Helmont, Starkey, and Thomson, calls "Seeds of the Plague" (233), "seeds of Infection" (167, 196), and "Seeds of the Contagion" (189). Air replete with seeds impels inwardness as a function of epistemology, not psychology. Indeed, to signal the different status of his personal musings from the introspection of a protagonist like Robinson Crusoe, H.F. denies the exemplarity or interest of his personal "intervals": "What I wrote of my private Meditations I reserve for private Use, and desire it may not be made publick on any Account whatever" (75). The interiority that occupies the *Journal* is not meditative or sentimental. Nor is it biographically or historically refined, as H.F. disclaims after recounting one family's escape from incarceration with their sick maid: "I cou'd give a great many such Stories as these . . . and which are very certain to be true, or very near the Truth; that is to say, true in the General, for no Man could at such a Time, learn all the Particulars" (51–52). Plague cuts characters' knowledge to the measure of the "General" rather than the "Particulars," as H.F. explains again: "It was for want of People conversing one with another . . . that it was impossible any particular Person cou'd come at the Knowledge of all the extraordinary Cases that occurr'd in different Families" (158). As the categorical impossibility of "particular" understanding affirms, Defoe's *Journal* excavates interiority that is not personal but, on the contrary, *formal*. While "form" refers most topically to statutes, norms, or patterns,[126] Defoe's "People" implicate the mutually empirical and structural composition of persons that possess unknowable or hidden recesses. Form also refers to stipulations, as Frances Ferguson and Sandra Macpherson suggest of tort or strict liability law, whose models of action displace individual intention or engender it impersonally, retroactively, and statutorily.[127] In *Journal*, the form of character is effected by neither psychology nor the radical foreclosure of volition, but the empirical limits of one's perception of one's own corpuscular interiority.

In Defoe's *Journal*, characters contain unknowable or secret things. But secrecy does not correlate, even from the side of its bearer, with subjective particulars. Instead, insensible seeds of plague determine character's form. This claim departs from histories of the novel that coordinate the progressive deepening of character with personalizing detail, because the *Journal* heralds a novel that does not equate interiority with the reader's access to psychology or a representative inner self.[128] The depth excavated by plague's insinuating physicality does not engender the early modern individual posited by Ernest Gilman, according to whom pestilence "would expose the perilously fragile boundaries upon which the bounded subject depends."[129] However precarious it may be, a "bounded subject" does not model persons who are already porous. Instead, penetrable persons render character's manifest attributes—for some indeterminate length of time, or for the remainder of a lifetime—signs the surety of whose reference to hidden contents cannot be plumbed, even in the medium of first-person narrative. In Defoe's *Journal*, interiority harbors impersonal and unverifiable danger.

H.F. reports of the plague's transmission to England: "the first Person had the Infection, was generally said to be, from a Parcel of Silks imported from *Holland*, and first opened" (196) in London. Contagion ensues when something is "opened," unleashing events that define persons, buildings, and the city as formal cognates. Offering a bird's-eye view of the progress of pestilence whose cumulative statistical impact is revealed by the weekly Bills of Mortality, H.F. remarks that "the thing began to shew it self" (8): even mathematically abstracted, "the thing" respects the topographic logic of a belatedly surfacing symptom, as when, in early summer 1665, "the Bills rise high, . . . [and] began to swell" (8). Just as the plague's mortal consummation—an endpoint whose chemical fatality H.F. stresses by likening pestilence to the London Fire of 1666—is imminent in Bills which "began to swell," the outcome harbored by history resembles a carbuncle that has not yet breached the surface of the patient's skin: "the Plague, was not, as I may say, yet broken out" (29). Of buboes whose matter must be expelled, H.F. notes that "the Surgeons had, it seems, hopes to break them" (165); incarcerated victims, themselves a kind of morbific agency that likewise cannot be held in, "would break out at all Adventures" (53). By summer 1665, when "the Distemper spread openly, and could not be conceal'd" (198), pestilence occasions unprecedented excrescences: of dissenting preachers able to sermonize without fear of persecution, H.F. writes that "those who they called silenced Ministers, had their Mouths open'd on this Occasion" (169); of magicians, astrologers, and fortunetellers peddling supernatural charms, "this Trade grew so open . . . that it became common to have Signs and Inscriptions set up at Doors" (28). During the plague, London succumbs to an efflorescence of "Signs," some of which persist after the crisis is past: of one burial ground whose mandated shallowest depth of six feet renders London's past as topographically imminent as its future, H.F. remarks that "the Mark of it also was many Years to be seen in the Church-Yard on the Surface" (59). As a participant in the appearance to empirical perception of contents that "would break out," H.F. advances a rare stake in his present tense, venturing the persistence of plague's formal impact on the city: "I believe I could go to the very Place and dig their Bones up still" (97).

Defoe's *Journal* recapitulates the formal imperative of plague medicine, which accommodates chemical physicians like Starkey and Thomson as well as the empiricists Sydenham and Locke. Boyle anticipates this ecumenical stance when he explicates the "Eye-witness" testimony of a "Doctour (whose name I am sorry I have forgotten)" who during the plague contrives a way to save patients after their buboes rise to the surface:

> [U]pon a little refrigeration of the Body by the Air, and oftentimes by the very fear that disheartened the Patient, the Tumours would suddenly subside, and the Pestiliential Matter recoiling upon the Vital Parts, would quickly dispatch the fatal work . . . [This doctor] usually did . . . enable and excite Nature to expell the peccant Matter into a Tumour; for then he presently clapp'd on an appropriated drawing Plaster, which would never suffer the Tumour to subside; but break it, or make it fit for opening, and thereby give Nature a convenient vent, at which to discharge the matter that oppress'd her.[130]

Motivated by "*Philanthropy*," Boyle includes the Latin text of the recipe for this doctor's "drawing Plaster." He also cites the far simpler "Antidote" used daily by another physician who survives employment at "a great Pesthouse": "a little Sea-salt dissolv'd in a few spoonfulls of fair Water; which . . . kept his Body soluble without purging or weakning it."[131] These case histories recombine Helmontian and empirical strategies. By citing "the fear that disheartened the Patient," Boyle recurs to the consentaneousness of a low-spirited Archeus; when he refers to toxic matter "recoiling upon the Vital Parts," Boyle confirms Thomson's and Sydenham's claims for subtle rays of disease which, deflected deep into the patient, thwart the directives of chemical and empirical medicine alike. The persistence of Boyle's first cure into Defoe's novel is evident: to "expell" inwardly amassed "peccant Matter," the doctor draws corpuscles from center to surface, which exit through the patient's now salutary "opening." The mechanism of the second prophylaxis, salt-water that makes the physician "soluble," is echoed by Defoe when, in an infrequent reference to his own health, H.F. recovers from a random malaise three days after he "rested well, [and] sweated a little" (15). By sweating, H.F. enacts the formal and chemical mechanism that makes pervious persons vent.

Defoe was neither a doctor nor a chemist, but *Journal* reflects his education at Charles Morton's dissenting academy from 1674 to "late 1679 or early 1680," where, Paula Backscheider recounts, Defoe received Morton's "revolutionary . . . science instruction." This included laboratory training and the study of current work by "Harvey, Newton, and Hooke" as well as "Boyle, Gilbert, Gassendi, and Descartes."[132] Published after their author moved to Harvard University, Morton's science lectures, the *Compendium Physicae* (1687; composed around 1680 but imparted to his students before then), mobilize Boyle's corpuscular chemistry to refute scholasticism, as when Morton approves modern "men . . . [who] do not allow the term (Substantial) to any form but the Soul of Man. See Mr. Boyles treatise of essential forms" [*The Origin of Forms and Qualities*].[133] Among other topics deemed "more fully handled by Mr Boyle," Morton includes "a transmutation of all mettals into Gold, by curing the Leprosities of them."[134] Defoe's *The History Of the Principal Discoveries and Improvements, In the Several Arts and Sciences* (1727) contains a chapter, "*Of the Discovery of the Magnet or Loadstone*," whose synopsis, Defoe writes, is "sum'd up out of the Learned Mr. *Boyl* and others."[135] Most tellingly, Defoe cites "the continual passage of the subtile magnetical Particles *through*" bars of metal to justify their transformation into "a kind of *Magnet* themselves."[136] Corpuscular science, also evident in H.F.'s references to seeds of plague, comprises the *Journal*'s epistemic backdrop. Empirical medicine's ongoing recourse not only to "magnetical" powers but to consentaneous patients persists when H.F. refers to those who "fell presently ill by the Fright" (71) as well as to plague-time debates over whether Helmontian "Fermentation and Heat already in the Blood" (210) is relieved by warmth or cold.

Boyle's doctrine of qualities redresses a stubborn mistake: as Locke repeats in the *Essay*, the assumption that perceived qualities are resident as such in the isolated object. In his interview with an exceptionally intrepid survivor, Boyle claims the occasion of plague to expose the reliance of sensed endowments on the health or disorder of the observer's perceptual apparatus:

> Being a while since in a Town, where the Plague had made great havock, and inquiring of an ingenious man, that was so bold, as without much scruple to visit those that were sick of it . . . he told mee . . . that he was able to tell divers Patients . . . before they . . . had any evident symptomes of the Plague, that they were indeed infected, upon peculiar observations, that being asked, they would tell him that the neighbouring objects . . . appear'd to them beautifi'd with most glorious Colours, like those of the Rainbow, oftentimes succeeding one another; and this he affirm'd to be one of the most usual, as well as the most early symptomes, by which this odd Pestilence disclos'd it self.[137]

In this scene, the discovery of "evident symptomes" is possible neither for the "ingenious man," to whom such evidence is invisible, nor for the infected but apparently sound victim, who cannot perceive the infinitesimal matter he harbors. What shows this person to be "indeed infected" is not the vehicle of the illness, whether an effluvium or a toxic particle, but rather his altered apprehension of "neighbouring objects." To hazard a mechanical explanation that Boyle does not undertake, upon infection his eyes seem to act like prisms. One of this plague's crucial "symptomes" is thus not how patients appear to the world but how the world "appear'd to them": if, in the perceptual regime espoused by Boyle, particulate excitation of human organs provokes the same sensations in everybody, then plague propels the victim's fall into a skeptical universe where any object can produce ideas of any color, or a rainbow of colors in succession.

In Defoe's *Journal*, H.F. recounts a story similar to Boyle's. It involves an observer who, like Boyle's "ingenious man," is armed with a supplemental method of apprehending infection:

> I knew a Man who conversed freely in *London* all the Season of the Plague in 1665 . . . and he had such a Rule to know, or have warning of the Danger by, as indeed I never met with before or

> since . . . He had a Wound in his Leg, and whenever he came among any People that were not sound, and the Infection began to affect him, he said he could know it by that Signal, (*viz.*) That his Wound in his Leg would smart, and look pale and white; so as soon as ever he felt it smart, it was time for him to withdraw . . . Now it seems he found his Wound would smart many Times when he was in Company with such, who thought themselves to be sound, and who appear'd so to one another. (184–85)

Defoe adapts Boyle's *New Experiments Physico-Mechanicall, Touching the Spring of the Air* (1660), which supports the likelihood of "strange Ebbings and Flowings, as it were, in the Atmosphere" by evidencing "the Pains and Aches that are often complain'd of by those that have had great Wounds or Bruises, and that doe presage great Mutations in the Air . . . whilst to strong and healthy Persons no sign of any such thing appears."[138] For Boyle, "healthy Persons" cannot apprehend a "sign" elicited only by those whose bruises make them susceptible. For Defoe as well, the problem is how to apprehend "Danger" or, rather, how to conjure an operational symptom from "such, who thought themselves to be sound, and who appear'd so." Whereas Boyle's ingenious observer has hit upon the right question to ask of victims in whom any given object triggers a rainbow of colors, Defoe's "Man" actually possesses an extra sensory orifice, his wound.

Just as Locke declares that secondary qualities exist only because there are organs to, in effect, create them—"let not the Eyes see Light, or Colours, nor the Ears hear Sounds; let the Palate not Taste, nor the Nose Smell, and all Colours, Tastes, Odors, and Sounds . . . vanish and cease"[139]—so Defoe's man's wound functions to convert impalpable matter into a sensible sign or, as Defoe writes, a "Signal." Registered by this supplemental sense, his companion's still undiscerned infection becomes the wound's secondary sensation of "smart[ing]," as well as "look[ing] pale and white." In the manner of the ingenious man who ascertains that infected people see rainbows, Defoe's wound converts impalpable matter into sensory ideas the observer can "know." Such sensitivity may seem improbable, but it respects the premise of Boyle's doctrine of qualities. To know who is infected, one must find a body part stimulated by impalpable danger. What portends such danger is not immateriality but primary materiality for which there exists no receptive organ. In normal historical times, Boyle repeats, heightened awareness aggravates a sick person's debility: "those subtile Steames that wander through the Air . . . are wont to be painfully felt . . . more constantly by men that have had great Bruises or Wounds."[140] But as H.F. makes clear, in plague-time the sensitivity to effluvial impressions sustained by "Wounds" affords a life-saving opening, the experience of a quality—pain—triggered by matter whose powers no one else yet knows.

Defoe's man's wound adjudicates between empirical perception and sickness that eludes human sense. But it also intervenes in the medium of thought, because it is in this medium that the infected person misapprehends himself. Such persons, H.F. writes, "thought themselves to be sound," because without the enhanced impressionability afforded by a wound, plague in H.F.'s London is not initially perceived by those who contain it. To demonstrate the empirical limits of internal self-understanding, H.F. relays a dialogue driven by an "untouch'd" "Citizen," who, in the "Month of *September* [1665], . . . was mighty chearful, and something too bold" (188):

> Says another Citizen, a Neighbour of his to him, one Day, *Do not be too confident Mr.—it is hard to say who is sick and who is well; for we see Men alive, and well to outward Appearance one Hour, and dead the next. That is true*, says the first Man . . . *I hope I have not been in Company with any Person that there has been any Danger in.* No! Says his Neighbour, *was not you at the* Bull-head *Tavern in* Gracechurch Street *with* Mr. — *the Night before last*: YES, says the first, *I was*, but *there was no Body there, that we had any Reason to think dangerous*: Upon which his Neighbour said no more, being unwilling to surprize him; but this made him more inquisitive, and as his Neighbour appear'd backward, he was the more impatient, and in a kind of Warmth, says he aloud, *why he is not dead, is he!* upon which his Neighbour still was silent, but cast up his Eyes . . . at which the first Citizen turned pale, and said no more but this, *then I am a dead Man too*, and went Home immediately, and sent for a neighbouring Apothecary to give him something preventive, for he had not yet found himself ill; but the Apothecary opening his Breast, fetch'd a Sigh, and said no more, but this, *look up to God*; and the Man died in a few Hours. (189; emphasis Defoe's)

This dialogue marks one of the most affectively nuanced scenes in the *Journal*. But it is the intuition of illness from the vantage of external relations that propels the "unwilling," "backward," "inquisitive," "impatient," and "Warm[]" escalation of the exchange. Rather than refining either man's personality, their dialogue dramatizes the citizen's discovery of a fate apprehended not from the inside out but from the outside in, a discovery that implicates the knowledge of others. While the plague rages, "*I am a dead Man*" does not signal insight that originates from the inside but empirical knowledge gleaned in conversation with one's neighbors. The citizen's encounter with a "Person that there has been any danger in" must be confirmed by an observer with a better understanding of what is now "*in*" the citizen himself. The still healthy citizen responds to his neighbor's insight by "turn[ing] pale": like the wound, he manifests fatal inevitability, even though he cannot yet know it from within.

When this doomed man futilely calls for "something preventive," he echoes the therapeutic advice promoted by Defoe's *Due Preparations for the Plague* (1722), a text written, with the *Journal*, in response to the contemporary threat of a plague outbreak in Marseilles. In *Due Preparations*, Defoe attests that "I have known that some have preserved themselves in the last plague, 1665, by only having a bottle of vinegar in their hands, and being continually smelling to it. I myself have rid through a town infected with the plague with a bunch of rue in my mouth, and have been secured."[141] Vinegar and rue are odorous prophylaxes that block, *Due Preparations* asserts, "poisonous or infectious effluvia,"[142] affirming the causality cited in the *Journal* to ratify the Lord Mayor's policy of killing "all the Dogs and Cats" in London: "because as they were domestick Animals, and are apt to run from House to House . . . so they are capable of carrying the Effluvia or Infectious Steams of Bodies infected, even in their Furrs and Hair" (117). While human characters harbor plague in their depths, dogs and cats trap "Steams" in their fur. In this sense, Defoe's novel is not precisely Cartesian: animal-human difference registers not as the presence or absence of soul but rather as the contrasting formal levels at which bodies are penetrated by the effluvia they pass "from House to House."

Defoe concurs with Boyle on the infectious agency of effluvia, but *Due Preparations* registers a shift of emphasis from Boyle's attribution of plague to underground belches. For Defoe, "Infectious Steams" must pass between persons: "the infection is taken from one to another by the infected bodies emitting poisonous particles either from the pores of the body or from the breath, or from some malignant effluvia which pass from the body infected and are received by the body at that time to be infected."[143] Defoe pens *Due Preparations* to promote strategies guided by "the common hypothesis, namely, that the distemper is what we call catching or contagious . . . and that conversing with those who are infected gives the infection."[144] The aspect of pestilence that compels Defoe to intervene is the prospect that its transmission is "catching." Against

Boyle and later proponents of a miasmist theory of plague,[145] Defoe insists that human communication is requisite to the distemper's spread: "plague is carried from one to another by infected persons conversing with one another . . . and not by any general stagnation of air, or noxious fumes infecting the air, or poisonous particles carried by the winds."[146] Rather than atmospheric malaise, what perpetuates and worsens plague, Defoe repeats, is the necessity of persons "conversing." To return to the citizen's discovery that "I am a dead Man too," this man's passage to the grave from "*the* Bull-head *Tavern in* Gracechurch Street *with* Mr.—" maps a blueprint of the course that such conversing might take as well as an enactment of the man's necessarily belated comprehension of what he has caught. The convergence of his self-understanding and his death stages the fatal revelation of form: only with the discovery that "*I am a dead Man*" does his "I" coincide with the knowledge of what he contains. As conversation in which two characters demonstrate how one's inner knowledge must be externally induced by the other, this scene represents the form of a person whose depths cannot be apprehended from within. *Journal* represents the conversing entailed by this form as well as the form entailed by this conversing.

H.F. declares of persons who do not know whether they "had received the Contagion": "These were the dangerous People, these were the People of whom the well People ought to have been afraid; but then *on the other side* it was impossible to know them" (184; emphasis Defoe's). Defoe belabors the policy conundrum unleashed by the impossibility of knowing who harbors the plague: namely, "shutting up the WELL or removing the SICK will not do it," because "People have it when they know it not" (185). H.F. thereby mitigates the force of his monitory "ought": why ought well people fear the dangerous ones if, for all the well people "know," they are dangerous too? Indeed, the only people in London's plague who are not dangerous in this mutually formal and empirical sense, who do not confound apparent wellness and the deathly impossibility of knowing, are those who are, as H.F. writes, "openly visited" (193).

To distinguish the threat posed by openly visited people from the threat posed by "apparently well" (191) people, H.F. invents a third category composed of what might be called epistemologically dangerous people:

> [I]t was not the sick People only, from whom the Plague was immediately receiv'd by others that were sound, but THE WELL. *To explain my self*; by *the sick* People I mean those who were known

> to be sick, had taken their Beds, had been under Cure, or had Swellings and Tumours upon them, and the like; these every Body could beware of . . .
>
> By *the Well*, I mean such as had received the Contagion, and had it really upon them, and in their Blood, yet did not shew the Consequences of it in their Countenances, nay even were not sensible of it themselves, *as many were not* for several Days: These breathed Death in every Place, and upon every Body who came near them. (183–84)

According to H.F.'s refinement of plague vocabulary, "*the sick*" are manifestly "known to be sick"; their buboes, pain, fever, and distraction are even extravagantly perceptible, both to themselves and to others. Yet "*the Well*" are also, in H.F.'s revised taxonomy, "really" sick; however, neither they nor the persons with whom they converse possess empirical knowledge of their infection or, that is, are "sensible of it." The well harbor "Contagion" whose materiality H.F. underscores with the stressor "really"; but before it produces perceptible symptoms, this corpuscular reality cannot be empirically known. Although the plague exists "in their Blood," it has not yet produced "Consequences" that "shew." The lag between infection and sensible symptom obtains both on the outside, in "their Countenances," and on the inside, in the minds of persons oblivious to what they contain. Although he endorses preemptive self-incarceration on the part of both the well and the apparently well, H.F. vouches for the suspension of self-knowledge that, for the duration of their nescience, relieves these persons of liability for spreading contagion they cannot apprehend: "I suppose them to be utterly ignorant of their own Condition" (192).[147] In Defoe's novel, "*the Well*" internalize the gap between insensible matter and its power to be humanly felt. That is, the well internalize primary-secondary difference to assume a form of character defined by the unknowable contents of all not openly sick persons.

While he narrates his own infection, the plague doctor Thomson situates himself at a unique textual juncture. As a result of his autopsy of the servant boy, he attests, "I entered into more than ordinary familiarity with a dead body." Because the pores of his hand are, in turn, entered by the corpse's pestilential seeds, Thomson acquires empirical knowledge not from one "body" but two: "what the Cadaver could not teach me of it self, was infused into me to my sad Experience." The source of Thomson's medical understanding is thereby peculiarly confused: "I shall now deliver to you the Physical Observations I made, resulting from both dead, and my own living Body."[148] Thomson's use of the first person plays upon the narrative opportunity that his infection by an opened corpse affords. Experimental knowledge is "infused into me," but it is seeds of plague, not sensible ideas, to which "more than ordinary familiarity" with a cadaver refers. Thomson deftly leverages the prospects his porosity presents for narrative: once he has "entered into . . . familiarity" with a carcass, its fatal contents penetrate deep inside him. Only belatedly does "sad Experience" ratify Thomson's knowledge of his perviousness.

Defoe does not exploit the narrative possibilities of the plague doctor's position. H.F. passes through the Great Plague unvisited, and his resistance to infection correlates with an explanatory stance that resembles Sydenham's more than Thomson's. After refusing to detail "the Physick or Preparations we ordinarily made use of on this terrible Occasion" (228), H.F. distills plague medicine into the formal imperative that patients' buboes be "broke and kept open" (230). Unlike those who intuit microcauses, H.F. affirms that "from the whole I found, that the Nature of this Contagion was such, that it was impossible to discover it at all" (195). Yet Defoe is committed to a corpuscular etiology of plague, a commitment motivated by both theological and disciplinary concerns. Reflecting a strain of piety that mitigates against vitalist speculation into infinitesimally animate causes, H.F. rejects as "a Discourse full of learned Simplicity" the agency of microscopic or "invisible Creatures" (73), "strange monstrous and frightful Shapes, such as Dragons, Snakes, Serpents, and Devils, horrible to behold: But this I very much question the Truth of" (195). Whereas miniature monsters constitute a literal embodiment of superstition, the belief that plague is divinely enjoined on individual sinners dissolves physical causes into enthusiastic fears of supernatural vengeance. H.F. refuses to surrender the agency of infection to "an immediate Stroke from Heaven . . . having Commission to strike this and that particular Person," because such a supposition is both "the Effect of manifest Ignorance and Enthusiasm" (73) and an incitement to the fatalistic disregard of public safety, since enthusiasts abdicate liability for movement Defoe is at pains to preserve: "if it pleas'd God to strike them, it was all one whether they went Abroad or staid at Home" (185). Fanatics who attribute plague to an "immediate" supervening divinity mimic the "profess'd predestinating Notions" H.F. routinely ascribes to "*Mahometans*" (13) but whose more topical historical referent is seventeenth-century British strains of hard-line Calvinist predestinarianism.[149]

Following Boyle, Defoe ascribes plague to "natural Causes" (21) whose first source is divine. Boyle "confess[es]" himself "at a loss about it's Origine" to warn against explanations that "recur without an absolute necessity to Supernatural Causes, for such Effects as do not manifestly exceed the power of Natural ones: though the particular manner of their being produc'd, is perchance more than we are yet able clearly to explicate."[150] H.F. likewise accommodates plague to divine "Judgment" of the "crying Vices" (17) of Charles II and his court while affirming that "it was really propagated by natural Means" (186). Indeed, for Defoe, the plague signals an exemplary instrument of divine punishment precisely because its spread is physical and insensible at once: "Now 'tis evident, that in the Case of an Infection, there is no apparent extraordinary occasion for supernatural Operation, but the ordinary Course of Things appears sufficiently arm'd, and made capable of all the Effects that Heaven usually directs by a Contagion. Among these Causes and Effects this of the secret Conveyance of Infection imperceptible, and unavoidable, is more than sufficient to execute the Fierceness of divine Vengeance, without putting it upon Supernaturals and Miracle" (187). For Defoe, the empirical fallout of contagion, the conversion of everybody—except the openly visited—into dangerous people, vindicates the mutually "natural" and "divine" agency of the plague. Plague's "secret Conveyance" is "imperceptible" and yet "really" material, neither the retaliation of a vengeful deity nor a microscopic devil. Corpuscular causation consummates "unavoidable" "Vengeance" as the very form of empirical knowledge: as Defoe's *Journal* shows, conversing entails the fatality of character that eludes human understanding. For Defoe, as for Boyle and Morton, plague seeds enable divine judgment. This proposition does not idealize infectious matter, as tens of thousands of Londoners attest, but rather mobilizes the deathly power of secondary qualities in the form of characters who must, of necessity, converse.

To show that minuscule particles cause plague, Boyle and Defoe insist that air can be made experimentally manifest. In a historically long-standing proof of the existence of imperceptible parts that volatilize and then accumulate in the chemist's apparatus, Boyle claims that when heated, "very agile Corpuscles" of "inflammable *Sulphur* . . . fasten themselves all about to the inside of the Receiver, and there compos'd divers thin Coats, or Films, as 'twere, of Sulphureous matter sticking to one another."[151] Layered "Coats, or Films" induce the visibility of singly insensible particles of airborne "matter." Defoe's *Due Preparations* derives the imminent perceptibility of what air may carry directly from plague victims themselves: "in the houses which were infected, and had been shut up, and where several persons, or perhaps the whole family, had died, there was a strange clammy or dewy sweat on the inside of the glass of the window . . . That this was the poisonous air breathed out of the infected people's bodies . . . was not doubted."[152] In this calamitous experimental scene, the quarantined house becomes something of a supersized "Receiver," its panes of glass rendering sensible contagion that proceeds directly from the depths of the incarcerated family. While Boyle resists attributing the plague to the "Internal Putrefaction"[153] of infected persons, Defoe insists: "We receive poison one of another, and we emit poison one to another."[154] For Defoe, conversing persons insensibly "emit poison." In the *Journal*, plague realizes divine judgment not because of ambient miasma but because of the inescapability of conversation with others. People who breathe poison are not, in Defoe's novel, trapped under glass that corrects for empirical perception's limits. While the quarantined house fixes poisonous air as palpable "sweat," persons who converse receive and emit what can never be seen.

H.F. retraces contagion's path: the plague "secretly, and unperceiv'd by others, or by themselves, communicated Death to those they convers'd with, the penetrating Poison insinuating it self into their Blood in a Manner, which it is impossible to describe, or indeed conceive" (193). At the same time that plague's vehicle "is impossible to describe," it possesses a physical reality and a formal destination that H.F. invokes with confidence. Particles that remain "unperceiv'd" enter conversing persons to make innerness unknown: "The acute penetrating Nature of the Disease it self was such, and the Infection was receiv'd so imperceptibly, that the most exact Caution could not secure us while in the Place" (187). Persons by whom plague "was receiv'd so imperceptibly" internalize primary-secondary difference as the form of character in Defoe's novel. To return to Locke's claim for the simultaneity of impression and idea, H.F. elaborates the paradox of a thing that is received yet not felt, "penetrating" and yet imperceptibly taken in.

Locke's influence on Defoe's *Journal* is evident in the ideas that fill H.F.'s mind. But just as profound a determinant of this novel are corpuscles that excavate depth not as psychology, not as bounded selfhood or transparent self-knowledge, but as the fatality that attends empirical understanding. If, in the history of the novel, Defoe's *Journal* exemplifies nascent narrative realism, then its remit is not the transparent and immediate representation of external objects, but seeds of plague "unperceiv'd" even by the persons they enter. In Defoe's *Journal*, character proceeds from matter whose powers can be only belatedly apprehended, and plot accrues in conversation with the unknowable innerness of others.

CHAPTER 4

Race and the Corpuscle

The Stranger view'd the Ladys with wonder, their strange Habit and tauny Complexions ill agreeing with the sweetness of their Features, and delicate Hands and Limbs.

—Penelope Aubin, *The Life of Charlotta Du Pont, An English Lady* (1723)

SERAGLIO, a Bawdy-house; so called from the Great Turk's Palace.

—*A New Canting Dictionary* (1725)

Touching Color

The playwright, theater historian, publisher, and promoter of Eliza Haywood's *Love in Excess* (1719) William Rufus Chetwood (d. 1766) wrote four travel novels, including *The Voyages, Dangerous Adventures And imminent Escapes of Captain Richard Falconer* (1720).[1] In *Falconer*, Chetwood restages the scene in which *Love*'s coquette Melantha tricks the Count D'elmont by switching bedrooms with the object of his desire. Taking Melantha for his coveted ward Melliora, the Count ravishes the coquette, unaware of the exchange until his spurned wife Alovysa enters the room: "Melantha slunk under the [bed]cloathes, and the Count started up in the bed at the first appearance of the light, which Alovysa had in her hand."[2] Roused from sleep by Alovysa's rage, Melliora herself then appears at the door. Haywood's syntax captures the empirical impossibility experienced by the Count at this moment: "He beheld the person, whom he thought had lain in his arms, whom he had enjoyed, whose bulk and proportion he still saw in the bed . . . come from a distant chamber." At this bewildering instant, the person the Count "thought had lain in his arms," the person he "had enjoyed," is at the door and in the bed at once. Primary endowments of her body experienced as "bulk and proportion" split off from secondary attributes whose ideational status Haywood exacerbates when the baffled Count wonders: "is every thing I see and hear illusion?"[3]

While sheets play a role in Melantha's ruse, Alovysa's interruption points to the condition that allows the coquette to perpetrate her amorous fulfillment. Melantha hides beneath the bedclothes because "of the light, which Alovysa had in her hand": before then, her identifying features are obscured by darkness. (An enabling corollary of this episode is Haywood's presumption of the dullness of rakish touch.) In *Falconer*, Chetwood replicates Haywood's scene to place pressure on the sensible quality occluded by light's absence: color.

Left ashore by his shipmates on the island of Dominica, Chetwood's protagonist Falconer is captured by a group of Indians. Rescued from death by "one of the Women," Falconer is compelled by the governors of the tribe to choose a female Indian "to be my Mate, or Bedfellow, or suffer Death."[4] Selecting the woman who saved him, Falconer eats as part of "the Ceremony" that makes her "my Bride": "I must confess I was so hungry, that I had a very good Appetite to my Victuals, being I had not eat any Flesh for four Days: But I had no great Stomach to my Bride, although a young well-featur'd Woman, yet her Complexion did not like me. When I had done Eating, my Bride and I were put into a Hutt, and shut close without any Light; but the old Proverb, *Joan's as good as my Lady in the Dark*, had like to have prov'd no Proverb to me."[5] With Haywood, Chetwood distinguishes "well-featur'd" contours from endowments rendered imperceptible by a bedchamber "without any Light." But in Chetwood's version, darkness does not operate to conceal traits that particularize Haywood's two flesh-colored—to repeat the Lockean condensation that insinuates geographical bias into simple ideas—Frenchwomen. Rather, Falconer's unlit bedroom obviates the universalizing specification of phenotypic whiteness itself. Because the secondary quality annulled by darkness is hue or "Complexion," Chetwood's text recapitulates the ontological instability endemic to literary and natural philosophical constructions of racial difference over the long eighteenth century.[6] Does "a Hutt shut close without any Light" restore Falconer's bride to the presumption of normative whiteness confirmed by the irrelevance of complexion to Melantha's and Melliora's swap? (Distinguished by this would-be primary attribute, they *are* the same.) The outcome of Chetwood's consummation scene, which accommodates complexion to the indifference of servant ("*Joan*") and lady, remains unclear. Darkness may not extinguish Falconer's failure to "like" a bride in whom complexion somehow persists even when he cannot perceive it.

As *Love*'s advertiser, Chetwood must have registered the similarity of these two bedroom scenes.[7] *Falconer* echoes *Love* to transpose domestic amorous tropes into a global frame. But because *melas* signifies black or dark, Melantha anticipates the crux of Falconer's situation: enjoyed in the absence of light, she may unmoor the regulating presumption of her own whiteness.[8] "Melantha" thereby transforms the one quality *Love*'s manifest referential range seems to neutralize into its opposite. More concretely, she animates a conundrum central to the corpuscular and empirical doctrine of qualities. Haywood and Chetwood stage the critical question posed by Robert Boyle and Isaac Newton: Are bodies colored in the dark?

To argue that scientific claims for color's relation to matter are restaged in eighteenth-century narrative treatments of racial difference, this chapter first delineates Newton's disagreement with Boyle over the prospective residence of color in bodies. The next two sections examine texts that grapple with the capacity of corpuscles to incarnate colorific identity. John Arbuthnot's *Essay Concerning the Effects of Air on Human Bodies* (1733) motivates national difference from the outside in, difference that, as a mechanical effect, ultimately exploits the sameness of particles that obey Boyle's Law. John Mitchell's *Essay upon the Causes of the different Colours of People in different Climates* (1744) refers human color to fragments of skin too tiny to bear it. My chapter's final section suggests that Boyle and Newton map the options engaged by Chetwood and his contemporary, the novelist Penelope Aubin, in their variant representations of the ontology of racial difference. I turn now to Newton's break from Boyle.

In *Experiments and Considerations Touching Colours* (1664), Boyle argues that color is produced when an object's particulate superficies disturb incident white light. Boyle offers a hyper-refined vision of the topographic access to color promised by microscopy:

> [I]f we were Sharp sighted enough, or had such perfect *Microscopes* . . . our promoted Sense might discern in the Physical Surfaces of Bodies, both a great many latent Ruggidnesses, and the particular Sizes, Shapes, and Situations of the extremely little Bodies that cause them, and perhaps might perceive . . . how those little Protuberances and Cavities do Interrupt and Dilate the Light, by

> mingling with it a multitude of little and singly indiscernable Shades, though some of them more, and some of them less Minute, some less, and some more Numerous; according to the Nature and Degree of the particular Colour we attribute to the Visible Object.[9]

This passage stands as an antecedent to the episode in Locke's *Essay* that speculates on the epistemological capacities of microscopical eyes. While for Locke such data is not humanly intelligible, for Boyle it schematizes the doctrine of qualities that defines empirical ideas as perceptual effects of corpuscular texture. Boyle's viewer does not see color itself, but the underlying particulate surface—composed of "latent Ruggidnesses" and "little Protuberances and Cavities"—that alters white light. His microscope promises to elucidate the causal link between primary matter and sensed qualities, between minuscule "Physical Surfaces" and the hue they trouble light to produce. The ability of "promoted Sense" to trace this link exceeds the limits of empiricism projected by Locke: Boyle's observer would justify "*how*" the primary affections of particles, "Sizes, Shapes, and Situations of the extremely little Bodies," intercept incident white light to engender the observer's experience of color.

Boyle's scene of magnified perception does not overwhelm the viewer with alien objects, because the intelligibility of blown-up protuberances and cavities is secured by the likeness of tiny particles and bigger things. Alongside proofs, like sublimation or distillation, that vindicate the reality of minuscule parts as the result of a reagent's disappearance and return to sense, the fundamental lucidity of microscopical discernment is assured by the analogical consistency of matter. To concretize this assurance with another exceptional instance of color's apprehension, Boyle offers his next percipient, "a Man . . . who at certain times can *distinguish Colours by the Touch* with his Fingers."[10] This person, who is "Blind," collapses analogically familiarized bumps and the haptic immediacy of color. When asked "what kind of Discrimination he had of Colours by his Touch," the blind man replies: "all the difference was more or less Asperity, for . . . Black feels as if you were feeling Needles points, or some harsh Sand, and Red feels very Smooth."[11] For this witness, matter holds color securely; indeed, color is so deeply embedded in matter that it circumvents the necessity of vision. For Boyle's blind man, bodies *are* always colored in the dark.

As we saw in Chapter 2, Locke's own blind man compares the color scarlet to the sound of a trumpet and, his vision suddenly restored, cannot reconcile the sight and touch of a cube. Locke's refusal to translate simple ideas into the corpuscular micro-explanation through which they might otherwise be figured—"Black feels as if you were feeling Needles points . . . and Red feels very Smooth"—foregrounds Boyle's willingness, in this instance, to collapse primary texture and secondary sensation, to conflate the simple idea "Red" and the palpable smoothness of the particles that effect it. The mechanical affection experienced as smoothness *is*, in the case of the man who apprehends color as texture, continuous with sensation that enters through the fingers. Literalized as well in this limit scenario is haptic "Discrimination" also encountered as texture. By distinguishing red from black by touch, the blind man broaches and resolves a basic problem: How does particulate matter secure difference? *Can* particles which do not instantiate elemental substance entrench difference at all? Pointiness anchors the singularity of each color in its engendering texture. But the corpuscle will not provide human difference so secure a home.

Despite the confidence with which his blind man locates color in the object, Boyle broaches a nagging semantic dilemma: "Colour may be considered, either as it is a quality residing in the body that is said to be coloured, or to modifie the light after such and such a manner; or else as the Light it self, which so modifi'd, strikes upon the organ of sight."[12] Boyle's doctrine of qualities asserts that as sensation triggered when "modifi'd" light "strikes" the eye, color really subsists in "the organ of sight." But color foregrounds the difficulty, physical as well as semantic, implicit in the divorce of corpuscles from the ideas they stimulate in persons. Boyle refines this difficulty in a caveat: "yet, because there is in the body that is said to be coloured, a certain disposition of the superficial particles, whereby it sends the Light . . . to our eyes thus and thus alter'd . . . it may also in some sense be said, that Colour depends upon the visible body."[13] The verb "depend" signifies a causal relation that recurs to the particulate infrastructure of the thing colloquially "said to be coloured." Color is not in bodies as such, but it relies on a light-altering "disposition of the superficial particles" continuous with the physical reality of the object. Color subsists neither in any freestanding hue nor in palpable smoothness or pointiness but rather in an imperceptible "disposition" assigned objects only after the viewer's sensory encounter.

The retroactive intuition of how primary texture is disposed to elicit human sensation provokes Boyle to revisit the question of color in the dark, a question he frames by invoking "that famous Controversie which was of Old disputed betwixt the *Epicureans* and other *Atomists* on the one side, and most other [peripatetic] *Philosophers* on the other side. The former Denying Bodies to be Colour'd in the Dark, and the Latter making Colour to be an Inherent quality."[14] Boyle argues that his answer—the body's corpuscular disposition to trouble light—exposes the failure of Epicurean and Aristotelian alternatives effectively to frame the stakes of the debate:

> [Y]et I doubt whether it be not in great part a Nominal dispute, and therefore let us . . . Distinguish the Acceptations of the word Colour, and say, that if it be taken in the Stricter Sense, the *Epicureans* seem to be in the Right, for if Colour be . . . but Light Modify'd, how can we conceive that it can Subsist in the Dark . . . but on the other side, if Colour be consider'd as a certain Constant Disposition of the Superficial parts of the Object to Trouble the Light . . . this Constant, and, if I may so speak, Modifying disposition persevering in the Object . . . in this Sense, Bodies retain their Colour as well in the Night as Day.[15]

As what is really a "Nominal dispute," this controversy exposes the constitutive ambiguity "the word Colour" sustains. As "Light Modify'd," color vanishes in the dark, but as a "Modifying disposition persevering in the Object," color is impervious to any extrinsic circumstance that leaves the object's texture intact. Of course, the identification of color and particulate disposition collapses the same distinction confounded by Boyle's blind man. Just as for him red and smoothness feel the same, so texture's "Constant Disposition" suggests that objects in the dark "retain their Colour," even if the latter idea is not equivalent to the visual intake of particulate superficies achieved by Boyle's microscopically eyed viewer. To specify the constancy of texture that is not a scholastic "Inherent quality," Boyle asserts the abiding power of the object mechanically to produce color in the sensoria of persons. Whether as smoothness or an infinitesimal panorama comprised of protuberances and caves, this power might itself somehow be apprehended.

A close reader of *Touching Colours*, Newton responds to Boyle in *Opticks; or A Treatise of the Reflections, Refractions, Inflections & Colours of Light* (1704; thirty-one total queries added, 1730).[16] In a stark repudiation of Boyle's "modificationist"[17] thesis, the first proposition to begin Part 2 of the first book of *Opticks* proceeds: "The Phænomena of Colours in refracted or reflected Light are not caused by new Modifications of the Light variously impress'd, according to the various Terminations of the Light and Shadow."[18] For Newton, color does not proceed from the object's modifying disposition. *Opticks* extirpates color from matter by denying light's capacity to be "impress'd"—in a signally Lockean restatement of Boyle's claim—by the object that reflects it. In a letter published by the Royal Society's *Philosophical Transactions* in 1672, Newton establishes the experimental grounds for his rejection of the modificationist theory endorsed by Boyle and others like Robert Hooke. (Indeed, Hooke's hostile response to the letter may have retarded Newton's publication of *Opticks*.)[19] The "*Experimentum Crucis*"[20] that propels Newton to overturn the modificationist theory employs two prisms: Prism 1, which separates a beam of incident white light to prove its compositeness, is succeeded by two fixed boards with apertures that permit only a selective expanse of colored rays to pass through. By rotating Prism 1, Newton selects which rays in the red-to-violet spectrum will enter the apertures to pass through a fixed Prism 2 and hit a screen. Crucially, a colored ray—for example, violet—that passes through Prism 2 exhibits the same degree of refrangibility (tendency to be refracted) as does the violet ray that exits Prism 1. As evinced by where it falls on the screen, its passage through Prism 2 does not alter the refrangibility of the incident violet light: violet, the spectral color refracted to the largest degree by Prism 1, is also refracted to the largest degree by Prism 2.[21] Newton concludes that "Light it self is a *Heterogeneous mixture of differently refrangible Rays*."[22]

Newton's first prism shows that white light can be separated into a continuous spectrum of rays each of which is refracted by glass to a different degree (or, each yields a different number for Snell's law or the law of refraction, the ratio of the sine of the incident angle over the sine of the refracted angle when it exits a medium like glass or water). If each sine proportion in the spectrum is affiliated with a color, then the double prism experiment leads Newton to stress the primacy of that affiliation for each separated ray: "As the Rays of light differ in degrees of Refrangibility, so they also differ in their disposition to exhibit this or that particular colour. Colours are not *Qualifications of Light*, derived from Refractions, or Reflections of natural Bodies (as 'tis generally believed,) but *Original* and *connate properties*, which in divers Rays are divers."[23] Rather than a modification of white light, color is, so to speak, welded to the substance of each separated ray.[24] As "*Original* and *connate properties*," colors are inalienable from the differently refrangible rays that "exhibit" them. Newton asserts every color's affiliation with one separated ray as well as that color's resistance to subsequent change: "The species of colour, and degree of Refrangibility proper to any particular sort of Rays, is not mutable by Refraction, nor by Reflection . . . When any one sort of Rays hath been well parted from those of other kinds, it hath afterwards obstinately retained its colour, notwithstanding my utmost endeavours to change it. I have refracted it with Prismes, and reflected it with Bodies . . . transmitted it through coloured Mediums . . . and diversly terminated it; and yet could never produce any new colour out of it."[25] To illustrate the claim that color is not a "*Qualification*[] *of Light*," Newton elaborates a proof whose frustrated seriality may recall Boyle's inverse stress on the capacity of selfsame particles to acquire new attributes in relation to other bodies. The narrative momentum of Newton's proof of the "obstinately" rooted property of color proceeds in the other direction. Despite multiple efforts—"I have refracted it with Prisms, and reflected it . . . transmitted it . . . and diversly terminated it"—Newton cannot change the defining attribute of his separated ray. The failure of Newton's "utmost endeavours" defines a quality not contingent on the topographic contours that effect it: color is, rather, an obstinate and "*Original*" entity.

Newton's color departs from the primary-secondary schema advanced by Boyle and refined by Locke. Because it inheres in a separated ray, color exists anterior to the corpuscular texture of other bodies. Whereas Boyle refers the ontology of color to minuscule bumps that trouble light, Newton inverts the relation between color and the physical object on which that quality may supervene:

> [T]he Colours of all natural Bodies have no other origin than this, that they are variously qualified to reflect one sort of light in greater plenty then another. And this I have experimented in a dark Room by illuminating those bodies with uncompounded light of divers colours. For by that means any body may be made to appear of any colour. They have there no appropriate colour, but ever appear of the colour of the light cast upon them, but yet with this difference, that they are most brisk and vivid in the light of their own day-light colour. *Minium* [red lead] appeareth there of any colour indifferently . . . but yet most luminous in red, and so *Bise* [or bice, a deep blue pigment obtained from smalt] appeareth indifferently of any colour with which 'tis illustrated, but yet most luminous in blew . . . And that this is the intire and adequate cause of their colours, is manifest, because they have no power to change or alter the colours of any sort of Rays incident apart, but put on all colours indifferently, with which they are inlightned.[26]

Newton divests matter of the colorific disposition sustained by corpuscular texture. Instead, color "illustrate[s]" the less constant bodies on which it impinges. We can thus read Newton's "no power" in proleptic foreclosure of Locke: unlike the object whose secondary qualities are specified as its capacity to trouble light, Newton's "natural Bodies" are powerless to refuse to "put on" whatever separated rays are imposed on them. The "body . . . made to appear of any colour" testifies to the radical integrity of each spectral ray separated by Newton's prism, because an object susceptible to the hue with which it is "inlightned" transfers the ontology of this attribute from the thing to "the colour of any sort of Rays incident apart."

There is a critical stipulation attached to Newton's proof that objects yield to the extrinsic integrity of uncompounded light: this proof transpires in a "dark Room." To the question posed by Falconer's Indian bride, Newton answers in the negative: "it can be no longer be disputed, whether there be colours in the dark, nor whether they be the qualities of the objects we see."[27] Bodies better able to reflect some blended rays rather than others cannot claim a denominating color without incident white light. Unlike the surface Boyle's blind man touches, Newtonian objects retain the endowment of hue only for the circumscribed duration of "their own day-light colour." In the dark, unreflecting things can be made to assume "any colour indifferently." In Newton's version of *Falconer*'s lightless chamber, the ontology of human color is suspended. We return to Chetwood's and Aubin's bedrooms below.

"Impressions of Air": Boyle's Law and Human Difference

Arbuthnot's *Effects of Air* and Mitchell's *Colours of People* invite appraisal as answers to a more particular question: How do corpuscles anchor human difference? Indeed, *can* corpuscles anchor human difference? This query is not posed by two leading accounts of British and French eighteenth-century racial thought, Roxann Wheeler's *Complexion of Race* and Andrew Curran's *Anatomy of Blackness*, neither of which analyzes the construction of race at the level of particles.[28] Wheeler downplays the rise of the corpuscle in eighteenth-century science by asserting the confluence of humoral physiology and environmental determinism: "Climate and humoral theory, in one form or another, provided the most important rubric for thinking about human differences in the eighteenth century." But Wheeler's "humoral/climate theory"[29] rubric conjoins historically competing explanatory modes to elide the anachronism of

Galenic medicine's commitment to four essentially qualified bodily fluids. Affiliated by the time of the Great Plague with the incompetent dogmatism of doctrinally bound operators, humors were denounced by chymical physicians like George Thomson and George Starkey. What Wheeler designates "humoral and climatic sensibility"[30] tends to obscure the difference of iatrochemists and empirical doctors (like Starkey, Starkey's emulator Newton, Thomas Sydenham, and Locke) from practitioners whose substance-reliant etiology of human illness posits four irreducible humors. The absence of corpuscular science from *Complexion of Race* may be due to the Foucauldian genealogy of knowledge Wheeler cites to justify an apparent "eighteenth-century interest in surface."[31] Michel Foucault overlooks the pivotal role of micromatter in seventeenth- and eighteenth-century constructions of phenotypic identity, constructions that reflect not a science restricted to surfaces but the centrality of the corpuscle to non-essential formulations of human difference. The particle that resists elemental identity in Boyle's *Skeptical Chymist* is the same particle that, according to Arbuthnot, claims the equivocal power to volatilize and re-instantiate the national peculiarities of persons.

Arbuthnot's *Effects of Air* marks a signal effort to justify how particles communicate the defining properties of geography. Because it plays out the paradox unleashed when Boylean and Lockean anti-innatism operate on a global scale, *Effects of Air* may be introduced with a symptomatic contradiction. The preface remarks that "tho' Abstinence from Air is not, the Sort of Air which they use, is in the Power of a great many People";[32] but Arbuthnot's final chapter revises this estimate without noting the change: "Mankind . . . have in some measure the Power of defending themselves from the Injuries of the Air; but few have the Choice of the Air in which they live" (205). While he wittily dispenses with "Abstinence" as an escape from air's influence, Arbuthnot does not conclusively gauge the extent to which pervious persons can elude determination by the particles that environ them. The question of whether "many" or "few" exert volitional control over what air makes of them illuminates, I will suggest, the inverse redundancy of human will facilitated as one historical outcome of corpuscular science: its anti-essentialist endorsement of the enslavement of African, Caribbean, and South American persons.

What is in the air? Arbuthnot affirms the epistemological limit which opens persons to the insensible contents of what they imbibe: "The Particles of Air are not discernable by a Microscope" (24). For Arbuthnot, the discontinuity of "Particles" stipulates non-identity that admits geographical difference at the scale of the corpuscle: "The Air near the Surface of the Earth, in which all Animals live and breathe, contains the Steams, Effluvia, and all the Abrasions of Bodies on the Surface of the Earth, when they are so small and light as to float in it; from whence it is evident, that the Contents of it must be different in different Places of the Surface of the Earth" (3). Air is "different" from itself because its difference is particulate. Air then perpetuates difference because growing and breathing creatures absorb discrepant submicroscopic parts. Yet as the bottom line of the empirical animal's extrinsically induced difference, the derivation of that difference from "the Steams, Effluvia, and all the Abrasions of Bodies" presents an obvious tautology. Where is difference rooted, if dissimilar bodies are produced by steams and abrasions of the same bodies? Tiny extrusions of dissimilar things "float" in the air to produce things whose steams, effluvia, and abrasions rise again to repeat the same difference. How is the circularity of Arbuthnot's attribution of geographical difference to the airborne constituents of the same difference not evident?

Effects of Air defers the recurrence of difference to difference in part by elaborating a more refined roster of air's "Ingredient[s]" (6). Air contains "a great Quantity of Water" (4) as well as dew, which "is not mere Water, but a Composition of all the watery, volatile, oily, saline Vapours, which exhale from the Earth" (6). Floating in air are "Exhalations of Vegetables" (8), "Oils, Salts, Seeds, and the insensible Abrasions of Vegetables" (9), "Earth, calcin'd" (9), "Particles of all Minerals" (10), "the perspirable Matter of Animals" (11), and "Eggs of Insects" (11). Arbuthnot cites Newtonian chemical attraction to galvanize his proof that water may be sucked from the air: "Fix'd alkaline dry Salts attract, and are dissolv'd by the watery Particles of the Air . . . 1 Ounce of Salt of *Tartar* making 4 Ounces of Oil of *Tartar, per deliquium*, only by Attraction of Water from the Air" (4). And in a bracingly Swiftian flip between macro and micro, Arbuthnot affirms that "all Excrements and all the Carcases of Animals vanish into Air" (11).

The conversion of "Salt of *Tartar*" to "Oil of *Tartar*" proves that the hygroscopic or deliquescent salt potassium carbonate (K_2CO_3) absorbs molecules of water from ambient air. This empirical demonstration of the physical reality of air's insensible parts is consonant with Arbuthnot's claim for subliming excrements and carcasses: just as salt of tartar's liquefaction and fourfold increase in weight show that water particles insensibly exit the air, so excrements that "vanish" enter the atmosphere part by invisible part. Yet the juxtaposition of these two proofs amplifies the circularity they would skirt. At what point in the breakdown of carcasses into smaller parts do they cease to retain the identity of the whole animal? Because Arbuthnot's "Salt of *Tartar*" does not entrench elemental identity—corpuscular components of things instantiate only mechanical affections and changeable texture—the micromatter into which a carcass dissolves does not metamorphose into elements. If discrepancies between bodies are not secured on the side of the corpuscle, at what level do the tiny parts of Arbuthnot's disintegrated body retain the difference that defines the animal—or, indeed, any difference at all? The invisibility that masks airborne difference demands particles whose tininess undoes their affiliation with qualitatively identifiable bodies.

What links persons and geography is not what air contains but how it pushes back. In this capacity Arbuthnot deploys Boyle's claim, made in *Spring of the Air* (1660), "That there is a Spring, or Elastical power in the Air we live in."[33] Air's elasticity is vindicated by analogy to "a Fleece of Wooll. For this . . . consists of many slender and flexible Hairs; each of which, may indeed, like a little Spring, be easily bent or rouled up; but will also, like a Spring, be still endeavouring to stretch it self out again."[34] Boyle's "Aerial Corpuscles," each of which springs "(by vertue of its structure)," are compressible but will expand to a volume inversely proportional to "the weight of the incumbent part of the Atmosphere."[35] Footnoting "*Mr.* [Stephen] Hales" (35), the author of vivisectionist experiments on internal arterial and venous pressure compiled in *Hæmastaticks* (1733), Arbuthnot invokes "Spring or Elasticity" (30) as the decisive variable contained inside the blood and fluids of living creatures: "the Variation of the Gravity and Elasticity of the Air, which proportionally expand and dilate the Liquors . . . must have sensible Effects upon animal Fluids" (36). Internalized air renders human bodies contingent on climactic "Variation," because the elasticity of the air contained inside persons, just like that of air inside air-pump receivers, "proportionally" distends or contracts the fluids in their vessels. Shaped from within by the reciprocal springiness of fluid-borne air, animals are shaped from without by the atmospheric weight that constrains them for the duration of their growth: "The Pressure of the Atmosphere keeps both Vegetables and animal Fibres within certain Limits of Accretion; it being always Fluid, the Pressure is equal upon every Part of their Surfaces" (23). From the inside, Boyle's Law of Gases governs the fluid-borne constituents of bodies; from the outside, Boyle's Law regulates vegetable and animal "Accretion."

Enforced by the atmosphere in tandem with internalized air's inversely proportional spring, "Limits of Accretion" acquire a startling level of resolution when Arbuthnot defends air's power to distinguish persons by nation. He recruits Boyle's Law to justify morphological particulars that vary according to place:

> [T]he Air operates sensibly in forming the Constitutions of Mankind, the Specialities of Features, Complexion, Temper . . . There are Faces not only individual, but gentilitious and national . . . This Diversity of National Features and Shapes is not altogether the Effect of Propagation from the same original Stock; for it is known by Experience, that Transplantation changeth the Stature and outward Shape, both of Plants and Animals . . . An Animal growing, expands its Fibres in the Air as a Fluid . . . and tho' the Fibres of several Animals shoot, as it were, in this Fluid, according to their original Shapes, yet such a Fluid resisting by its Pressure, is, in respect to the Animal, like a soft Mold, in which the Body is form'd. (146–47)

It is no overstatement to credit globally dissimilar "Constitutions of Mankind" to the homely resilience of Boyle's fleece of wool. Indeed, Arbuthnot opposes the overweening influence of air to a conservative internal mandate that would make accreting creatures "shoot . . . according to their original Shapes." Operating not as an atomic element but as springy texture that, in Boyle's words, may be bent or rolled, air discriminates between persons at the level of "gentilitious and national" difference. Although Arbuthnot invokes the monogenetic guarantor of species uniformity that credits human identity to "the same original Stock," this original cedes its unity before the countervailing mechanics of the air. "Transplantation changeth" because ancestral beginnings claim no mechanical priority over the extrinsic power of air. As pressure that environs an entire body, environmental elasticity trumps a static internal dictate to reproduce the same original. Air is not one environmental contingency among others but a "soft Mold" whose reciprocal compressibility belies its all-encompassing power to communicate geographical particularity as "outward Shape."

Still writing of the leading morphological influence imparted by a person's place on the globe, Arbuthnot notes that "Variations of the Pressure of the Air . . . must produce proportional oscillatory Motions in the Fluids and Solids of Human Bodies" (42). Here he shifts from outward figure to the far more pervasive and metaphysically inclusive domain of human physiology. Boyle's Law governs the outcome of variable atmospheric pressure by justifying "proportional oscillatory Motions" driven by air whose springiness is unimpaired by its incorporation into humans, because, Arbuthnot stresses, porous persons provide air "constant Admittance into the Body": "it not only operates by outward Contact, but we constantly imbibe it at all the Pores of the

Body" (64). Air that acts like air whether it is outward or imbibed licenses geophysiological difference whose corpuscular antecedent is inversely reactive particulate motion: "Changes in the Gravity of the Air, affecting Human Bodies with oscillatory Motions of the Solids and Fluids, the more frequent and great those Changes are, the greater Effects they will have upon the Nerves and Spirits; but so it is, that within the Tropicks, where there are no variable Winds, the Variations of the Height of the Mercury are but small" (75). Echoing Hales, for whom human vitality corresponds to a "vibrating State"[36] of the blood, Arbuthnot places ambient corpuscular motion in third-degree relation to human physiology: first, local air pressure changes; then, those changes are more or less "frequent and great" as a function of geographical place; finally, more frequency and greatness induce "oscillatory" responses whose "greater Effects . . . upon the Nerves and Spirits" internalize globally discrepant observances of Boyle's Law as human physiology.

At a fourth remove from barometric change, Arbuthnot's oscillating nerves and fibers exert an impression upon the medium of human will:

> In Northern Countries, where the Alterations of the Height of the Barometer . . . are frequent and great, the Fibres of Human Bodies are in a continual oscillatory Motion . . . tho' this . . . is insensible, and not dolorifick, it is a sort of Exercise which the Inhabitants of Countries where the Variation of the Height of the *Mercury* is small or nothing [do not feel]. By the Difference of the Tension of the Fibres, the whole nervous System, and the animal Spirits, are in some measure affected . . . [T]he Temper, both of their Body and Mind, must be different, and . . . greater Variety in the oscillatory Motion of the Fibres of Northern People must produce the same in their Spirits, and therefore a proportional Inequality in their Passions, and consequently greater Activity and Courage . . . [T]he Inhabitants of the Climates, where the Difference . . . is but small . . . and the Motions of their Fibres and Spirits being more uniform, they may be for that Reason, and from excessive Heats, lazy and indolent: From Inactivity and Indolence there will follow naturally a slavish Disposition, or an Aversion to contend with such as have got the Mastery of them. (151–52)

The arc of this proof is sustained by subtle but still physical "animal Spirits," which occupy the interface between "Body and Mind." Because oscillatory movement in the nerves and fibers effects "the same in their Spirits," Arbuthnot finally transposes the same variety perpetuated by particles into volitional terms, a move canonically sanctioned by René Descartes's mechanical physiology of affect *The Passions of the Soul* (1649). Critically, a wide range of barometric variance dictates "Inequality in their Passions": northern persons who contain outside air translate its oscillations into greater permutations of feeling, while lower barometric pressure dictates affective neutrality that cramps the faculty of volition. Slavishness—or, the inability to resist "Mastery" by others—thus results from internalized air that fails to promote resistant will. Because barometric motion is conserved in the medium of animal spirits, the same atmospheric difference vindicates a historical dispensation that denies political humanity to southern persons.

It is tempting to read Arbuthnot's deployment of Boyle's Law as a transparently bad-faith recourse to climate, one that vindicates the instrumentality of human bondage to imperial commerce regardless of climate's ontological hold on persons. Yet Arbuthnot defends the primacy of environing air in calibrating the differential claim to property in one's labor he calls "Degrees of Slavery" (153).[37] His strongest appeal to the atmospheric determination of national character asserts the irrelevance of its genetic original: "In perusing the Accounts of the Temper and Genius of the Inhabitants of different Countries, we discover in them a great Uniformity, even tho' the Race has been chang'd" (149). As the guarantor of consistent difference, geographical place trumps starting familial stock or "Race." But Arbuthnot's corpuscular derivation of degrees of slavery does not preclude the generic proneness of southern persons to subjugation. *Effects of Air* enacts the signal failure of empiricist anti-essentialism to provide a scientific, philosophical, or ideological counterweight to the historical institution of chattel slavery and the plantation economy.

The capacity of anti-essentialist science to justify "slavish Disposition" as a function of climate does not herald the deflection of racist determinism from bloodline to culture.[38] Indeed, corpuscular texture that rationalizes slavery is formally cognate with the bilateral logic Anthony Appiah deconstructs by showing that racial phenotypes recur not to the difference but to the overwhelming sameness of an underlying human genotype.[39] As with the genotype-phenotype schema, *Effects of Air* predicates macroscopic effects on the physicality of an empirically inscrutable productive agent. But in Arbuthnot's account, human geomorphological difference is produced not by genes, nor even floating fragments of carcasses, but by the variably springy action of the same internalized air. For Arbuthnot, the insensible substrate that engenders human difference is not matter that genetically holds the essence of race, but identical corpuscles whose imposition of pressure varies with the barometric discrepancy between north and south. The production of phenotypic variation from corpuscles that indifferently obey Boyle's Law thereby enacts one concrete historical outcome of Boyle's doctrine of qualities: particles that share an identical repugnance to elemental substance effect the denominating qualities of perceptible bodies. Anti-essentialism operates as an historical justification for the bondage of southern person *because* it resists qualitative identity all the way down to the indifferent springiness of Boyle's aerial corpuscle.

Returning to the question of who can elect to move, Arbuthnot never decides whether persons possess the "Power" to displace themselves. This is the same aptitude, movement, on which Locke predicates the *Essay*'s revisionary definition of human liberty—instead of, Locke argues, the nonsensical freedom of a will that is unable to displace body in any case.[40] Rather than an innate endowment, the briefly projected capacity of southern persons to choose "the Sort of Air . . . they use" resembles Locke's secondary quality, an object's imminently perceptible demonstration of the agency its texture possesses. Indeed, Locke uses the same word, "power," to signify both texture's disposition and a person's manifest capacity to move himself. Southern people who muster the motion to avoid air that makes them slavish would thus exercise power insulated from an atmosphere that robs their fluids and spirits of it. Arbuthnot's retrenchment of such exertion from "many" to "few" tracks his text's arrival at the disposition that shores up the historical inexorability of southern persons' subjugation, an outcome enforced by the insensibility of atmospheric particles that wreak their effects as soon as a body begins to accrete. Volition does not interrupt air's insinuation of slavishness into the many persons who find themselves unable to move. A vindication of slavery mobilized by corpuscles, Arbuthnot's text postulates persons who lack a secondary power that would be evidenced as resistance to air itself.

Blackness's Body: Newton's *Opticks* and Mitchell's *Colours of People*

Among the features shaped by air's soft mold, Arbuthnot lists complexion. To account for this quality, *Effects of Air* does not adopt the expedient of some supplemental black fluid but rather cites as the effect of "*African*" (154) warmth blood or bile that congeals and rises upward: "The Blood too, in hot Countries, is thicker and blacker by the Dissipation of the serous Part, by sensible Perspiration . . . Great Heats exalt the Bile . . . it stops at the Surface of the Skin, and discolours it" (154–55). Arbuthnot's recourse to "Bile" exploits the mechanical propensities of this fluid while reifying its capacity to color: because, unlike the "serous" part of blood, "exalt[ed]" bile is "unperspirable" (155), it stops at the outer edge of the person whose pores it cannot exit, darkening her skin from below. As Curran shows, the assistance of tinted fluids to the production of phenotypic variation was cemented by Pierre Barrère's influential *Dissertation sur la cause physique de la couleur des nègres* (1741). Curran explains: "Barrère's text ushered in a new era, one in which anatomical research was claiming the ability to put forward an all-encompassing explanation of the *nègre*'s fundamental difference."[41] Rather than anatomy, it is Newton's color science that justifies the divergent explanatory strategy advanced by Mitchell's *Essay upon the Causes of the different Colours of People in different Climates*, read by proxy before the Royal Society from May to June 1744 while Mitchell was resident in Virginia.[42]

Designated by James Delbourgo "a locally knowledgeable Creole with a sure eye to powerful audiences in the imperial capital,"[43] the anatomist, cartographer, botanist, and Royal Society Fellow John Mitchell (1711–1768) was born in Virginia to slave-owning mercantile parents and educated in natural philosophy at the University of Edinburgh before returning to the Atlantic coast, where he amassed financial capital and natural philosophical expertise. With Curran, Delbourgo situates Mitchell's *Colours of People* in the thick of climactically inflected theories of human difference, but he notes the text's singularity: "Mitchell's account *was* unique, however, in its attempt to harness the enormous prestige of 'Newtonianism' to his project."[44] Read by Delbourgo as a bid for "rare public authority" in "a larger colonial matrix in which the meanings of Newton and his work were actively being fashioned,"[45] Mitchell's claim on the *Opticks* leverages the authority of Newton's most accessible publication, proffering a non-essential derivation of the skin's hue whose political inflections do not necessarily jibe with *Colours*'s subsequent enlistment by the abolitionist cause.[46]

Mitchell's text solicits attention as an effort to explicate color at the microparticulate level, for rather than exploiting the tingeing influence of black bile, Mitchell excavates color's corpuscular antecedents. As we have seen, in opposition to Hooke's and Boyle's modificationist theory, Newton declares that things are extrinsically colored. In the absence of white light, separated rays may be imposed on objects whose own texture poses no obstacle. Newton flags the alternative scenario as the object's "own day-light colour": bathed in compounded white light, things reflect some of its constituent rays rather than others. The latter predilection selectively to reflect, which assumes some material, and hence particulate, tendency in the object, preoccupies Mitchell. *Colours* will thereby recapitulate the key incoherence of Newton's theory of colored bodies, which attempts to approximate the daylight color of three-dimensional objects to Newton's computation of the corpuscular diameter of colored films or plates.[47] Because Newton deploys thin films to render materials like air or glass equivalents for all colored things, *Colours* amplifies the ontological precariousness of the human embodiment of blackness, whose definition in *Opticks* as both film and dimensional object precipitates Newtonian color theory's most insoluble contradiction.

Recapitulating Newton's theory, presented after multiple early iterations in Book 2 of the *Opticks*, demands a return to Boyle's *Touching Colours*. Newton's account of colored bodies is, to its core, a corpuscular theory, whose most topical antecedent is Boyle's coarticulation of particulate divisibility and the visual metamorphosis of opaque bodies, once magnified, into diaphanous ones. Opacity, it turns out, is a perceived quality whose primary constituents may be intuited under the microscope, a discovery established for Boyle by objects made of "Thin parts": "that Instance is remarkable, that is afforded us by *Muscovie* glass [mica] . . . for though plates of this Mineral . . . do often appear Opacous, yet if one of these be Dextrously split into the thinnest Leaves 'tis made up of . . . these Leaves will afford the most Transparent sort of consistent Bodies . . . and a single Leaf or Plate will be so far from being Opacous, that 'twill scarce be so much as Visible."[48] Newton's analogy of "Thin parts" and all colored bodies is enabled by the visible failure of smaller and smaller material units to sustain perceived opacity. Singly transparent plates which, stacked, turn into a solid body lead Boyle to speculate on the diaphanousness of every object's smaller parts: "And multitudes of Bodies there are, whose Fragments seem Opacous to the naked Eye, which yet, when I have included them in good *Microscopes*, appear'd Transparent . . . I am not yet sure that there are no Bodies, whose Minute Particles even in such a *Microscope* . . . will not appear Diaphanous."[49] Boyle's thinnest sheet of mica models the reversion of any opaque fragment, once magnified, to anterior transparency.

The discovery of opacity's constitutive evanescence seems like another instance of Boyle's doctrine of qualities: sensible attributes are produced by corpuscular microstructure that eludes human sense. But the transparency of thin sheets catalyzes Boyle's deeper incursion into that microstructure, an incursion which signals the latent incompatibility of his modificationist theory of troubled incident white light and the newly disclosed diaphanousness of "Minute Particles." Addressing *Touching Colours*'s designated reader, his nephew Richard Jones or "Pyrophilus" (lover of fire), Boyle concedes that the collapse of color and texture dramatized by his blind man cannot account for the event of opacity's dissolution into transparent layers:

> Hitherto, *Pyrophilus*, we have in discoursing of the Asperity of Bodies consider'd the little Protuberances of other Superficial particles . . . as if we took it for granted, that they must be perfectly Opacous and Impenetrable by the Beams of Light . . . But to deal Ingenuously with you . . . I have often thought it worth a Serious Enquiry, whether or no Particles of Matter, each of them singly Insensible, and therefore Small enough to be capable of being such Minute Particles, as the *Atomists* both of old and of late have (not absurdly) called *Corpuscula Coloris*, may not yet consist each of them of divers yet Minuter Particles, betwixt which we may conceive little Commissures where they Adhere to one another, and, however, may not be Porous enough to be, at least in some degree, Pervious to the unimaginably subtile Corpuscles that make up the Beams of Light, and consequently to be in such a degree Diaphanous. For, *Pyrophilus* . . . you'l [*sic*] easily grant . . . that whereas Perfectly Opacous bodies can but reflect the incident Beams of Light, those that are Diaphanous are qualified to refract them too.[50]

Again we encounter a corpuscle that is not irreducible. Embodied by the hypothetical "*Corpuscula Coloris*," the quality of color implicates matter in hierarchy. To account for diaphaneity, Boyle does not switch larger for smaller indivisible parts: he posits the recurrence of one "singly Insensible" corpuscle to the yet "Minuter Particles" of which it is composed. Tiny atoms do not simply vanish below the threshold of sense, because Boyle's corpuscle itself effects transparency as structure. This microstructure entails "little Commissures" between subcorpuscular parts that render each body "Porous." Rather than a solid bump that reflects incident light to produce color, as Boyle's modificationist theory claims, the porous *Corpuscula Coloris* harbors infrastructure able to make a single particle "refract." Color galvanizes the corpuscle's subsidiary porosity because Boyle "Ingenuously" acknowledges the discrepancy between light that bounces off hard points and light that amplifies a magnified body's diaphanousness.

Citing Boyle's *Corpuscula Coloris*, Alan Shapiro remarks: "This is perhaps the first time that Newton encountered the compositional theory of matter."[51] Newton's theory of colored objects, which Newton will analogically yoke to his account of color in "thin transparent Bodies,"[52] claims a debt to Boyle's *Touching Colours* as well as Hooke's *Micrographia*, where, a year after Boyle, Hooke examines mica's "indefinite number of plain and smooth Plates, heaped up" to query whether from them "may not be deduced the true causes of the production of all kind of Colours." Newton's corpuscular theory is decisively influenced by *Micrographia*'s claim that colorific variety is the structural effect "of an infinite number of thin flakes joyned . . . one upon another":[53] Hooke observes that "if you take any small piece of the *Muscovy-glass*, and . . . cleave it oftentimes into thinner and thinner *Laminæ* . . . each Plate, after it comes to such a determinate thickness, shall appear most lovely ting'd or imbued with a determinate colour." He cites the vanishingly thin plate's "determinate colour" to explicate the inverse case of a thin film of air trapped between two glass surfaces, which, like denser "*Laminæ*" surrounded by rarer air, produces the appearance of "several *Irises* or coloured Lines."[54] (The phenomenon of color in near-invisible plates is replicated, Hooke reports, by bubbles of "Soap-Water," "*Pitch*," "*Turpentine*," "*Wort*," "*Wine*," and "*Glare of Snails*,"[55] among other fluids.) Since my focus is color and the porous corpuscle, I will not trace Newton's quantification of what are now named Newton's rings—Newton's critical departure from Boyle and Hooke inheres in the compositeness of white light, whose colors are separated, not produced, by the thin plate or film—in whose seminal experimental reenactment Newton traps air under a convex glass lens pressed against a glass plane. Book 2 of the *Opticks* describes the periodic "Rings" of discrete colors extending outwards in radiating "Circles" from the glasses' point of contact, a point whose singular blackness Newton evokes as follows: "[Light] seemed in that place of contact to be wholly transmitted, insomuch that when look'd upon, it appeared like a black or dark spot, by reason that little or no sensible Light was reflected from thence . . . and when looked through it seemed (as it were) a hole in that Air which was formed into a thin Plate, by being compress'd between the Glasses."[56]

Newton's "black . . . spot" which is also a "hole" anticipates the difficult embodiment of blackness in Mitchell's *Colours of People*. But first we must track the radical recurrence to primary matter enabled by Newton's mathematical calculation of the thickness of the film of air at each periodic band of color his rings exhibit (at the defining points violet, indigo, blue, green, yellow, orange, and red). Because he can geometrically derive the width of the space between the curved lens and plane at any given color as the rings radiate outward, Newton constructs a predictive chart or nomograph which fills in the values for the film or plate at submicroscopic scale: "if it be desired to know the diameter of a Corpuscle, which being of equal density with Glass shall reflect green of the third Order; the Number 16¼ shews it to be 16¼/10000 parts of an Inch."[57] By visually identifying a given ring's color and incidence of periodic recurrence—for example, "green of the third Order"—Newton defines a key physical criterion of the unit of matter that reflects it, the particle's "diameter." It is worth pausing over the significance of this intervention in the ordering of primary-secondary difference asserted by Boyle and Locke. The infinitesimal "Corpuscle" whose core physicality "shall reflect" a particular colored ray stimulates green of the third order in the eyes of the viewer, but the sensible idea green is now a precondition for Newton's measurement of the primary attribute of particulate size. The idea green refers not to the haptic immediacy so seductively analogized as texture—the varying degrees of asperity Boyle's blind man professes to feel—but to a coordinate that concretizes the primary reality of the particle. Color not only demands Boyle's, Hooke's, and Newton's acknowledgment of corpuscular matter's diaphaneity: in *Opticks*, color is fused to the quantitative specification of each particle's actual size.

Newton's physiology of ocular sensation still locates color's occurrence in human organs. But his nomograph implicates the observer's visual assessment of whether any given color, like yellow, is "pure and intense," "pretty good," "faint," or "more perfect"[58] as a necessary variable in the computation of its engendering corpuscles' diameter, because this variable indicates to which of yellow's serial rings, or orders, any given incidence of the perceived color must be referred. That is, because the viewer's secondary sensation of the yellow object's colorific "Intenseness"[59] dictates how far away from the glasses' central contact point the periodic yellow band that best resembles the perceived object is located, Newton confounds the mathematical specification of the corpuscle's primary features and the viewer's experience of the quality of color. Sensation becomes integral to the quantitative approximation of particulate physicality. As a result, Newton erodes primary-secondary difference from the side of the corpuscle whose size proceeds from the viewer's determination of the order to which her experience of color belongs. As Newton concedes of the insinuation of human sense into the mathematical calculation of corpuscular reality, "[t]he greatest difficulty is here to know of what Order the Colour of any Body is."[60]

The primary attributes of corpuscles pivot on the viewer's empirical and sensational capacity "to know": to calculate the size of the particle that reflects the hue she sees, she must match her sense of its intensity to the order of Newton's rings. This exercise confounds primary and secondary qualities because Newton assimilates corpuscle-thin films of air, water, and glass to the colorific disposition of the particles of three-dimensional opaque and transparent objects, a premise he formulates as an analogical link: "*Of the permanent Colours of natural Bodies, and the Analogy between them and the Colours of thin transparent Plates*."[61] The same analogy sustains the refusal that motivates Mitchell's *Colours of Persons*: "The Colour of Negroes does not proceed from any black Humour, or fluid Parts contained in their Skins" (114; italics in original). Mitchell, who instead aspires to "deduce the Colour of the Skin from its Structure" (103) invokes the analogical likeness of "*thin . . . Plates*" and "*natural Bodies*" signally deployed in the *Opticks*, Book 2, by Newton's Proposition 7, to which Mitchell refers: "The bigness of the component parts of natural Bodies may be conjectured by their Colours."[62] But as the color whose body Mitchell approximates to "thin transparent Plates," the blackness at the center of Newton's rings provokes the same revisionary incursion into singly unified parts heralded by Boyle.

With the *Opticks*, Book 2, Proposition 7, which affiliates the phenomenological experience of color—when the viewer assigns her sense of a hue to the order of its periodically recurring ring—and the mathematical determination of the size of each selectively reflecting corpuscle, Mitchell cites Book 2, Proposition 4: "The Parts of Bodies and their Interstices must not be less than of some definite bigness, to render them opake and colour'd."[63] Here Newton assigns a lower limit applicable only to "opake" bodies while acknowledging the proneness to invisibility established for all objects by a seminal chymical proof: "For the opakest Bodies, if their parts be subtilly divided, (as Metals, by being dissolved in acid Menstruums, &c.) become perfectly transparent."[64] As we have seen in Chapter 1 with the *reductio* experiment, Newton theorizes transparency as a result of the dispersion of insensibly "subtil[e]" parts in a corrosive medium. The constituents of the acid menstruum that occupy the "Interstices" between metallic particles, like the particles themselves, fall below the "definite bigness" requisite to reflect incident light. Newton rationalizes the transparency of dissolved metals in acidic solution by referring to the rainbow-colored bubbles noted by Hooke: "the Reflexion of the

Water-bubble where it became thinnest was almost insensible, so as to cause very black Spots to appear on the top of the Bubble, by the want of reflected Light."[65] Whether dissolved by mineral acids or spread out in the "thinnest" extension of a bubble's surface, Newton's subtile parts—as well as the air or menstruum in the spaces between them—are "too small to cause Reflexions in their common Surfaces."[66] Regardless of how dense the starting body seems, its chymical or mechanical separation into tiny parts undoes its perceived solidity. Because they are too small to reflect, these divided parts, as well as the air or fluid that sustains the extremely porous object in "almost insensible" state, inversely establish the particle's "definite bigness" as Newton's criterion for the apparition of colored opacity.

Mitchell refers to the minimal bigness undergirding colorific opacity when he rejects the "general received Opinion, since [the Italian physiologist Marcello] *Malpighi*'s Time, that the Cause of the Colour of Negroes is a Juice or a Fluid of a black Colour, which lies between the *Epidermis* and *Cutis*" (114). Because it colors persons, Mitchell ascribes Malpighi's fluid another attribute: "whatever this supposed black Humour may be . . . it must be opaque" (117). To disprove "this supposed black Humour," Mitchell thus cites Proposition 4's stipulation of definite bigness, which accompanies the anatomical contention that dark persons contain "rather subtiler . . . pellucid aqueous Juices" (117) than pale persons do. But Mitchell's mutually experimental and microstructural defense of the transparency of bodily fluids for white and nonwhite persons alike is not unequivocally supported by *Opticks*'s Proposition 4. Newton exemplifies his defense of the definite bigness requisite to induce opacity with the countervailing qualities of objects whose parts are too tiny to reflect light. Chief among these is the bubble whose vanishing thinness renders it "almost insensible, so as to cause very black Spots to appear on the top." With their homologue, the "black or dark spot" perceived at the center of Newton's rings, these very black spots define *blackness* as the perceived result of parts too tiny to reflect. Soap bubbles stretched almost to bursting and glasses whose contact precludes even a corpuscle-thin film of air pose next to nothing between the sides of an environing medium: the blackness of vanishingly thin interstitial layers registers the maximum dispersion of their parts. By invoking the impossibility of an opaque colored fluid, Mitchell aims to divest physiological "Juices" of a quality whose apparition is not, for Newton, foreclosed by the tininess of its suspended parts. The apparition of blackness is not proscribed by corpuscles whose determinate bigness Newton stipulates only for every other colored body.

Newton's spot triggers the *Opticks*'s overdetermination of blackness. The area where the convex lens and glass plane are contiguous precludes the color-inducing break in density otherwise introduced by the thin film: as Newton postulates, light at this zone of contact is "wholly transmitted" to produce "a black or dark spot" when the touching surfaces are "look'd upon." The same area of contact "seemed (as it were) a hole in that Air which was formed into a thin Plate" when the glasses are "looked through." A spot when looked upon and a hole when looked through, the contact zone marks the contiguous glasses' unbroken transmission of light as well as the gap in the film of air that otherwise divides them. Because Newton establishes empirically that "Reflexion" (the observer looks on) and "transmission" (the observer looks through) at the same thickness of air produce "opposite" colors, blackness—or, the glasses' failure to reflect light of any color when they are looked upon—becomes its contrary when the glasses are looked through: "The central Spot was now white."[67]

As black and white, a hole in the trapped air and the singular contiguity of lens and plane, blackness has no body while, as whiteness, it "is" the nonexistence of the film that otherwise interrupts the density of an ambient medium. The spot that is black and white, no body and all body, anticipates the difficulties of Newton's theory of reflection and its inverse, transparency.[68] Newton cites mechanical and artisanal objections to a billiard-ball schematic in which light would simply bounce off particles: for example, their corpuscular facets would have to line up exactly to produce any consistent color, and "in polishing Glass with Sand, Putty, or Tripoly, it is not to be imagined that those Substances can, by grating and fretting the Glass, bring all its least Particles to an accurate Polish."[69] Newton's reference to glass polishers extends to divine agency which has not planed the microsurfaces of worldly things to "an accurate Polish" either, leading to *Opticks*'s Proposition 8: "The Cause of Reflexion is not the impinging of Light on the solid or impervious parts of Bodies, as is commonly believed."[70] With the *Opticks*'s denial that reflection—or, in other words, color itself—is produced when light hits "impervious parts," we may recall Boyle's *Corpuscula Coloris* and the diaphaneity of his unified corpuscle. As proof of the penetrability of their engendering particles, Newton offers the qualities of transparency and opaque blackness:

> Now if Light be reflected, not by impinging on the solid Parts of Bodies, but by some other principle; it's probable that as many of its Rays as impinge on the solid parts of Bodies are not reflected

> but stifled and lost in the Bodies . . . Should all the Rays be reflected which impinge on the internal parts of clear Water or Crystal, those Substances would rather have a cloudy Colour than a clear Transparency. To make Bodies look black, it's necessary that many Rays be stopp'd, retained, and lost in them; and it seems not probable that any Rays can be stopp'd and stifled in them which do not impinge on their parts.[71]

If rays simply "impinge[d]" and "reflected," made contact with hard atoms and bounced off, then neither clear nor black bodies would be produced. "Water or Crystal" would instead be "cloudy" because their solid interior parts would jumble inwardly directed spectral rays to make the object appear whitish. Just as parts of totally clear things must do something other than reflect, bodies that "look black" demand corpuscles able to take incident light "in." As the perceived effect that most urgently recruits parts able to "stop" and "stifle" rather than transmit or reflect light, opaque blackness demands a corpuscle in which incoming rays are "lost." Rather than diaphaneity, it is, for Newton, opaque blackness that necessitates the stifling porosity facilitated by subcorpuscular parts: "And hence we may understand that Bodies are much more rare and porous than is commonly believed."[72]

Yet by refining the corpuscle's structure to admit not only light's transmission and reflection but also its absorption by wholly black bodies, Newton appears to violate his Proposition 2, which echoes Boyle and Hooke to posit as a basic presupposition "easily granted by them that have been conversant with Microscopes" that "The least parts of almost all natural Bodies are in some measure transparent."[73] As Shapiro concludes: "we have encountered a genuine contradiction. The least parts cannot be both transparent and impervious."[74] Shapiro cites Newton's revised attribution of absorptiveness to the corpuscle as a source of perplexity registered by the flame emission spectographer Thomas Melvill (1726–1753) in his *Observations on Light and Colours* (1756): "Since the cause of blackness in bodies is the smallness of their *transparent* parts, which renders them incapable of reflecting any colour; how can black bodies . . . be at the same time opaque?"[75] As Melvill's query affirms, the embodiment of blackness entails further contradiction: while Newton's black spot models the transparency of the corpuscle, his opaque black object marshals an absorptive or light-deadening—and hence, itself also black—corpuscle.[76] As the site of transparency *and* opacity, transmission and absorption, blackness is keyed to irreconcilable microstructures. It proceeds from the dispersion of parts so finely suspended that they enable the continuity of the surrounding medium, but it also propels the deeper porosity of parts that stifle light inside the three-dimensional object. Blackness is deep "in" the latter while it heralds the dematerialization of the former. As the spot at the center of Newton's glasses, the secondary sensation of blackness precludes the calculation of corpuscular diameter because there "is" no intervening body where the glasses touch. But perceived as three-dimensional opacity, blackness proves the power of corpuscular parts to deaden light. The analogy of bigger to smaller bodies that sustains corpuscular philosophy propels discrepant incarnations of blackness's engendering texture: blackness marks both the corpuscle's absence and its normative attribute.

Colours of People recapitulates this paradox. Human blackness appears to acquire a physiological rationale in Mitchell's text: perhaps most emphatically, in his anatomy of the "many black Fibres pervading the inner *Lamella* [of the epidermis or top layer of skin], and perforating the upper one, which appear like so many black Spots on these two Surfaces, when separated from one another" (110). Mitchell's "black Fibres," which terminate in "black Spots" or "small Scale[s]" (113), court the ultimate impossibility anticipated by the human incarnation of Newton's thin plates, the analogical likeness of "small Scales" that coat persons to the contact point at which blackness has no body at all. Indeed, Mitchell proceeds to erode the corporeality of human blackness. Citing the lens-maker and microbiologist Anthony van Leeuwenhoeck (1632–1723), Mitchell compares Leeuwenhoeck's computation of the porosity of human skin to the data provided by Newton's nomograph:

> Sir *Isaac Newton* informs us, that the Particles of Bodies, on which their Colours depend, are about 600 times less than those which can be discerned with the naked Eye . . . But *Leeuwenhoeck* shews, that a Portion of the *Epidermis*, no bigger than what can be discerned with the naked Eye, is divided into 125000 Pores; which Pores must divide such a Portion of the Skin . . . into 125,000 Particles; therefore each of these Parts of the Skin, between its Pores, must be about two hundred times less than those Particles, on which the Colours of Bodies depend. (119)

As *Colours*'s sole alignment of Newton's reflective corpuscle and contemporary epidermal anatomy, Mitchell calculates that skin's particles are "about two hundred times" smaller than the bodies that occupy the space between Newton's lens and glass plate. This claim marks a dissolution of skin color's undergirding texture as reciprocally literal as its instantiation by the blind man who feels red as smoothness: skin particles whose diameter renders them, Mitchell writes, "too small to reflect any Rays of Light from their common Surfaces, or to appear coloured from such reflected Rays" (120) elude any colorific identity whatsoever. Because the appearance of color depends on corpuscles, as Newton stipulates, of a determinate bigness, the "125,000 Particles" comprising the least visible "Portion of the *Epidermis*" lack the capacity to produce it. The nomograph whose smallest numbers exceed the diameter of skin's parts deprives those parts of secondary power. But if epidermal particles are divested of the power to produce color, from where does this human endowment proceed?

If color cannot be produced by thin films that coat the person, it must come, Mitchell writes, from "below" (121). At this juncture, Mitchell echoes Arbuthnot's displacement of difference from earthbound bodies to airborne fragments of those same bodies, because Mitchell transposes color from the epidermis to "the pure White of the Membranes below it" (121).[77] Because too-tiny epidermal particles are not colored but "transparent" (120) in all persons, deviations from the epidermis's underlying "pure White" occur as a result of another "Texture": "those who have such thick and coarse Skins, are never of so perfect and pure a White, as they who have a thin and fine Skin (as (*a*) *Cowper* observes)" (121). Mitchell footnotes the anatomist William Cowper's (c. 1666–1709) *Anatomy of Humane Bodies* (1698), signaling his debt to Cowper's claim to "doubt[] of the Existence of . . . Aqueous Vessels, between the *Cuticula* and *Cutis*; in which some have placed the Seat of that Tawny Tincture of the *Ægyptians*, and that Black one of the *Æthiopians*."[78] Before Mitchell, Cowper denies that the so-called Malpighian layer—located "between" the outer epidermis (or "*Cuticula* or Scarf-skin")[79] and the epidermis's bottom or "*Cutis*"—contains either "Tawny" or "Black" "Tincture." (We can note the congruent roles played by pigmenting and alchemical tinctures, alienable parts that color metals or persons.) But Cowper does not propose density as an alternative "Seat" of the skin's shade; rather, he stresses skin's accretion of separable layers as a mechanical effect: "With the Assistance of the Microscope, the *Cuticula* appears composed of divers *Strata* or Beds of Scales . . . According to the Number of these *Strata* or Beds of Scales, the Skin appears to be more, or less Fair, and the Person is commonly said to have a thicker or thinner Skin . . . [O]n the Lips not above two *Strata* appear; on other parts more, seldom less; in the Bottoms of the Feet of those who walk much, and the Palms of the Hands of Laborious Mechanicks, these *Strata* are not only very numerous, but each

Scale is thickned."[80] Like Hooke's laminae of mica, "*Strata* or Beds of Scales" prove the skin's discontinuity: "the Skin can no more be esteem'd a Similar or Simple Part, than any of those call'd Dissimular [*sic*] or Compounded Parts."[81] By deriving the skin's imminent dissimilarity from layers of identical "*Strata*," Cowper anticipates Mitchell's ascription of epidermal particularity to the same structural difference: "altho' the Particles, of which white and dark skinn'd People are composed, may not be very different from one another, as they seem not to be; yet . . . more *Strata* of them, in thick Skins, and the Smallness of their Intervals in Skins of a dense Texture, will increase . . . the Opacity of them" (123). As an effect of "Texture" and not ontologically discrete particles of blackness, skin color, Mitchell affirms, accommodates the monogenetic origin of humanity's starting body: "black and white People . . . might very naturally be both descended from one and the same Parents" (145).[82] Racial difference proceeds not from innately qualified parents or "Stock" (131) but from the impeded transparency that comes to distinguish some persons' epidermal texture.

As testimony to contingent variation in the skin's amassment of identical strata, Cowper cites "the Palms and the Hands of Laborious Mechanicks." This evidence implicates a corollary proof Cowper does not pursue: repetitive use makes the epidermis coarse. Extending Cowper's claim for the local thickening of workers' skin, Mitchell claims another extrinsic determinant of texture: "we see in Smiths, *&c.* constantly used to handle hot and hard Things, who have the Skin of their Hands become so thick and hard or cartilaginous by it . . . And thus it is . . . with the Skins of Negroes, *Indians*, &c. constantly exposed, and generally naked, to the scorching Heat of the Sun in a perpetual Summer" (133). While the immediate source of southern skin's density is the sun, Mitchell's most proximate exemplar of solar effects is not a tanned body but a working part, "Hands . . . constantly used to handle hot and hard Things." An extreme historical circularity is sealed by this enjambment of examples. Because the smith's forge stands in for the sun, the entire epidermal surface of "constantly exposed, and generally naked" southern persons is represented by the worker's calloused palm. Cowper's laborers demonstrate the local capacity of mechanical duress to thicken select portions of skin, as is not the case, for example, with "Lips," where "not above two *Strata* appear." Mitchell, however, elides Cowper's articulation of epidermal anatomy to posit an entire person coarsened by exposure to "perpetual Summer." Callosity modeled by the locally dense strata of English laborers thereby incarnates blackness as anticipatory fitness for mechanical use. Epidermal thickening may be effected by climate, but it sustains the preemptive vindication of its bearer's capacity to labor—or, be "constantly used." Although they descend from the same stock, some persons' texture is adequate to fix color as holistic mechanical aptitude.

We might pause to consider recent critical alignment of racial non-essence and physiological or ontological superficiality. That is, an eighteenth-century episteme without a biological conception of race has tended to be denied any counterweight that entrenches human difference in, for example, texture conjured from identical parts.[83] Wheeler, for example, defines climate theory as "superficial and malleable beliefs about skin color and race" or "superficial and changeable notions of human difference."[84] For Wheeler, the word "superficial" connotes meaning in three registers: in a formal register, as a body's surface or outer layer; in a political register, as the form's prospective compatibility with an abolitionist or humanitarian agenda; and in an ontological register, as the form's capacity to anchor difference. In the second register, Wheeler denies that the shallowness of color was historically coincident with anti-racist institutions and politics. But in the third, she aligns eighteenth-century color's "exterior" placement and a "malleable" or "changeable" ontology of phenotypic difference. I have argued that while human blackness is not secured by particulate matter—indeed, it is not irreducibly embodied by single corpuscles at all—neither Arbuthnot nor Mitchell projects a superficial racial episteme in this third sense.

For Arbuthnot, varying levels of micro-oscillation proceed from air pressure outside and inside the body's fluids; for Mitchell, callousness that seals the epidermal surface anticipates southern persons already mechanically made to labor. If inner depths harbor particles that obey Boyle's Law to engender slavishness at low pressure, so, according to Mitchell, superficies built from the same starting stock render some persons' fitness for work irreversible: southern skins are "rendered so thick and hard, or tough and callous, as not to be so easily affected, or readily wrought upon, to render them again of their original swarthy or pale Colour" (148). For Arbuthnot and Mitchell, depth does not harbor essence, but neither can surface assure malleability. As Delbourgo cogently observes: "an environmental account of skin color and the naturalization of slave labor that emphasized the physical hardiness of African bodies were not mutually exclusive positions."[85] By revising his text to restrict the number of persons able to move from environing air, Arbuthnot affirms an inevitability dictated not by particularizing substance but by the mathematical outcome of Boyle's Law. Humanity's identical starting particles are more than adequate to effect qualities—like slavishness or fitness for mechanical work—whose tenacity reflects one of the powers of corpuscular texture.

By referring light's suffocation not to a single deadening corpuscle but to epidermal strata that conflate somatic hue and the capacity to labor, Mitchell endorses human bondage legitimized by science as susceptibility to mechanical use. Both Arbuthnot and Mitchell collapse blackness and personal forfeiture of ownership in labor: as proneness to enslavement induced before persons can muster the volition to move, corpuscles that refuse racial essence engender texture exempted from Lockean personhood. Arbuthnot and Mitchell thus demand some refinement of the critical tendency to affiliate eighteenth-century matter—whether indivisible atoms or bigger things—with facticity. Dror Wahrman, for whom the permutations of the British eighteenth-century self are unencumbered by its physicality, nonetheless claims the body as the repository of a surge of racism emergent in Britain in the 1760s. Wahrman does not question the readiness of eighteenth-century persons to receive racist essence.[86] Yet whether Arbuthnot's identical airy particles or Mitchell's too-tiny plates of skin, the corpuscle itself does not secure difference. Instead, corpuscles situate a scientific rationale for the subjugation of southern persons in the secondary powers of texture. This power amplifies Catherine Molineux's claim that phenotypic color after Boyle "was a secondary rather than a primary quality,"[87] because thin films hardened by sun or resistant movement attenuated by low pressure sustain an inexorable historical end, corpuscular science's vindication of the forced labor of southern persons.

As I read Chetwood's and Aubin's travel narratives, I pursue an ontology of color effected by relations, including seminal relations. First, I address epistemology in Chetwood's novels, which motivates my turn to Aubin's imagined entrenchment of European purity in displaced noblewomen's bodies.

Mutilating Innocence: White Femininity's Virtue

In *Falconer*, Chetwood's protagonist underscores the signal epistemological affiliation of the travel narrative. Averring that he witnessed a caught shark holding "in his Paunch the Collar-bone of a Man, and a Boatswain's silver Whistle, with a red Ribbon in it, entire," Falconer informs his reader that the latter item "may be seen at my Bookseller's, if any one has the Curiosity to ask for it."[88] With this assurance of the referential immanence of the tale his narrator recounts, Chetwood enters *Falconer* to vouch for its truth. Cited in the preface as "*Honest* Chetwood, *my Bookseller*,"[89] Chetwood really did publish and promote his own novel. A curious reader could have tracked Falconer's reference, since the "Whistle" is linked to a real address—"W. CHETWOOD, at *Cato*'s *Head*, in *Russel-Street*, *Covent-Garden*"—whose location in the apparatus framing Falconer's plot promises to cement its empirical integrity. In advance of Swift's *Gulliver's Travels*, Chetwood deploys extradiegetic zones of the printed book to authorize its narrator's honesty. (Like Samuel Richardson's, Chetwood's avocation as a writer-printer may have enhanced his sensitivity to the prospective novelization of the physical artifact.)

By invoking his publisher, Falconer assumes the posture Michael McKeon designates "naive empiricism,"[90] the alignment of truth claims and rigorously referential sensory experience. In a move that flags the debt of naive empiricism to Locke's decrials of abstract or ornamental verbiage, Falconer, writing of Jamaica, declares his failure to describe one of Locke's own prized objects: "But of all the Fruit this Island produces, the Pine-Apple is the best . . . the Taste is so excellent, that I know not what to compare it to, it having the Relish of all fine-flavour'd Fruits."[91] Falconer cedes his inability to relate the pineapple's "Taste" to ideas acquired from restricted contact with other fruits. While the proximity of the object would prove the silver whistle is real, Falconer's inimitable sensory experience is communicated by the fact that it exceeds the words at his command. As Locke stipulates of renovated language, Chetwood's signifiers yield to the pressure of their referents, whether the nearness of a traveler's relics or the extralinguistic quiddity of exotic sensory ideas.

As Falconer catalogs Jamaica's edibles, he repeatedly collates old and new ideas. His reluctance to compare the pineapple gives way to a complexly conjoined set of resembling impressions:

> *Plantins* are a very good Fruit, which is, when bak'd in the Ashes, us'd by the Slaves instead of Bread: 'Tis a Fruit with a Skin on it, like our Beans, which is taken off, and then the Fruit appears, about the size of a *Bolognia* Sausage . . . The *Guana* is another Creature as amphibious as the *Alligator*, but nothing nigh so large: There's an Island near *Jamaica*, call'd *Guana* Island, inhabited by nothing else; our Seamen eat of these latter, but much Good may it do 'em, for the Flesh looks like a piece of a *Black-a-more*'s Arm; but how it tastes I can't tell . . . The *Cocoa* Nut is a Fruit that is both

> Meat, Drink, and Cloathing to the Natives, (I mean the Blacks) the Rind serves for weaving of Cloaths, nay, and rigging their Canoes, before they knew the *Europeans*.[92]

As a not rigorously systematized inventory of eatables, Falconer's survey veers from fruits to slaves to reptiles to sailors and back to fruit again. He concludes this episode with a rare meta-narrative nod to the scene of its composition: "I don't doubt but my Readers will laugh at me for thus running from one Thing to another, but I relate 'em just as my Memory prompts me."[93] By defending a variant of natural history whose impetus is how "my Memory prompts me," Falconer's justification of his method harks as directly to Lockean empiricism as the vindicable whistle or ineffable pineapple. Dictated by the impulses of memory, Falconer's "running from one Thing to another" claims its idiosyncratic order as the result of what Locke names "*Association*," the "strong Combination of *Ideas*, not ally'd by Nature."[94] While Falconer's excursus into Jamaican natural history draws from past experience or, as Locke writes, "*Memory* . . . the Store-house of our *Ideas*,"[95] Falconer's narrative style is more topically shaped by "habits of Thinking in the Understanding," chief of which, Locke posits, is "the accidental Connexion of two *Ideas*."[96] In this passage, "strong" and "accidental Connexion" advances a formal articulation of qualitatively novel bodies.

The ideas comprising the lattice of associational reference in Falconer's catalog of foodstuffs recur to "Slaves" and to "Natives, (I mean the Blacks)." The alliance between the former and the latter is not, as Locke suggests, sanctioned "by Nature." Just as striking in this arbitrary but strong combination is the insistence with which Chetwood's ostensible objects of empirical scrutiny—plantains, iguanas, and coconuts—occasion semantic resonance that lends structure to neighboring ideas. In each instance, the connotative blurring of foods and persons—including the explicitly cannibalistic confusion of reptile and human "Flesh"—entails violence at the same time that it precipitates form. While the manifest empirical strategy of natural history is evidenced by domestic ideas, like that of the Bologna sausage, which vivify the narrator's reference to unfamiliar foods like plantains, the submerged pull of other referents propels a formal resemblance that entails the prospective alienation of surface from body. Chetwood's history deploys associational coordinates to motivate a form which—while its central, looming instance threatens the spectacle of dismemberment—figures skin as rind and rind as clothes. This form pivots on qualitative difference, but solicits ideas of such difference stripped from the body that bears it. Chetwood's travel narratives project a *formal* ontology of human color, which assumes both the contingency and the weightiness of an attribute whose variable attachment to persons embroils British percipients in relations.

Chetwood's narrators do not encounter phenotypic difference as static substance. The epistemic prerogative of British observers is not consistently enabled by the facticity of skin color, as Chetwood's second novel, *The Voyages and Adventures of Captain Robert Boyle* (1726), makes clear. Chetwood gestures toward his protagonist's namesake (designated "Robert" to avoid confusion) when Robert declares that, after his merchant father dies in a wreck, he "chose [the calling] of a Watchmaker, as imagining my self to have a good mechanical Head."[97] Upon going to sea and being captured by "an *Irish Renegado*"[98] turned Muslim, Robert's "mechanical Head" qualifies him to serve as his captor's gardener in Salé, where he works elaborate Cartesian repairs on a disused fountain; when he falls in love with a captive Englishwoman whom his master desires, Robert buys time to scheme their flight by pledging to concoct a "chymical Liquid" or love potion of "calcin'd Gold"[99] for his master's use. Robert deploys the same competencies to conceal his lover, Mrs. Villars, before they embark on their escape: "I inform'd her she must submit to another Disguise. She ask'd me what that was: I took out a Paper of *Ombre*, and told her she must permit me to rub it over her Face and Hands; which I did: But the Pleasure of touching her Flesh in that gentle manner perfectly put me in an Ecstasy, which she observ'd, for I could not help softly squeezing her Hand."[100] Robert's use of umber pigment on his beloved's "Face and Hands" exploits skin color as a reversible apparition of chymical and mechanical "Disguise." With *Falconer* and Chetwood's third novel, *The Voyages, Travels and Adventures, of William Owen Gwin Vaughan, Esq.* (1736), *Boyle* coordinates vertiginous shifts from white mastery to white slavery with the loosening of somatic color's fixity.[101] The "*Ombre*" Robert spreads over his lover's skin illustrates the schematics of color that is not attributable to the disposition of engendering parts but rather, as Newton writes of objects made to display any illuminating hue, may be put on.

Even though the extrinsic application of Mrs. Villars's tint suffices mechanically to excite another person, Chetwood does not rigorously distinguish between inner and outer sources of colorific effects. As Falconer remarks when he is cast away en route to Jamaica, "I had been here now a Month, by my Reckoning, and in that Time my Skin look'd as if it had been rubb'd over with Walnut-shells."[102] This suntan appears no more entrenched than extrinsic pigment or sartorial disguise, as Robert affirms when he crosses paths in Morocco with an Englishman he met while enslaved: "I could not perceive . . . he knew me; but that might be from my Change of Habit, and the Sun's tarnishing my Complexion."[103] The susceptibility of whiteness to both "tarnishing" and rubbing extends to the recovery of filial likeness: settled in Somersetshire after his sea voyages, Robert encounters "twenty Gypsies . . . with a Child in the midst stark naked, which they were rubbing over with Walnut-Shells . . . that made it look like one of their Fraternity."[104] After a series of trials reunite him with Mrs. Villars, Robert learns that "the Infant I had so marvellously sav'd, was our own Child."[105] At a generation's remove, Robert recovers his own occluded whiteness, whose estrangement briefly enables alternative fraternal affinity. Chetwood's protagonist William Vaughan, having "ombred my Face, and appear'd as a Gypsy," murders the kidnapper of his beloved; liability hinges on sensible ideas that split him from the doer of the act: "Pursuit, I know, has been made after the Gypsy, but no Enquiry after me."[106]

The mitigation of whiteness by rubbing, tarnishing, or "Habit" does not cancel domestic mnemonics for fairness like those invoked by Vaughn to an English paramour: "The Roses and Lilies, were Remembrancers of your amiable Face; the Down of Swans, of the Whiteness and Softness of your Skin."[107] Yet the engine driving Chetwood's travel narratives is not the "Down of Swans" but the distilled imperative emblematized by Mrs. Villars's rubbing. As Robert recounts, "touching her Flesh in that gentle manner perfectly put me in an Ecstasy": the mechanics of qualitative disguise, the motion with which Robert applies umber particles to Villars's "Flesh," trigger his feeling. As a corpuscular determinant of character's always subsidiary affective response, the rubbing scene offers insight into Chetwood's travel plot. For its European actors, this plot is galvanized by slippage in the alignment of climate and qualities—by persons who, displaced from the atmosphere that makes them, cannot conserve their native attributes without many intervening, even countervailing, compensatory motions. Robert's application of umber to his mistress's skin affirms as one determinant of plot the suspension of qualities' anchoring liaison to place.

Like all of Chetwood's novels, *Boyle* contains inset stories which sketch further experimental lines for the deracination of qualities to take, as when Robert's fellow traveler Don Pedro Aquilio discovers his fiancée Isabella's infidelity. In revenge, Aquilio impersonates her lover and is admitted into her chamber at night: "I soon discover'd my false Fair-one, (tho' in the dark.) She was undress'd and disincumber'd of every Lett to Enjoyment."[108] Soon after,

Aquilio murders his disguised rival only to learn that the man is his father; he decides to leave his home in Seville "for the City of *Mexico*."[109] Waiting for a fair wind, Aquilio receives a letter asking him to serve as its anonymous writer's "*Protector*" to Mexico City; Aquilio's servant informs him that its author is "a young Negro . . . [whose] Father, tho' a Negro, was a Man of Substance, and had sent him in his Infancy to be educated at *Sevil*." Aquilio "was mightily pleas'd with the Person of the black Gentleman,"[110] who proceeds to request a private audience with him. The "young Negro" then confesses: "This Veil of Night would not hide my Blushes, if I were not convinc'd in your Knowledge of my Frailty . . . Be not surpriz'd to discover in this Disguise" Isabella, Aquilio's estranged fiancée. Vowing a renewal of her tenderness, she declares, "I do not mean the tye of Wedlock, but if you will accept of me as I am, I'll be as subservient to your Commands, as your meanest Slave."[111]

Like Falconer's Indian bride, "in the dark" Aquilio embodies enjoyment so effectively that the secondary refinement of his identity is irrelevant. As he embarks on his voyage, however, the capacity of darkness to defer a person's sensible endowments migrates to his discarded fiancée. Transmuted into the "young Negro" with whom Aquilio finds himself "pleas'd," Isabella collapses the absence of light and her subsequent impersonation of blackness: "This Veil of Night would not hide my Blushes." A prototypical inset story, this episode condenses the displacements that propel Chetwood's novels: the inset plot feigns an interest in denominating Isabella's lover only to shift to the overlay of ambient lack of light, somatic blackness, pleasure, and slavishness entertained by his object. Promoted by the temporal density of this subgenre, which admits increments of chronological regress, secondary qualities move from one person to another, insinuating difference into a chamber where bodies, as Newton argues, may be made to put on any color.

Marking the ambivalent restitution of Isabella's identity on board the ship to Mexico, "we agreed to dress her in the Habit of a Man, and let her wear her own Complexion."[112] Complexion Europeans can elect to "wear" loses its permanence by virtue of its distance from the air whose barometric specificity, Robert assures a fellow northerner, upholds the evidence that "one Climate gave us Birth."[113] To broach the more particularized mandate of travel novels penned by Chetwood's contemporary Aubin, Isabella's example elicits the question: Does dress that makes her a "Man" also signal femininity's uprooting by travel? As one episode in *William Vaughan* suggests, Chetwood may concede the radical outcome of plots that unsettle the qualitative determinants of sex as well as whiteness. Finally released from slavery in the Ottoman city of Tunis, Vaughan's brother John relates how he accepted his fate at the hands of his new master: "It wou'd avail me nothing to reason with him, therefore I was forc'd, through a fatal Necessity, to accept of my Chains as chearfully as I could."[114] Remarkably, John's statement of resignation iterates the resolve of wives housed in the English domestic novel. The collapse of public and private despotism central to eighteenth-century British orientalist thought compels displaced Englishmen "chearfully" to endure the conjugal tyranny borne at home by wives.[115] Chetwood's captive man staves off abject slavishness by mimicking wifely cheer, for John conserves his native—or Lockean—disposition by simulating willed duty. The likeness of Ottoman captivity and wifely compliance includes an intimation of conjugal forbearance threatened by his master's licentious son, to which John responds: "I would rather suffer Death, than comply with his infamous Desires. I can hardly mention it without Blushing."[116] At the intersection of phenotypic whiteness and coverture that extends to the wife's own status as sexual property, blushing marks the premier authenticator of eighteenth-century British women's patrilineal use-value. By pledging death in exchange for chastity empirically proven by his change of tint, Vaughan safeguards British women's sexed and raced contribution to modern patriarchy.

Aubin authored, among other titles, *The Strange Adventures of the Count de Vinevil And his Family* (1721), *The Life of Madam de Beaumount, a French Lady* (1721), *The Noble Slaves: or, the Lives and Adventures of Two Lords and two Ladies* (1722), *The Life and Amorous Adventures of Lucinda, an English Lady* (1722), and *The Life of Charlotta Du Pont, An English Lady* (1723). These plots link whiteness and sexed virtue to adjudicate between the options polarized by Newton and Boyle as color's alienation from or incarnation as body. Decisively and violently, Aubin vindicates whiteness as embodiment, but this whiteness is crossed, like it is for John Vaughan, with a stipulation of feminine reproductive service also endemic to European climes. Boyle's blind man, who identifies color by touching its engendering microtexture, stands for the prospective obduracy of feminine value whose physicality is likewise thrown into relief by contiguous duress. However, to mitigate the incarnation of color as reflective surface, Boyle's *Touching Colours* admits agents other than the refracted light that renders hard particles diaphanous. When he considers "The Cause of the Blackness of those . . . we are wont to call *Negroes*,"[117] Boyle deviates from the collapse of blackness and asperity to hazard other occasions for human shade. By recapitulating this effort, I anticipate modalities of patriarchal value realized by Aubin's women in concert with the North African and African actors who threaten their ruin.

Touching Colours denies that climate engenders racial difference: "Experience doth not Evince . . . That that Heat alone can produce a Discolouring that shall amount to a true Blackness, like that of *Negroes*."[118] Most markedly in a text that otherwise insists on the relational production of hue, Boyle argues: "the Principal Cause . . . of the Blackness of *Negroes* is some Peculiar and Seminal Impression, for . . . we see that *Blackmore* boyes brought over into these Colder Climates lose not their Colour."[119] As I suggest in Chapter 3, semina represent one expedient by which ends may be installed inside particles. In the case of "*Blackmore* boyes," seminal impression conserves a projected outcome impervious to climatic influence, the extrinsic standard of blackness that remains "true." But seminal impression promises to communicate an ontology of true blackness only for individuals, because Boyle acknowledges reports of "Twins, whereof one had a White skin, the other a Black."[120] Although, Boyle concedes, blackness and whiteness are isolated in discrete persons, siblings and families cannot reproduce that difference: "I see not why it should not be at least as possible, that White Parents may sometimes have Black Children, as that *African Negroes* should sometimes have lastingly White ones."[121] Boyle insulates the distinction of black and white from environmental encroachment, but *semina* foreclose the contribution of other, intimately propagating bodies, including parents. The irresolvable ontological outcome of seeds that Boyle opposes to climate will surface in Aubin's fiction.

In anticipation of Chetwood's skins and rinds, Boyle, writing in the register of natural history, evinces "Cherries, Plums, and I know not how many other Bodies, wherein the skin is of one Colour, and what it hides of another." The figuration of skin as a surface that "hides" extends to trials of solvents that change the outer texture of metals, as when invisible corpuscles of dissolved gold return to sense by leaving a sheen on some mercury placed in the solution, "that [gold] Mettall after a little while Cloathing the Surface of the *Quick-silver*, with a Thin Film of its own Livery."[122] Indeed, metal's dissolved particles may fuse directly to human skin: "the purer Crystals of fine Silver made with *Aqua fortis* [nitric acid], though they appear White, will presently Dye the Skin and Nails, with a Black, or at least a very Dark Colour."[123] Color manifest to the eye as "Dye," "Cloathing," "Livery," and "Thin Film" once more lends form to humans: "the Seat of that Colour seems to be but the thin *Epidermes*, or outward Skin, for I knew a young *Negroe*, who having been lightly Sick of the Small Pox or Measles . . . in those places where the little Tumors had broke their passage through the Skin, when they were gone, they left Whitish specks behind them."[124] "Measles" condense corpuscular and seminal ontologies of phenotypic color to articulate a paradoxically individuated person: the seat of true blackness is a vanishingly fine outer surface which, like dye comprised of silver crystals, may be ruptured to disclose "Whitish specks." The truth of Boyle's blackness correlates with neither familial continuance nor the formal assistance offered ontology by depth.

Writing after Boyle, Aubin grapples with additional ontological predicates for human color, because her overwhelming concern is British women's whiteness. To argue that her fiction envisions an instantiation of this tint inextricable from gentlewomen's sexual utility to European patriarchy is to make a corollary claim for Aubin's conservative mandate.[125] This is first evident as the misalignment of talents and places—which McKeon designates "status inconsistency"[126]—glossed in her Tory appeal for the sensory refreshment of aristocracy: "I heartily wish . . . our Nobility [would] make themselves distinguish'd from the Crowd, by shining Qualities, for which their Ancestors became so honour'd, and for Reward of which obtain'd those Titles they inherit."[127] Aubin's call for the sensory revivification of nobility's pedigree is impelled by a representative "lustful *Londoner*" who inhabits "that curs'd Town, where Vice takes place of Virtue, where Men rise by Villainy and Fraud."[128] In *Count de Vinevil*, Aubin asserts an intertextual warrant for the ensuing proof: "As for the Truth of what this Narrative contains, since *Robinson Cruso* has been so well receiv'd, which is more improbable, I know no reason why this should be thought a Fiction."[129] The "this" whose probability Aubin champions is not the likelihood of castaway survival but the staunchness of noblewomen's preservation of their chastity under orientalist threat. Marking Aubin's particularized deployment of a travel plot that tries nobility's "shining Qualities," she depicts "Women [who] are really vertuous, and such as we ought to imitate."[130] Her Tory animus precludes her interrogation of alliances sketched, for example, in the amorous destiny of *Vinevil*'s protagonist Ardelisa and the Count of Longueville: "in fine, her Soul was capable of every thing that was Noble. There needed nothing more than this Sympathy of Souls, to create the strongest and most lasting Affection betwixt this young Nobleman and Lady; they loved so tenderly, and agreed so well, that they seem'd only born for one another."[131]

Soliciting only notational treatment, the marriage of Ardelisa and Longueville precedes their arrival in Constantinople, where the couple follows

Ardelisa's father the Count de Vinevil, who, disgusted by the rampant status inconsistency in France, has liquefied his estate and turned merchant. Events unfurl rapidly after they embark: Vinevil is killed in Constantinople, Longueville is separated from his wife in a storm, and Ardelisa hides in disguise as a man. While in drag, she is seized by a Turkish general, who proceeds to elucidate how Aubin's travel plot exploits despotism to sustain both universal and particular trials of British gentry. The general addresses his disguised captive: "Lovely Boy, or Maid, I know not which as yet to call you . . . If you are a Man, renounce your Faith, adore our Prophet, and my Great Emperor, and I will give you Honors and Wealth exceeding your Imagination: If you're a Woman, here are Apartments, where Painting, Downy Beds, and Habits fit for to cover that soft Frame, Gardens to walk in, and Food delicious, with faithful Slaves to wait upon you, invite your Stay; where I will feast each Sense, and make you happy as Mortality can be."[132] The alternate pathways whereby a "Boy, or Maid" abdicates Englishness underscore Aubin's foreclosure of feminine political personhood. While the hypothetical man is enjoined to forfeit his Christianity and British citizenship, the woman's theological and governmental ties rank below her carnal value as an asset of Western patriarchy. Whereas the Englishman's estrangement merits the abstracted enticements of "Honors and Wealth," the maid must withstand sensual titillations like "Painting, Downy Beds, and Habits." If Aubin's man is political, the woman is already domestic: before she enters the seraglio, she is poised to betray not enlightened nationhood but her role as reproductive transmitter of European patrilineal privilege. As the protagonist of *Amorous Adventures of Lucinda* remarks when she is captured by Barbary pirates: "My only Consolation was, that they had not discover'd my Sex, being more willing to undergo any Slavery they should enjoin me as a Man, than be forc'd to submit my self as a Woman to their libidinous Desires."[133] At this juncture Lucinda, like Ardelisa, could become either "Sex." According to the calculus driving Aubin's orientalist plot, as a slave Lucinda will be a "Man." But as an object of "libidinous Desires," the novel will test her virtue as a "Woman."

Aubin refuses the inverse determination whereby captives who are treated like women might include John Vaughan. The discrimination of male slaves from female objects of "libidinous Desire" isolates the warrant of a travel plot that musters an ontology of race as the accessory to European women's proof of their corporeal virtue. It is not the fitness of nobles who are "born for one another" that travel threatens to disturb, but the imperative of upper-class white women's reproductive instrumentality. Ardelisa's husband stipulates the link between that instrumentality and an ontological predicate for whiteness: "Permit not a vile Infidel to dishonour you, resist to death, and let me not be so completely curs'd, to hear you live, and are debauch'd." Before he dies, Vinevil likewise testifies to his daughter's preemptive obligation: "My Daughter . . . will, I know, prefer a noble Death to such Dishonour."[134] In Aubin's seraglio, the latter dishonor is no discrete act, but the unremitting imminence of feminine debasement, opening an experimental—and pornographic—space in which Western patriarchy's would-be essential guarantor is repeatedly undone. But, in a signal of the racialized security this threat yields Aubin's novel, it is only female Venetians who are thus "debauch'd": Violetta, for example, includes herself among "noble Virgins, who, like me, have liv'd too long, being now made Slaves to the wild Lusts of cruel Infidels."[135]

Debauched virgins threaten the redundancy of women who have "liv'd too long"—that is, women who have survived the ruin of their marital use-value. But Venetian women's contiguity to the East admits a finer distinction communicated to Aubin's English reader at the conclusion of *Charlotta Du Pont*, which cites the exemplarity of a subset of more fiercely resistant heroines:

> These Examples should convince us, how possible it is for us to behave our selves as we ought in our Conditions, since Ladys, whose Sex and tender manner of Breeding, render them much less able than Men to support such Hardships, bravely endured Shipwrecks, Want, Cold, and Slavery, and every Ill that human Nature could be try'd withal; yet we who never feel the Inclemency of foreign Climates, that never saw the Face of barbarous Pirates, or Savages, are impatient at a fit of Sickness, or a Disappointment, shake at a Storm, and are brave in nothing but in daring Heaven's Judgments. Let us blush when we read such Historys as these.[136]

Sustained by truth claims that rival *Crusoe*'s, Aubin's "tender" ladies endure trials which are exotic but real. Yet as her injunction to "we who never feel the Inclemency of foreign Climates" makes clear, the motivation for displaced noblewomen's endurance of "Shipwreck, Want, Cold, and Slavery" is patriarchy at home. The particularity of this commitment is conveyed by the reader's power to "blush," which links whiteness to patriarchal allegiance whose contradictory requirement—that ladies delimit rape from the specter of consent also insinuated by force—Aubin's oriental plot aspires to resolve.

In *Noble Slaves*, the Spanish virgin Maria, captured by a Turk and sold in "*Ispahan*" (Isfahan, Iran), pledges love to the Persian captain Tanganor after he agrees to "turn Christian."[137] Abducted by the state's emperor, Maria is later restored to Tanganor (who to retrieve her "black'd my Face and Hands, and changed my Dress for that of a Slave"[138]), but upon lifting her veil he discovers she is now blind. After declaring that "my Eyes are sacrificed to Virtue, with the loss of them I have procured your Happiness," Maria explains how she preserved her chastity:

> [Maria's abductor declares:] "Now I am resolved never to part with you: Force must I find procure me now what your Consent shall afterwards secure me of." 'Here he took me in his Arms, and carry'd me to a rich Bed, on which he threw me . . . "My Eyes shall never see my Shame, *said I*, nor more inflame Mankind: These I offer up to Vertue, and they shall weep no more in ought but Blood." At these Words I tore my Eyeballs out, and threw them at him. "I saw no more, but heard him say, *Ah cruel Maid, what have you done*? *Tanganor*, you are happy . . . [Maria,] you are more than Woman, and I will never presume to sue again for what you must deny."[139]

As the structural impasse Maria must defuse, the equivocation of masculine "Force" and feminine "Consent" may be catalyzed by the former in the absence of the latter, because Aubin approves the corrective intimation of consent conjured "afterwards" to salve the illegitimacy of already exercised violence. Indeed, the depth of Aubin's approval of the encroaching redundancy of force and consent can be gauged by the superhuman measures required to forestall it. On the brink of rape, Maria internalizes liability for her power to "inflame Mankind": this entails such a profound arrogation of agency that she must preemptively ruin her own body. A ruin inflicted not on her hymen—which, as the property of patriarchy, she cannot presume to touch—but on "Eyes" that would bear witness to the inaugural breach between force and consent, Maria's blindness appeases the emperor with a grisly surrogate embodiment of her irreplaceable "Vertue."

Aubin repeats the trial in *Noble Slaves* when the protagonist Teresa, running down stairs after throwing wine in her assailant's face, breaks her leg: her pursuer, "seeing the Blood running on the Floor, soon discover'd . . . the Shin-bone was shiver'd, so that it had cut thro the Skin and Sinews, and appear'd. This Sight dash'd his amorous Fires."[140] While not as spectacular a counter for feminine honor as Maria's eyes, the apparition of "Blood" and punctured "Skin" constitutes an adequately dashing—or phallic—manifestation of surrogate ruin. Teresa henceforth embodies the patriarchal value of her hymen as an incurably "lame"[141] leg. Blindness and lameness anchor chastity as durable disfigurement.

In a final appeal to the travel narrative's domestic reader, Aubin brings home the outcome leveraged by her seraglio: "the Ladies, I fear, will scarce find any here who will pull out their Eyes, break their Legs, starve, and chuse to die, to preserve their Virtues . . . The Nuns of *Glastenbury*, who parted with their Noses and Lips to preserve their Chastity, are, I think, the last the *English* Nation can boast of. 'Tis well in this Age if the fair Sex stand the Trial of soft Persuasions; a little Force will generally do to gain the proudest Maid."[142] Aubin's scornful reference to "a little Force" tips her hand on the resistibility of rape. The little force that "will generally do" indicts the excessive weakness—or, proneness to seduction by "soft Persuasions"—of Englishwomen who fail to trade permanent disfigurement for hymeneal loss.[143] In the absence of non-consent as an effective deterrent, since, for Aubin, refusal and consent converge when rape occurs, Aubin offers the surer expedient of mutilation. Concrete testimony to the will to resist, pulling out one's eyes typifies a standard of feminine intransigence that cannot be credited in the absence of sensible signs. Aubin's attempt to promote the exemplarity of "*English*" nuns "who parted with their Noses and Lips" corrects for an hymen whose theft leaves no perceptible trace. No longer an insensible and volitional attribute, susceptible to alienation by the despot's precarious likeness to any domestic exemplar of tyrannical government, scarring testifies to the virtue of women who ruin themselves in advance.

Samuel Richardson adopts Aubin's plot to amplify the despotism of his rake Lovelace.[144] As the portability of wild lust suggests, orientalist duress has its uses within Anglo-patriarchy. What role, then, is played by actors outside the harem, like the "barbarous Pirates, or Savages" whose cognate instrumentality Aubin also cites? To frame Aubin's explicit turn to an ontology of non-whiteness, I look ahead to Edward Long's *The History of Jamaica* (1774), a text whose vicious racism nonetheless sustains the same instability as Boyle's *Touching Colour*. Most infamously, Long posits polygenesis to affirm the primordial seminal ground of phenotypic difference: "For my own part, I think there are extremely potent reasons for believing, that the White and the Negroe are two distinct species."[145] But Long's chapter "*Of the Inhabitants*" includes a schema that delineates the "denominations" engendered by "intermixture"

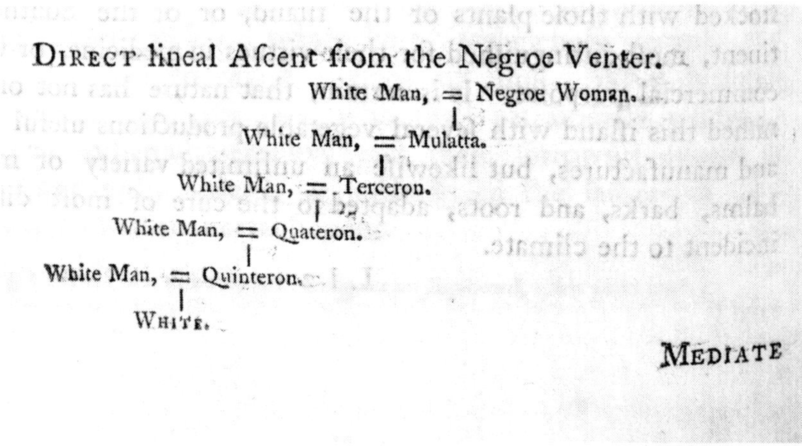

Figure 8. "Direct lineal Ascent from the Negroe Venter," Edward Long, *The History of Jamaica* (1774). Photo courtesy of the Newberry Library, Chicago. F 974.52 vol. 2.

(Fig. 8). This "table" of "different casts"[146] admits not only breaching but effacement of qualitative and genetic variance. Promulgated through the female "Negroe Venter," the fifth generation of progeny by a succession of European men claims the qualitative identifier "WHITE." Long's routing of mixture through black women reflects their historical sufferance of rape inflicted by white masters, as Angela Davis and bell hooks influentially argue, to produce a succeeding generation of enslaved persons.[147] In the plantation system schematized by Long, mixture proceeds through the racially qualified woman, whose serial seminal impression leaves intact the racial prerogative reincarnated by Long's "White Man." While the white master is impervious to change, the daughter produced from her mother's rape becomes the vehicle of ontological slippage that undoes Long's taxonomic separation of white from black.

Aubin's *Noble Slaves* and *Charlotta Du Pont*, whose publication spans the years 1722–23, represent an intermixture, and a qualitative terminus, rigorously eccentric to Long's table. In an intertextual relay that is neither explicit nor acknowledged, both novels sketch the relation between an African man and white woman by enlisting a black slave named Domingo. Aubin thus departs from the orientalist threat that finally secures both whiteness and feminine value, since those ends preclude exposure to foreign seminal influence. (Even orientalist debauchery, in Aubin's novel, operates as a simulacrum of domestic seduction and tyranny rather than a spur to childbearing.) Aubin's two Domingos augur qualitative difference that does not amplify Englishwomen's dedication to the patrilineal virtue they promise to transmit. Alienated from that service, Englishwomen produce bodies that cannot be marshaled by Aubin's conservative plot.

In *Noble Slaves* (unlike Aphra Behn's, Aubin's qualifier applies to white Europeans), the "faithful Slave" Domingo saves his Spanish mistress Teresa from death at sea on a pleasure boat, but falls sick on their desert island after eating from "a Tree with Fruit he had never seen before."[148] On the verge of death, he confesses:

> 'my Soul ador'd you, but Christianity restrain'd me from asking what my amorous Soul languished to possess. I brought you to the Wood with Thoughts my Soul now sinks at. I was born free as you, and thought I might with Honour ask your Love, since Heaven had singled me out to save your Life, and live your only Companion and Defender; but God has thought fit to disappoint me. May no other rob you of that Treasure which I no longer can protect. Angels guard you. Give me one Kiss, and send my Soul to rest.' Here he grasp'd her Hand, and strove to rise, but fell back and expir'd . . . she could not but applaud the Goodness of God, who had so wonderfully prevented her Ruin; for tho he had a Soul fair as his Face was black, yet *Domingo*, her Father's Slave, was not fit to enter her Bed.[149]

Despite Aubin's appeal to the fairness of Domingo's "Soul," *Noble Slaves* plots an opposing outcome. In this inset story, the ultimate stabilizing value is not an egalitarian pretense to human spirit but the Western patriarchal preserve of white female property. Unlike the "lustful *Turk*" appropriated to prosecute white ruin, Domingo solicits the "wonderfully prevented" allowance of "one Kiss." The alacrity with which Aubin authorizes divine vengeance flags her refusal to invest him with the despot's power to test English innocence. In a divinization of hymeneal chastity that leaves Teresa intact, Aubin enlists Domingo to enact the equivocation of white virginity and black personhood. Rather than maiming herself, Teresa watches her father's slave testify to her patriarchal value by dying in its stead.

In this instance, Aubin circumvents the equivocation of masculine violence and feminine consent by dispensing with Domingo first. *Noble Slaves*—or, more precisely, its God—denies Domingo the conjugal power that other men in Aubin's plots retroactively garner by deploying force. But in *Charlotta Du Pont*, Aubin repeats the experiment. Cast away with her company on the way to Virginia, Charlotta discovers "a Blackmoor Man" at "the Entrance of a Cavern"[150] on the island's rocky coast. Signaling this microplot's divergent outcome, he is espied "standing at the entrance of his Cave, with a white Woman who seem'd to be very young and very handsom; she had a *Molotta* Child in her Arms about a year old, her Gown and Petticoat was made of a fine Silk." Charlotta's party penetrates the family's retreat to hear the woman relate "her sad Story"[151]: born in Virginia to an "extremely rich" plantation owner, Isabinda, educated in England till age thirteen, is tricked away in a pleasure boat by her father's Angolan slave Domingo. At the instant when he declares his love, "he proceeded to kiss me, my Distraction was such, that I swooned; he took the advantage of those unhappy Minutes, when I was unable to resist, and, in fine, has kept me here two whole Years . . . I have had but this Child by him . . . He is a Christian, and would gladly marry me."[152] Like the first Domingo's story, this plot's endorsement of Isabinda's value—expressed as Domingo's eagerness to marry rather than his readiness to die—is not shaken by the inset narrative. Domingo takes "advantage" only to solicit the legitimizing sanction of "Christian" domestic authority. Yet though Isabinda "would now willingly consent to be his Wife,"[153] Aubin invokes the first Domingo's death as a prospective resolution of this plot as well: "Had he but once given me the least Intimation of his Passion, I should have acquainted my Father with his Insolence, and his Death would have prevented my Ruin."[154] But despite this reminder of the readiness of white patriarchy to defend its imperviousness to difference, "consent" and mixture instead transpire. Isabinda—or her "Father," or her God—fail to conserve her Anglo-patriarchal virtue.

To preclude Isabinda's return to that selfsame virtue, she and Domingo parent a "tauny Child."[155] This child marks a site of deep intersectional confusion in Aubin's oeuvre: Aubin's God intercedes to kill one Domingo before he transgresses a racial prohibition isolated by contrast with the serial violations enacted by random Turks. But Domingo number two sustains the collapse of masculine force and feminine consent retroactively sanctioned, Aubin's same God affirms, by Christian marriage. *Charlotta Du Pont* proves able to engineer only a local and provisional resolution. After Domingo and Isabinda are wed on the island of Saint Domingo, the narrator concludes: "the Governor sent him [Domingo] to a little Market-Town about twenty Miles from the City, to a House of Pleasure which he had there; and here he found a little

Paradise, a House so neat and richly furnish'd . . . that his Imagination could not have form'd a more beautiful Retreat . . . Here *Domingo* and his little Family liv'd happily the remainder of their days, having many Children."[156] *Charlotta Du Pont* fails to preserve white feminine use-value elsewhere salved by the reparatory powers of marriage, because Domingo's conjugal right is upheld by the same patriarchy that authorizes feminine non-consent only as manifest disfigurement. As a husband, Domingo thus does something other than prosecute the hardship Aubin enjoins her English readers to defuse. Together, he and Isabinda add "many" new bodies to a plot whose first iteration foreclosed them. These children do more than reverse the inaugural Domingo's providential decease, because their violation of Aubin's conservative mandate is intimated by their difference from and relation to their mother. However underspecified a seminal mandate she represents, Isabinda's "*Molotta*" child animates a counterplot that cannot deploy matter to recuperate white hymeneal value.

A creole educated in England, Isabinda is not licensed by Aubin's plot to send her mulatto children there. Fifty years later, Long's *History of Jamaica* imagines their future. Writing of creoles and Englishmen who engender "spurious offsprings of different complexions" with black Jamaican women, Long censors "the strange manner in which some of them [the children] are educated":[157]

> Master is sent to Westminster, or Eaton . . . whilst Miss is placed at Chelsea, or some other famed seminary . . . After much time and money bestowed on their education . . . at length they return . . . From this period, much of their future misery may be dated. Miss faints at the sight of her relations . . . for, however well this yellow brood may be received in England, yet here so great is the distinction kept up between white and mixed complexions, that very seldom are they seen together in a familiar way, though every advantage of dress or fortune should centre with the latter. Under this distinction, it is impossible but that a well-educated Mulatta must lead a very unpleasant kind of a life here.[158]

Because his vision of the "future misery" unleashed by these children's schooling in Britain testifies to anxiety not assuaged by the taxonomic barrier of species, Long's admonition to parents who fondly educate their mixed progeny is couched in an ontological register that clashes with his defense of polygenesis. It is not an innate guarantor of species difference but Jamaican observance of cultural "distinction" that Long invokes to enforce the imperviousness of race to education. To threaten "that a well-educated Mulatta must lead a very unpleasant kind of a life" upon her return to Jamaica, Long grounds this fate in an avowed particularity of geographical circumstance assured by the tautological violence of his own guarantee: "however well this yellow brood may be received in England, yet here so great is the distinction kept up." An outcome predicated on the keeping up of racism uninflected by endowments like "dress or fortune," the Jamaican daughter's portended misery erodes Long's defense of racial speciation. The effects of the quality of color will be produced by something else: neither genetic essence, nor indifferently springy corpuscles, but the historical contingency of these children's relations with other persons.

CHAPTER 5

Quality's Qualities

Fielding's Alchemical Imaginary

> The greatest Curiosity which I saw there, was the 300000000000th Part of a Louse, which, tho' it was so very small that I could not distinguish it by the naked Eye, one of their learned Men cut it in two, and gave me one Half, telling me, that little Bit of Matter would find me Employment all my Life-time: But, by the Extremity of ill Luck, I dropt it out of my Pocket somewhere in St. *James's* Park, and tho', since my Return home, I searched after it many Days, could never find it.
>
> —Henry Fielding, "*Extract from Mr.* JOB VINEGAR*'s Travels*," in *The Champion*, April 16, 1740

> Ah! Master, Master, (says the Host,) if you had travelled as far as I have, and conversed with the many Nations where I have traded, you would not give any Credit to a Man's Countenance. Symptoms in his Countenance, quotha! I would look there perhaps to see whether a Man had had the Small-Pox, but for nothing else!
>
> —Henry Fielding, *The History of the Adventures of Joseph Andrews* (1742)

Writing for the opposition periodical *The Champion* in the guise of the travel narrator Job Vinegar, Henry Fielding describes the land of the "PTFGHSIUMGSKI, *or the* INCONSTANTS."[1] This mock travel narrative cursorily defamiliarizes the targets of Fielding's commentary by capitalizing and removing vowels from key words, as when Vinegar observes that "The

Name of their Deity is MNEY."[2] His anecdote about "the 30000000000th Part of a Louse" that "would find me Employment all my Life-time" transmits Fielding's appraisal of "their RYL-SCTY,"[3] the rigorous futility of whose pursuits is encapsulated by the "learned man" who carefully cuts the specimen "in two." Empirical learning that engenders invisible curiosities reflects the epistemological inconsistency perpetrated by Vinegar's INCONSTANTS: the fractured insect literalizes the divisibility of scientific objects that can produce only smaller and smaller fragments of knowledge. While the partitioning of the louse is legible as Fielding's satire of modern as opposed to classical techniques of understanding, it also illuminates the comedy unleashed by an imperceptible particle that Vinegar reports he "saw" and then "dropt," "searched" for and failed to "find."[4] Vinegar's attempt to retrieve what he cannot see underscores the narrative absurdity of invisible things: unlike Swift's Gulliver, whose travels expose him to magnified lice, and unlike Comenius's child reader, who experiences the insect by analogy with a hog, Vinegar dramatizes the power of subdivided bodies to evacuate referential language of sense. The loss of this "little Bit of Matter" anticipates the place of the corpuscle in Fielding's novel: it is neither an analogically accessible object nor an epistemologically legitimate source of "Employment."

But even though he spoofs a particle that somehow more palpably disappears in St. James's Park, Fielding's prose fiction is not disengaged from the repercussions of corpuscular philosophy. The world-weary host who enters *Joseph Andrews* to chide Abraham Adams for "giv[ing] any Credit to a Man's Countenance" illuminates Fielding's concern with how empiricism plays out between characters as well as between novels and their readers. In *Joseph Andrews,* apparently felicitous signs in a stranger's face lead Parson Adams to vouch for "that Sweetness of Disposition which furnishes out a good Christian" (158). Indicators not of corpuscular contents but of still untested moral predilection, sensible qualities in *Joseph Andrews* would determine not whether a person is sick but whether he is good. Indeed, the host's rebuke transposes corpuscular science's doctrine of qualities to persons. Just as perceptible endowments do not prove the presence of elemental essence but rather are contingently produced by particulate structure or texture, no attribute in a person's face guarantees that he harbors the moral ingredient goodness. The host who berates Adams echoes Boyle's *Sceptical Chymist,* which bemoans "what sleight and easily producible qualities they are that suffice . . . to Denominate a Chymical Principle or an Element."[5] The host's own skepticism transmits Fielding's refusal to refer countenance to some abiding moral depth. Forensic testimony to whether he has been scarred by smallpox, a stranger's face cannot prove his untried virtue—except in the eyes of gullible Adams, who demonstrates, in Boyle's words, that superficial signs "suffice" to satisfy lax analysts.

In the novelistic environment Fielding's characters inhabit, sensible qualities are not unerring symptoms. If anything, personal features that would signify one inner disposition more reliably herald its opposite. In *Shamela* (1741), *Joseph Andrews* (1742), and *Jonathan Wild* (1743), Fielding characterizes persons by recourse to a medium whose sensible qualities more surely manifest latent disposition: the book itself or, more precisely, printed characters. By print, I do not mean semantic content, for Vinegar credits the dysfunction of the country he visits to governmental corruption so pervasive that it vitiates the descriptive capacities of language. Indeed, Vinegar employs Fielding's most abused superlative to evoke the subdivided louse: by naming it "the greatest Curiosity which I saw there," Vinegar participates in a debasement of meaning thrown into relief by the minuteness of what is literally *not* great. Vinegar thus isolates the instrumentality of particulate matter for Fielding after all: this hyperbolic instance of greatness's antithetical referent makes the word's aggrandizing claim *physically* ridiculous. However tiny a 30,000,000,000th part may be, the physicality of the particle trumps the inflated pretensions of Vinegar's modifier. But the real cannot as decisively adjudicate the scene of language use Vinegar offers to portray social interaction under the government of Prime Minister Robert Walpole:

> Perhaps the Reader will think these Persons are a Nation of Hypocrites; whereas, it is quite the contrary: For no one is at all deceived by any of these FRMS (in *English* Farces), nay, the Persons themselves very rarely intend to impose upon you: For the most known Pickpockets publicly cant of their Honesty, even to those whom they have cheated, and to whom they know their Roguery is no Secret; and the Women boast of their Virtue before the Faces of the Adulterers, who had, perhaps, a few Hours before, enjoy'd them: A Man tells you he hath not a Shilling, with 100*s*. in his Pocket; a Tradesman swears he loses by selling at such a Price; a Courtier, that he will give you his Interest; a Lover, that he will be constant; a Friend, that he will serve you; all these are look'd upon as GD BRDNG, and no one is deceiv'd.[6]

As Fielding repeats, "no one is deceiv'd": no one, that is, but simpletons like Abraham Adams, who fail to exchange the meaning of a word like "Honesty" for the opposing property possessed by "Pickpockets." Adams's incomprehension of the "FRMS" that render vaunted virtue a signifier of its opposite constitutes a litmus of his innocence that will persist in Fielding's novels. Yet for the majority of Fielding's characters—and, just as crucially, Fielding's readers—language does not only sustain honesty's opposite. Like Vinegar's own reference to GD BRDNG, it involves its users in games. The principles that regulate meaning for Vinegar's INCONSTANTS are not honesty and virtue, but the rules of the game.

In this introductory discussion, I turn from linguistic fraudulence practiced by modern doctors to the role of language in the quality of status or class. Nobility may be transmitted by blood, as the recovered parentage that resolves Fielding's *Tom Jones* (1749) affirms. But what perceptibly vindicates status for the reader of Fielding's novels, I argue, are the sensible qualities of print. Starting with his inaugural work of prose fiction *Shamela*, Fielding refutes hypocritical pretensions to merit by amplifying the interceding attributes of the medium of typography. While the severed louse marks an example of particulate matter whose physicality opposes Vinegar's superlative, Fielding's novels do not debunk fraudulently elevating words with bits of lice. Instead, they grant words qualities that are perceived in the act of seeing as well as reading. Whereas Fielding excerpts prose by fellow men of letters Colley Cibber and George Cheyne as obviously ludicrous instances of self-aggrandizement, Samuel Richardson's *Pamela* (1740) provokes him to author a debasing variant that adds perceptible qualities to the printed artifact.

This chapter's next sections argue that Fielding amplifies the perceptibility of print to arrest the power of language to elevate its users. For Fielding, print *is* the medium that adjudicates claims to status. But his second novel, *Joseph Andrews*, foregrounds a dilemma: if abused words buttress Vinegar's FRMS, then the virtue of a character like Abraham Adams is vindicated as his exemption from hypocrisy's game of GD BRDNG. The very rigor of Fielding's indictment of *Pamela* as hypocrisy promulgated in print leads him to posit virtue manifest in language only as a good man's incapacity to use it. In the land of Vinegar's INCONSTANTS, linguistic ineptitude is goodness's surest proof. But because, in Fielding's novel, print corrects print, Adams's simplicity marks a critical dead end. *Jonathan Wild* claims the remedial perceptibility of typographical characters to transmute greatness into its opposite, illiteracy.

The vice that most steeply devalues linguistic meaning, hypocrisy is not blocked by some countervailing real. Unlike the 30,000,000,000th part that refutes greatness nobody can see, material facticity does not prevent Fielding's nation of hypocrites from claiming qualities in language. Vinegar offers a graphic example of matter's failure to regulate what attributes persons profess: "women boast of their Virtue before the Faces of the Adulterers, who had, perhaps, a few Hours before, enjoy'd them." This episode distils an empirical crisis: adulterous hypocrites violate John Locke's founding derivation of linguistic signification, the reference of words to ideas triggered by the speaker's sensory encounter with an object. As Locke argues in the *Essay*'s primordial instance of words and the simple ideas to which they are bound, "a Child knows as certainly, before it can speak, the difference between the *Ideas* of Sweet and Bitter (*i.e.* That Sweet is not Bitter) as it knows afterwards (when it comes to speak) That Worm-wood and Sugar-plumbs, are not the same thing."[7] Sensible qualities like "Sweet and Bitter" enforce the child's subsequent certainty of linguistic difference. By applauding the virtue of women they have enjoyed, adulterers violate a liaison between perceived "*Ideas*" and names that should be no less emphatically felt. Fielding may mock the empirical impossibility of scientific forays into the microworld, but he deplores the devaluation of linguistic meaning in empirical terms. The problem is not that unchaste women lie—since no one is deceived—but that the word "virtue" accommodates flagrantly opposed sensible qualities.

Adulterers who betray their intimate sensory experience animate a disconnect of words from stimulating particles whose most egregious instance in the sphere of polite letters is, for Fielding, not the self-promotional excess of Poet Laureate Colley Cibber's (1671–1757) *An Apology for the Life of Colley Cibber* (1740), but medical speculation authored by the doctor, philosopher, and theologian George Cheyne (1671–1743). *The Champion*'s May 17, 1740, issue recounts the proceedings of "*a Court of Censorial Enquiry*" assembled to try the poet and dramatist Cibber, whose *Apology* "in and upon the *English* Language an Assault did make."[8] During the trial, a "*Critick*" testifies that "the *English* Language has had more Violence done it by a very great and eminent Physician." Fielding quotes an "Instance . . . of this Barbarous Treatment"[9] from Cheyne's *An Essay on Regimen: Together with Five Discourses, Medical, Moral, and Philosophical* (1740):

> "Perhaps the primitive animal Body might consist of the first pure, specific and hallowed[10] Elements, harmoniously combined, and elegantly ranged in their original Natures, of which our present patched gross Bodies, are only the confused dense Kind; as our

> present Globe of Earth, its Water, Salt, Air, Light and Earth, are but probably the purified [*sic*; "putrified" in original Fielding and in Cheyne] Carcase of the primitive Planet; but both may continue to have some remote Analogy to one another, as a Carcase hath to a living Beauty, or an *Egyptian* Mummy to a *Cleopatra*."[11]

To be sure, this does not look identical to the violence enacted on language by sanctimonious adulterers. Yet it claims a deep affinity, because Cheyne writes nonsense—or, Fielding states, he "bring[s] Sentences together without any Meaning"[12]—about "Body" itself. Cheyne's narrative is recognizable as a borrowing from Paracelsus, who allies man and nature's theological "fall" with the activity of chemical "separation," as Walter Pagel recounts: "All individuation and specialisation is seen as a breaking away from the original divine unity, simplicity, and homogeneity."[13] By calling Earth a "putrified Carcase," Cheyne cites the Paracelsian and Helmontian agency of putrefaction, which opens bodies to propel their declension into differently qualified versions of their former selves. An inextricably theological pursuit, alchemy aims to penetrate corruptible metals and restore their original integrity. But as an echo of Paracelsus's anti-dualist "tendency to spiritualise matter"[14] refined in what Betty Jo Teeter Dobbs designates "neoPlatonic"[15] terms, Cheyne's cosmogony is a garbled pastiche. Fielding indicts Cheyne for the haphazard deployment of Paracelsian keywords that devolve into jargon remote from any sensible—let alone simple—idea.

Fielding does not protest Cheyne's inadequately rigorous reference to his chymical source. Rather, because the profession of medicine ranks as Fielding's premier instance of the simultaneously nonsensical and mortal detachment of words from body, he indicts Paracelsus and his epigones as its instigators. Fielding probably wrote *A Project for the Advancement of Physic in this Island*, which with his "A Catalogue of Choice Books in the Art of Physic" is attached to Fielding's *The Charge to the Jury* (1745), a pamphlet satirizing the medical fecklessness hastening Walpole's death in March 1745.[16] The "Catalogue" includes such choice headings as "5 *De Podagra renovanda*, or, a Method of bringing on a Fit of the Gout."[17] But the critique transmitted by the first item in this list of books to "be sold cheap" is contained in its title alone: "1 *PARACELSI Opera omnia*."[18] Paracelsus sits at the head of a catalog of otherwise fabricated titles and glosses, like "the Art of Prescribing for a Patient without knowing his Distemper,"[19] whose survey of modern medical incompetence pivots on the chemical physicians' repudiation of classical

Galenic method. Fielding's second item spells the ascription out: "2 *Scriptores cont. Galen.* 6 Vols. in Fol."[20] In this capacity, Fielding's *Project* invokes "the Profound and Oracular *Paracelsus.*—'Twas he who laid the Foundation of modern Physick, and we ourselves have had the Happiness of seeing it perfected by the Labours of some celebrated Genius's of the present Age."[21] Paracelsus—or, the iteration of his oeuvre as incomprehensible gibberish by literary physicians like Cheyne—propels the fraudulence that alienates modern medicine from empirical things.

Fielding's catalog of faked (with the exception of Paracelsus's *Works*) Latinate titles concludes with the caveat: "*Note,* They are all translated into *English.*"[22] Along these lines, his *Project* introduces a bill to expurgate antique languages from the medical college so "that Physick may become entirely *English,* and be within the Reach of all Capacities."[23] Here Fielding offers a refinement central to his novelistic engagement with qualities. It is not simply Paracelsus's "*Opera omnia,*" but their availability in "*English,*" that licenses theologico-Paracelsian extravagances like Cheyne's *Essay.* By gesturing toward the vernacular diffusion of medicine spurred during the Civil War and Great Plague by chymical practitioners who wrested authority from the College of Physicians as well as those who translated the official Latin pharmacopoeia, Fielding suggests that the unintelligibility of Cheyne's text does not preclude the diffusion of its pretenses to knowledge. As William Newman argues of the Helmontian-Galenist controversy, the popularization of British medicine enabled not only chemical remedies but also promotional publicity disseminated by would-be healers of many persuasions and, in Fielding's words, "all Capacities."[24]

Writing in 1664, the Helmontian translator and doctor George Starkey denounces empirical (in this context, untaught) pretenders to chemical facility in terms that anticipate Fielding's denial of medical wisdom to persons whose formation is restricted to English: "Since I sent forth my Apology for *Helmont* . . . a man would bless himself to think what a number of Coleburners there are, and how they dayly multiply, many of whom one could scare imagine of any other extraction, but that they grew out of some dunghil: such heteroclyte pieces of Mortality, as Weavers, Taylors, Botchers, Coblers."[25] A Harvard-educated chemist who bemoans the fact that his translation of Helmont has fueled the endeavors of "Coleburners," Starkey enables competition in print that print cannot referee. By contrast with the democratizing reach of sense-based understanding promoted—at least for children—by Comenius and the Hartlib circle, Starkey refuses to dignify essays at iatrochemical proficiency enabled by the accessibility of his own text. By dismissing the competence of manual laborers who "grew out of some dunghil," he hostilely recapitulates the structural effect of professional competence evinced in vernacular English. At stake is not only the potency of cures advertised by "Weavers" but the mobility of a class of practitioners composed of "heteroclyte pieces of Mortality," mobility advanced by the expanded claim to medical competence forged by Starkey himself. Fielding's catalog of cheap books follows Starkey's lead by assimilating social pageantry into the corpus of medical practice: "*Ars Practica*, or the Method of getting Business and Reputation: Containing the Secret of Puffing."[26] For Fielding, Cheyne performs medical puffery to mask ineptitude furthered by duplicities of both countenance and print.

By attacking medical practitioners of low extraction, Starkey bemoans an unmooring of values whose enactment in language produces Fielding's nation of hypocrites. As facilitated by Starkey's translation of Helmont, print is the medium of elevating pretensions untrammeled, at least until the patient succumbs, by experimental trial. But because it is a Lockean standard to which Fielding holds linguistic signification, his treatment of Paracelsus betrays some ambivalence. Although Fielding spoofs chymical jargon, it is Paracelsus to whom Locke is indebted for the notion of *spagyria*, or analysis and synthesis, the experimental strategy deployed by Locke's mentor Boyle to refute scholastic (and indeed Paracelsian) elemental essence.[27] This chapter's final section pursues the *formal* influence of chymistry on Fielding. As the enactment of a textual game in which the reader is addressed as a decoder, as a formal avatar of hypocrisy whereby qualifying words point to their deeper antithesis, and as a representation of mobility that entails the total transmutation of value, *Jonathan Wild* reflects what I will call Fielding's alchemical imaginary. A satirical undoing of hagiographic biography, *Jonathan Wild* incarnates the form of the alchemical object and, in so doing, compels the reader to rectify print's pretension to greatness.

Shamela and the Secondary Qualities of Print

> The *Thought* is everywhere exactly *cloath'd* by the *Expression*: And becomes its Dress as roundly, and as close, as *Pamela* her Country-habit.
>
> —[Aaron Hill,] prefatory letter to second edition of Samuel Richardson's *Pamela: or, Virtue Rewarded* (1741)[28]

> *Joseph* was soon drest in the plainest Dress he could find . . . and as this Suit, which was rather too large for the Squire, exactly fitted him; so he became it so well, and looked so genteel, that no Person would have doubted its being as well adapted to his Quality as his Shape.
>
> —Fielding, *Joseph Andrews* (254)

In Fielding's version of *Pamela*'s prefatory puff, Pamela Andrews's unauthorized sibling realizes *Pamela*'s claim to virtue as his own ability, when dressed in his brother-in-law Mr. B's cast-off clothes, to look "so genteel." Because the footman Joseph Andrews's assumed gentility proceeds from his strapping "Shape," the satiric bite of this iteration of *Pamela* derives from its Swiftian reenactment of Aaron Hill's metaphor. As illustrated by *Gulliver's Travels*, in which islands incarnate the visual field afforded by microscopes, one trademark of Swift's satire is its rendition of concepts as things. In the case of feminine "*Thought* . . . everywhere exactly *cloath'd* by the *Expression*," Fielding's procedure is assisted by the salaciousness evident in the original. The inner virtue that redeems Pamela's marriage into a titled family is represented by Hill as antecedent thought that fills out its epistolary "Dress as roundly, and as close" as Pamela's body does the countrified outfit she dons to elicit her master's appraisal as "such a tight prim Lass."[29] Pamela's apparition as "such a tight prim Lass" models Richardson's pioneering deployment of the epistolary novel, the snug fit of letters and the "*Thought*" that impels them. Yet Hill's metaphor is prone to deflating misprision. A figure for the letter's proximity to its writer's inner thought, "Country-habit" locates obscene reference at *Pamela*'s core. In addition, the fact that Pamela "trick'd myself up as well as I could" (55) in her rustic garb threatens to define the obtrusive tightness of her textual costume as a contrived effect.

Joseph Andrews's accession to Mr. B's used suit plays on Hill's already bawdy endorsement of *Pamela*'s formal premise. If Pamela's pretense to elevating virtue hinges on the close fit between printed letters and motivating thought, Joseph's unwitting assumption of quality proceeds solely from the snugness of his clothes. The virtue of *Pamela*'s letters, become the appeal of Joseph's brawn, thereby launches a double critique. As "doubted" by "no Person" or, more precisely, none of the women inhabiting Fielding's satire, Joseph instantiates the declension of status into a predicate of feminine lust. Just as Pamela's exceptional moral loveliness serves, for Fielding, as a cover for her lascivious squire's intemperate passion, so Joseph's palpable merits vindicate his value even to the squire's snobbish aunt Lady Booby. For Fielding, poor people's ascensions to quality are adjudicated as a function of aristocratic desire. *Joseph Andrews* defines the working-class virtue extolled by *Pamela* as a highly selective projection of aristocratic appetite. But the novel also—given the abuse of language which, for Fielding, *Pamela* instantiates—offers unlettered pulchritude as a possibly superior attribute.

As provocation for Fielding's pseudonymously published *Apology for the Life of Mrs. Shamela Andrews* (April 1741), Hill's praise solicits further attention. At the level of its form, *Pamela* concretizes a textual crisis to which Fielding's novelistic career, beginning with his first turn to prose fiction *Shamela*, marks an equally concrete textual response. Hill's immanently lewd distinction of snugly fitting "*Expression*" and well-rounded "*Thought*" amplifies the problem, but even shorn of any attendant metaphor, an epistolary text that declares fidelity to its writer's psychological state operates to alienate the typographical letter—already alienated at one remove by the defining conceit of an inaccessible holograph—from the writer's mental content. By disaggregating thought and printed reference, and by puffing *Pamela* as the obscenely snug correction of that break, Richardson's novel divides first-person narrative from antecedent ideas that escape typographical expression. In Defoe's *Journal of the Plague Year*, particulate contents lend characters hidden depths. As historical controversy over her prospective hypocrisy suggests, what cannot be apprehended inside Pamela is not morbific matter but thought prior to the typographical apparition of the epistolary letter.[30] Because, for Fielding, epistolary print alienates thought from type in the very act of professing fidelity to it, he does not intervene to respecify always elusive mental content. *Shamela* instead locates the truth of character in the physicality of the typographical artifact.

Locke's *Essay* stipulates that words trigger past ideas but cannot, without the help of memory, elicit empirically inimitable sensation. As a prophylaxis against the spread, among other things, of religious enthusiasm driven by parroted marvels rather than personal revelation, Locke tries to assign language a rigorously referential function. Michael Ayers agrees that the obstacle to human understanding is not a so-called "veil of ideas" but rather what Locke himself represents as a veil of words: "[Words] interpose themselves so much between our Understandings, and the Truth, which it would contemplate and apprehend, that like the *Medium* through which visible Objects pass, their Obscurity and Disorder does not seldom cast a mist before our Eyes" (3.9.§21).[31] The *Essay* acknowledges the physical being of words—or, treats words as things—only to the degree that words "interpose themselves" between human understanding and "Truth." Locke's ultimate representative of compromised knowledge, the target of the *Essay*, is "He that . . . imposes on himself, and mistakes Words for Things" (3.10.§32). Granted, this person's avatar is Don Quixote, who imposes on himself not by registering the thingness of words but by believing romances are history. Yet Locke bemoans the residual physicality of language whose distortion of truth implicates "all the artificial and figurative application of Words Eloquence hath invented" (3.10.§34). That the superfluous bulk of these words demands real bibliographic redress becomes clear when the *Essay* projects its salutary effect: "several of those great Volumes, swollen with ambiguous Words . . . would shrink into a very narrow compass; and many of the Philosophers (to mention no other,) as well as Poets Works, might be contained in a Nut-shell" (3.11.§26). Locke's solution obviates language that is not only "swollen" but "loose" (3.5.§6 and passim), referential promiscuity also signaled when he denounces linguistic deviation from perceived ideas as "perverting the use of Words" (3.2.§5).

If for Locke the quiddity of simple ideas is elicited by things, not words, then Fielding's words possess sensible qualities whose significance exceeds their semantic meaning. Indeed, Fielding activates language's objecthood to meet the standard espoused by Locke, because words perceived as things claim the power to stimulate extralinguistic sensation. Words that trigger ideas not as signifiers but as objects resolve the problem *Pamela* poses for Fielding by contrast with the problem glossed by Locke: *Pamela*'s letters are not too obscure but, on the contrary, too transparent. *Shamela* corrects not looseness but a performance of innocence enabled by the epistolary novel's claim to transmit thought unadulterated by the interceding density of the printed book.

Shamela's most obtrusive typographical intervention is its name. A cognomen that revises only one character in Richardson's title, Fielding's neologism fuses "Pamela" and the emergent slang "sham," defined by *A New Canting Dictionary* (1725) as "a Cheat, or Trick. *Cut a Sham*; To play a Rogue's Trick."[32] Because "sham" connotes cheating, trickery, and imposture, the revised moniker is indelibly qualified. Both the neologized revision of Richardson's title and its eponymous protagonist's proper name—its referent both a fictional person and the impaired transparency of the prior word—*Shamela* operates as a signifier and an object at once. Its revised orthography knits qualities into "Shamela" at the same time that, as a depravation of the name it cites, it arrests the reader at the surface of the word. As a signifier, *Shamela* refers to the quality of hypocrisy; as an object, the word's depravity is real, entrenched in the orthographic corruption of Richardson's original.

Pamela provokes vertiginously intratextual remediation. Richardson's chronological original becomes the expurgated copy of Fielding's satire, whose most salient typographical strategy is described in Fielding's play *Tom Thumb. A Tragedy* (1730), which, spoofing Cibber, mocks the vacuity of the genre of the preface: "As *Charon* in *Lucian* suffers none to enter his Boat till stripped of every thing that they have about them, so should no Word by any means enter into a Preface till stripped of all its Ideas. Mr. *Lock* complains of confused Ideas in Words, which is entirely amended by suffering them to give none at all: This may be done by adding, diminishing, or changing a Letter, as instead of *Paraphernalia*, writing *Paraphonalia*: For a Man may turn *Greek* into Nonsense, who cannot turn Sense into either *Greek* or *Latin*."[33] As Glenn Hatfield explains, this "typographical or orthographic"[34] blunder really occurred in the preface to Cibber's hit play *The Provoked Husband* (1728). A fortuitously sophisticated recombination of "*Greek* or *Latin*," the nonsensicality of "*Paraphonalia*" models an obfuscating—or, perhaps, medical—strategy of word use whose aspirational aims diverge from Fielding's shammification of Richardson's vernacular fiction. But as his reference to "Mr. *Lock*" affirms, Fielding's *Shamela* does pursue the empirical effects of typographical and orthographic prestidigitation. These effects are experiential, for the reader of "*Paraphonalia*" is arrested by the alien physicality of a string of characters which do not reinvest the word with perceived ideas but rather render palpable its imperviousness to meaning. If, like the title *Shamela* a decade later, Cibber's incorrect orthography does not fail entirely to signify, it signifies in the extra-Lockean sense that its characters claim the opacity of objects.

The word whose corrupt orthography operates to greatest effect in *Shamela* is "vartue." At the start of Fielding's text, Shamela professes her refusal to submit to her master's advances for anything "under a regular taking into Keeping, a settled Settlement, for me, and all my Heirs, all my whole Life-time" (318). A base version of Pamela, this servant's highest aspiration is her exchange of sexual compliance for "a settled Settlement" restricted to the duration of her temporal "Life-time." After her master Squire Booby prematurely declares that "I cannot live without you" (325), however, Shamela raises her hopes. At this juncture, she insists that she has never been on the market for "Keeping," as she explains to Booby's housekeeper Mrs. Jewkes: "I would have you know, Madam, I would not be Mistress to the greatest King, no nor Lord in the Universe. I value my Vartue more than I do any thing my Master can give me; and so we talked a full Hour and a half, about my Vartue; and I was afraid at first, she had heard something about the Bantling, but I find she hath not; tho' she is as jealous, and suspicious, as old Scratch" (325–26). As Fielding's bifurcated rendition of *Pamela*'s plot makes clear, Shamela's "Vartue" is a belatedly adopted pretext for the non-negotiability of sexual favors she has proven all too willing to purvey. This revision transforms the staunch chastity of Richardson's serving-girl into the crude rhetoric of a pretender who only long after the fact discovers that chastity might have any exchange value at all. A confusion of "virtue" and "value" that marks the "adding, diminishing, or changing" of letters in two source words at once, "Vartue" disables the transparency of virtue's signifier as, Fielding intimates, it passes in *Pamela* for the ineffable innocence of its writer's thought. Even Fielding's most exact repetitions of *Pamela* thereby lend its utterances a contaminating thingness: "Yes, Sir, says I, I see your Honour intends to ruin me, that nothing but the Destruction of my Vartue will content you" (327). Since Shamela has previously produced, in her heartlessly cant approximation, "the Bantling," Fielding's iteration conflates bad orthography with anatomical "ruin" that has already happened: the ruin Booby intends is preempted not only by Shamela's body but, most immediately in the eyes of *Shamela*'s reader, by her text. Because "Vartue" is typographically ruined, what was *Pamela*'s appeal to an inextricably anatomical and ephemeral good becomes in *Shamela* a punning tautology that is, however, no lie: "Sir, said I, I value my Vartue more than all the World" (329). Shamela values her ruined virtue's value, a formulation whose redundancy foregrounds the argument evident in the brevity of Fielding's satire. Because Shamela's "Vartue" finds no external referent other than Booby's misapprehension of an obviously vicious servant's innocence, "Vartue" telescopes the bulk of Richardson's novel into *Shamela*'s skeletal references to the specious prolixity of *Pamela*'s language use: "and so we sat down and talked about my Vartue until Dinner-time" (328).

I have intermittently used the verb "iterate" to refer to Fielding's repetition and revision of Richardson's original. This verb echoes Jacques Derrida's definition of the linguistic sign as terminally "deferred presence,"[35] a deferral stemming from the provisional determination of any given word's meaning, according to the linguist Ferdinand de Saussure, by the synchronic or structural array of all the words around it. Because words signify not by recourse to some transcendent or primordial referent but rather by means of "the playing movement" of intralinguistic "differences,"[36] the verb "to iterate" entails as the effect of repetition in language the possible noncoincidence of the repeated word with itself. Iteration thus entails the playfulness or, to employ another Derridean verb, the slippage of linguistic meaning produced as the effect of one word's physical dissimilarity from others. *Shamela* slips from the pretended virtue of Richardson's original as a function of Fielding's orthographically imperfect iteration of the word "virtue" itself. This slip connotes both the quintessentially ruinous act of Shamela's impregnation as well as her text's failure correctly to mimic the virtue assumed by *Pamela*. But Fielding does not participate in a process of ludic repetition that affirms the signifier's indefinite deferral of meaning. Rather, *Shamela* constitutes a singularly opaque reapparition of the original book. Fielding offers a single iteration of *Pamela* which would make perceptible hypocrisy sustained by that novel's defining formal conceit.

In the compressed prelude to Shamela's marriage, Fielding spoofs the enactment of virtue that dupes Booby: "I . . . put on my prettiest round-ear'd Cap, and pulled down my Stays, to shew as much as I could of my Bosom . . . and then I practiced over all my Airs before the Glass, and then I sat down and read a Chapter in the whole Duty of Man" (327). Clearly, Fielding suggests, it is the prurience of Shamela's countrified persona, and not any emanation of native innocence, that excites Booby. Passages like these make it easy to indict Shamela as an imitator who reconverts artlessness into the regressive simulation of *Pamela* itself. But while Shamela tricks herself up with a deliberateness no less marked than that of Richardson's original, she also "sat down and read a Chapter in the whole Duty of Man." Insofar as her appeal to her master is perceived, Fielding suggests, as her "Bosom," reading seems to depart from the medium of sartorial artifice that affirms Booby's lack of interest in his servant's mind. In fact, Fielding takes care to specify that Shamela's "Vartue" involves her consumption of books: "I went and hid myself in the Coal-hole, where I lay all Night; and comforted myself with repeating over some Psalms, and other good things, which I had got by heart" (326). With "the whole duty of Man" and the Bible, Shamela advertises her facility with moral discourse. Fielding's incrimination of *Pamela*'s intertextual affinities therefore includes not just the pornography contained in Shamela's assortment of books (most egregiously, "*Venus in the Cloyster: Or, the Nun in her Smock*") but also "*The Whole Duty of Man*, with only the Duty to one's Neighbour, torn out" (332) and "some Psalms." The indifference of Psalms and *The Nun in her Smock*, of showing one's bosom and reading *The Whole Duty of Man*, points to the deepest crisis that *Pamela*, for Fielding, realizes: moralizing language does not defuse vice but, on the contrary, assists it.

Pamela's moral climax occurs when Mr. B declares to his servant that "I love you with a purer Flame than ever I knew in my whole Life! . . . And I know more sincere Joy and Satisfaction in this sweet Hour's Conversation with you, than all the guilty Tumults of my former Passion ever did, or (had even my Attempts succeeded) ever could have afforded me" (265–66). Richardson predicates Mr. B's achievement of "a purer Flame" on *Pamela*'s discursive influence. With an assurance whose fundamental lewdness Fielding may have savored, B assures *Pamela*'s reader that verbal "Conversation" now satisfies him even more than the projected success of his "Attempts" at Pamela's rape. For Fielding, the surety of a rake's reform might well be eroded by his still carnal endorsement of virtue's pleasures. Whereas Richardson, however unevenly, predicates his novel's beneficent influence on its power to soothe lustful tumults, *Shamela* denies words any privileged effect as a result of their abstraction. For Fielding, the assumed transparency of words is no metaphysical guarantor of their potency but an harbinger of the ease with which they can be used. As Shamela confirms with an unwittingly apt synopsis of a sermon preached by her paramour Parson Williams, "those People who talk of Vartue and Morality, are the wickedest of all Persons" (324).

Writing in *The Champion* about a year before *Shamela*, Fielding analyzes the referential difficulties posed by the term "good-nature":

> No Virtue or Quality in the Mind of Man hath met with so various a Reception as this, whilst some few have held it in the most sacred Esteem, several wise Men have considered it as a Mark of Folly and Weakness, and several brave Men have despised it as a certain Indication of Cowardice or Pusillanimity.
>
> I am apt to suspect when I see sensible Men totally differ in Opinion concerning any general Word, that the complex Idea in their several Minds which this Word represents is compounded of very different Simples. 'Those gross and confused Conceptions (says Mr. *Lock*) which Men ordinarily have, and to which they apply the common Words of their Language may serve them well enough in their ordinary Discourses and Affairs; but this is not sufficient for philosophical Enquiries.' And a little lower in the same Page, this great Man declares, that 'the most he can find in all the Volumes and Varieties of Controversies, with which the World is distracted, is, that the contending of learned Men of different Parties do, in their Arguings with one another, speak different Languages.' I will venture to illustrate this by a familiar Instance: Suppose an Apothecary (as perhaps they often do) after

> mixing up a most pleasant Cordial, and a most nauseous Potion for different Patients, should write the same hard Word (*Haustipotiferous Draught* for Example) on each of the Bottles, would not these two Patients ever after conceive very different Ideas of *Haustifpotiferous*, and would not they stare equally at each other, when one commended the Pleasantness, and the other exclaimed against the Nauseousness of the Draught?
>
> This I apprehend may be the Case in Good-nature; for the common Use of the Words, without certain and fixed Ideas annexed to them, will not at all mend the Matter.[37]

Fielding's quotations closely approximate sentences from Book 3, Chapter 10 of Locke's *Essay*, "Of the Abuse of Words." Hatfield invokes Fielding's citation to confirm its overall agreement with Locke's diagnosis of "how words and their ideas . . . often became separated and confused."[38] But Fielding departs from Locke at this juncture, because Fielding appropriates the *Essay* to illustrate a scenario sharply discordant from Locke's discussion of "complex *Ideas*" (3.10.§22). The scene Fielding portrays may seem to offer a more emphatic example of the confusion sustained by speakers who use the same words while entertaining different ideas of their referents. But the divergence of Fielding's "*Haustipotiferous Draught*" from Locke's consideration of complex ideas is amplified by the illustration Locke affords in the text before Fielding's first excerpted sentence. Locke exemplifies the difficulty caused by words that are, he writes, "no more but the voluntary and unsteady signs of their [men's] own *Ideas*" (3.10.§22) as follows: "And yet if it comes in Question, whether a Plant, that lies ready formed in the Seed, have Life; whether the Embrio in an Egg before Incubation, or a Man in a Swound without Sense or Motion, be alive, or no, it is easy to perceive, that a clear distinct settled *Idea* does not always accompany the Use of so known a Word, as that of *Life* is" (3.10.§22). With the word "*Life*," Locke illustrates difficulties posed exclusively by words that stand for complex ideas. Out of their contingent experience of entities whose defining criteria are not rigorously delimited, no two speakers will compose "the same just precise Collection" (3.10.§22). The historical persistence of "Noise, and Wrangling" over the claim to life of embryos and persons in swoons proceeds from the mistaken presumption that the word refers to "the same precise *Ideas*" (3.10.§22) in the mind of every speaker. This is not a problem, as Fielding puts it, of "different Simples," because the conjoined ideas each speaker summons refer not to one impression but to differently curated sets of them. An example specific to complex ideas, argument over the life of embryos diverges from dispute triggered by the qualities of "Pleasantness" and "Nauseousness." Locke illustrates wrangling induced by names of compounded abstract ideas, whereas Fielding imagines a deadlock contrary to the foundational linguistic experience consummated when a child tastes the difference between wormwood and sugarplums. For Locke, no sentient children could fight over these words. But Fielding conjures a situation in which the cornerstone of empirical understanding—everybody's apprehension of simple sensory ideas like sweet or bitter, pleasantness or nauseousness—fails to institute linguistic certainty.

Fielding's "*Haustipotiferous Draught*" is not Fielding's more dramatic staging of discrepant congeries of ideas, but his vision of a word which fails stably to recur to one simple sensation. Because the word's referent is either a "pleasant Cordial" or a "most nauseous Potion," two men disagree over simple sensory experience. Fielding thus summons the agency of particulate matter once more: these men fail to settle the meaning of "*Haustipotiferous*" because a doctor places identical labels on objects that stimulate opposing qualities. Fielding does not skeptically insinuate that the same draught provokes opposing ideas; rather, he respects the corpuscular basis of simple ideas even in this maximally conflicted case. To motivate the crisis enacted by Vinegar's INCONSTANTS, Fielding deploys different particulate textures to produce opposing qualities.

Fielding exemplifies the referential dilemma posed by "Good-nature" with opposing simple ideas rather than variously combined sets of them. The specter of moral language unmoored from any regulating sensation recalls his play *The Author's Farce* (1730) when Witmore, an aspiring author who laments the lack of dramatic merit in the popular marketplace, voices the complaint: "I have heard Sense run down, and seen Idiotism, downright Idiotism triumph so often, that I cou'd almost think of Wit and Folly as Mr. *Hobbes* does of Moral Good and Evil, that there are no such Things."[39] Insofar as Hobbes defines morality as a contractual artifact that does not exist anterior to institutional coercion,[40] Witmore's grievance is real: there really are "no such Things" behind the names. Like "Good and Evil," "Wit and Folly" signify by virtue of Hobbesian obligation or lowbrow public predilection. *Shamela* aims to restore order, blocking *Pamela*'s pretenses with a preemptively ruined signifier. Unlike the *Haustipotiferous* draught, however, *Shamela* does not summon the reality of a corpuscular referent that undergirds words, because Fielding embraces the corrective powers inherent in print as print. In a literary sphere where signifiers fail to stimulate simple ideas, morality finds its justification in "Mr. *Hobbes*." But if Hobbes finds no guarantor for morality outside of language, then in Fielding's novel, language itself might fight back.

Pamela, Shamela recounts, is written at Booby's whim by a hack "Parson . . . who can make my Husband, and me, and Parson *Williams*, to be all great People; for he *can make black white*, it seems" (341). For readers of *Gulliver's Travels*, the exchange of "*white*" for "*black*" recalls Gulliver's testimony to his master Houyhnhnm on language use in England: "there was a Society of Men among us, bred up from their Youth in the Art of proving by Words multiplied for the Purpose that White is Black, and Black is White, according as they are paid."[41] Lawyers "are paid" to weaken the link between words and simple sensory ideas. Gulliver's Houyhnhnm gathers with difficulty that by such practices "I am led to believe a Thing *Black* when it is *White*, and *Short* when it is *Long*."[42] Satire that rectifies not the "Thing" but the assumed innocence of epistolary print, *Shamela* corrects for the whitened blackness of the word itself. But what, *Shamela*'s readers might ask, about Shamela? Her activities as an object of self-reference surely augment her depravity: "sitting down to Supper on our Wedding Night . . . I behaved with as much Bashfulness as the purest Virgin in the World could have done. The most difficult Task for me was to blush; however, by holding my Breath, and squeezing my Cheeks with my Handkerchief, I did pretty well" (334). With this canonically baseless enactment of feminine chastity—a set piece whose emasculating import is not blunted by ruined orthography—*Shamela* frames a question crucial to the novel's engagement with the qualitative apparition of virtue. Can Shamela mimic sensible attributes that proceed from an anterior reality she does not, in fact, possess? By blushing, Shamela fakes a secondary quality detached from the truth she actually embodies. Can secondary qualities, like words, be severed from the corpuscular texture by which they should be produced?

But Shamela's "Bashfulness" is a red herring, for only Booby experiences the success of her trick. When she avers that "I acted my Part in such a manner, that no Bridegroom was ever better satisfied with his Bride's Virginity" (335), the reader has every reason to believe her. Her husband has remained unshaken by cant of which *Shamela* provides ample dialogic evidence, since this Booby does not read but only hears it. At the end of Fielding's satire, Booby intuits his new wife's character when she defends her escalating financial rapacity, to which he responds: "this is a Spirit which I did not expect in you, nor did I ever see any Symptoms of it before" (336). Booby's obliviousness to "Symptoms" does not vindicate his servant's assumed purity, because "Vartue" has already impaired the transparency of *Pamela*'s self-representation. Because *Shamela* exposes its letter-writer's speciousness in the medium of typography, the printed book drains credence from an "acted . . . Part" apprehended as truth only by Booby. Shamela's power to evince qualities that contravene her "Vartue" is restricted to her appeal to a licentious squire whose own moniker concretizes his eagerness to be fooled.

As a satire of printed self-reference, *Shamela* does not assign Pamela a corrective psychological reality, as some readers suggest, so much as a corrective typographical reality.[43] *Shamela* does, however, implicate things other than print in its rendition of a servant girl's "Vartue." One of these is marriage. Wedded, Shamela mimics the ambivalently assimilationist logic that justifies Pamela's betterment: "Truly, says I, Sir, I shall live like other Ladies of my Fashion; and if you think, because I was a Servant, that I shall be contented to be governed as you please, I will shew you, you are mistaken" (335–36). In this declaration of equality with "other Ladies of my Fashion," Shamela exploits the leveling mechanism built into *Pamela*'s form, which promotes the virtuous immediacy of letters that adjudicate Pamela's title to Mr. B's hand as a function of her pious text. *Pamela* circles widely around the competing retrenchments of status driven by the novel's form: while Mr. B pledges to "protect your Gentleness to the utmost, as if you were a Princess by Descent" (329), Pamela counters with self-abasing decrials of "what is it for such a Worm as I to be exalted!" (363). But behind the fait accompli that reconsolidates social class as patriarchal tautology—as B affirms when they are wed, "my Wife *is* my Wife" (445)—lies a demand that the servant is better prepared to meet.

Mr. B explains his reluctance to marry within his own class by remarking that "We People of Fortune . . . of both Sexes, are generally educated wrong . . . We are usually so headstrong, so violent in our Wills, that we very little bear Controul." The servant Pamela, however, has been "educated" in a manner that propels her spontaneous pledge: "I will endeavour to conform myself, in all things, to your Will" (443). Shamela's redaction of this promise thus marks more than a grasping defense of her postmarital likeness to ladies of her fashion. It exposes the unevenly obfuscated anti-egalitarianism of *Pamela*'s endorsement of its protagonist's fitness to wed B. In this capacity, Pamela makes a better wife than a lady because she has been trained to "bear Controul." Shamela, however, turns the pliancy of Richardson's servant girl into an expedient jettisoned as soon as her livelihood no longer depends on it. *Pamela*'s real referent, Fielding threatens, is a wife whose refusal to internalize servility renders her not "contented to be governed." If, for Richardson, marriage contains the threat *Pamela* poses to aristocratic status, then Fielding exposes the precariousness of a sexed claim to domestic superiority represented by a Booby.

Pamela's upending of status is rectified by the sensible qualities of *Shamela*'s printed words. Fielding's next novel, *Joseph Andrews*, plays out the inverse expedient of servant virtue proven by Joseph's exclusion from the field of letters.

Debasing Games: *Joseph Andrews*

> [I]f we could by a Kind of Chymical Operation, separate those Parts of our ordinary Conversation, which either leave any Idea in the Mind of the Speaker, or convey any to that of the Hearer, from those which do not, the former would be found scarce to bear the Proportion of a tenth Part to the latter.
>
> —Fielding, *The Champion*, January 17, 1740[44]

> [*C*]*oncerning the* Marriages *of the* PTFGHSIUMGSKI, *or the* INCONSTANTS.
>
> . . . the young Couple are left alone together, when the Following *Formulary* is made use of . . .

O	MAD	DAM.
TU	HAD	DAM.
TOL	LOL	DOL.
MY	GOL	MOL.
HOC	POC	US.
ALL	JOC	US.
NO	HOB	NAIL.
BUT	BOB	NAIL.
MY	ALL	GAL.
FAL	LAL	DAL.
[She answers,]		
SKUR	HUSH	MUSH.
RUB	UP	BLUSH.

> —Fielding, *The Champion*, August 5, 1740[45]

In *The Champion*, Fielding cites "Mr. *Lock*, in his Chapter of the Remedies of the Abuse of Words"[46] and posits an experimental answer to the problem: "a Kind of Chymical Operation" to "separate" empirically bankrupt "Parts" of speech from dialogue that still stimulates an "Idea in the Mind of the Speaker, or . . . Hearer." Fielding's punch line resides in the mock quantification of his results, for at least 90 percent of "ordinary Conversation" would fail to meet Locke's standard. After Fielding's imagined chymical separation, nine-tenths of everyday chatter would be *caput mortuum*, useless residue at the bottom of the operator's flask. Hatfield claims this passage as a blueprint for Fielding's strategy of linguistic remediation: Fielding assays "a personal decontamination of the English language" that works "to *insulate* truth from the corruptions of medium and agent."[47] *The Champion*'s vision of chemical analysis entails void words that fall away from still stimulating signifiers, leaving debris—like punctilio or the figurative ornament decried by Locke—to equate junk language with the worthless density of waste matter. For Hatfield, Fielding attempts "a surgical separation of the diseased growth of corruption from the healthy tissue of 'original idea'" by deploying irony and, more vividly, showing the immediate value of "dramatic action."[48] For Sandra Macpherson, men in Fielding's novels take action to avoid the distorting circumlocutions of legal discourse: "In Fielding, a commitment to the humanity of persons is one and the same as a commitment to fighting as an alternative to legal remediation."[49] Whether the unimpeachable virtue of "charity"[50] or the extra-juridical expedience of "fighting," for Hatfield and Macpherson action marks the novelistic reenactment of Fielding's proposed chemical procedure, obviating redundant language with the more direct impact of a pound or a punch.

Yet action in Fielding's novels is recounted in words. Hatfield addresses this proviso by suggesting that Fielding's episodes of "burlesque diction" serve to neutralize his spells of "straightforward narration," thus enabling the identity of "plainness of language with bedrock truth."[51] While for Hatfield, Fielding's narrative burlesque "mak[es] his subjectivity itself a dramatic fact which the reader can apprehend objectively,"[52] Michael McKeon argues that Fielding's "modes of self-conscious narration work . . . to subjectify the objective historicity of the narrative line."[53] For both Hatfield and McKeon, the "reflexive narration"[54] that defines Fielding's third-person novelistic style after *Shamela* sustains a dialectic of objectivizing and subjectivizing effects, adulterating the author's capacity to transmit omniscient history while transforming authorial agency into a historical artifact.

Hatfield's and McKeon's claims for the objectifying influence of burlesque narrative do not, however, engage Fielding's novel as a perceptible object. Hatfield glosses Fielding's imagined chymical separation as "a serious extension of the Lockean position"[55] on the abuse of words. But Fielding breaks from Locke at this juncture, because rather than pruning superfluous volumes into nutshells, Fielding mobilizes the sensible thingness of even the most redundant, meaningless, and formulaic signifiers. These are not meaningless waste but instruments of intralinguistic correction that effect Fielding's restoration of status by attaching critically debilitating sensible qualities to the printed text. As evinced by the "Marriages *of the* PTFGHSIUMGSKI," Fielding forces the perceptibility of text as thing, for his residually semantic derivation of satire from lines like "RUB UP BLUSH" is subsumed by the rigorous consistency of meter, spacing, capitalization, and rhyming couplets. Phonemes resistant to sentiment, "FAL LAL DAL" embody a premarital "*Formulary*" evacuated of internal propulsion. This is not language whose near total thingness Fielding deplores, but palpable absurdity which situates Fielding's reader within the same game that prompts the rote users of his "*Formulary.*"

In *Joseph Andrews*, the game is class status sustained not by scripted formulary but lexical competence evidenced in the medium of novelistic speech. The character who most decisively enacts social aspirations as a game is Mrs. Slipslop, "the Waiting-Gentlewoman" to Mr. B's widowed aunt Lady Booby. "[T]he daughter of a Curate," Slipslop "insisted on a Deference to be paid to her Understanding": as *Joseph Andrews*'s narrator explains, "she was a mighty Affecter of hard Words" (21). Slipslop involves this novel's reader in a scenario whose rules Fielding proceeds to specify: Parson Adams fails to promote a higher station for Lady Booby's footman Joseph Andrews, of whose "Industry and Application" (20) he approves, "after a pretty long Discourse with her [Slipslop] on the *Essence*, (or, as she pleased to term it, the *Incense*) of Matter" (21). This game is propelled not by misspelling but by catachresis, because Slipslop's claim to "Understanding" is arrested by the discrepancy between the word she employs ("*Incense*") and the near homophone required by the context of her speech ("*Essence*"). At this introductory moment, Fielding stipulates what role the reader must perform, since only here does he specify the difference between Slipslop's homophone and the appropriate replacement. For the remainder of *Joseph Andrews*, Slipslop's hard words solicit the corrective necessity of play.

If, as Saussure argues, the physical difference between signifiers engenders their meaning, then Fielding wrongly mobilizes small units of type to aggravate the physicality of Slipslop's speech. Its perceptibility amplified by the few characters that divorce sound from sense, the typographical organization of Slipslop's words precludes their transparent transmission of meaning. Because homophony and not education dictates her choice of signifiers, Slipslop makes them things without the assistance of italics. But Fielding augments their visibility by lending them supplemental typographical stress. In this, he mimics the strategy he explicates in *The Champion* when, recapitulating Cibber's *Apology*, he notes that "as I have endeavoured to use my Author's own Words in the same Sense which he hath attributed to them, as often however as I am capable of finding it out, so I have distinguished all these Words so used in *Italicks*."[56] His mock deference to the merit of Cibber's inscrutability flags Fielding's appropriation of "*Italicks*" as a critical instrument. Italics exert an ancillary satirical force by altering the look of words Fielding claims to repeat verbatim, lending sensible opacity to what is already semantically opaque. Italicized, Fielding's citation undermines the physical medium of Cibber's pretended greatness as surely as Vinegar's 30,000,000,000th part of a louse deflates the greatness of the spectacle he reports. While Slipslop's discourse again spoofs the empirical nonsensicality of disputes over the "*Essence* (or . . . *Incense*) of Matter," Fielding nonetheless deploys matter—typographical characters—to expose Cibber's and Slipslop's baseless self-promotion. Italicized, words undermine pretensions that would be effected in the very medium of the printed page.

As they are "used" by Fielding, Cibber's words signal the groundlessness of his claim to literary fame, perhaps provoking the reader to embrace the alternative merit of sense. Slipslop's terminological blunders dictate more narrowly specified corrective action. Both italics and homophonic near misses, Slipslop's catachreses extract a ludic imperative from language itself. This is not to say that Slipslop's botched display of understanding claims no external correlate: "having made a small Slip in her Youth," Fielding narrates, the "not at this time remarkably handsome . . . short . . . rather too corpulent" (27) Slipslop bears a name that links botched linguistic self-aggrandizement to some queasily portended erotic proclivity. But the more immediately deflating referents of her hard words are their proper surrogates. Because the act of reading implicates all literate persons in her game, language itself corrects Slipslop. While "Vartue" adulterates the professions of *Pamela*, the adjustment of Slipslop's claims is an intranovelistic phenomenon. Despite—and, in Fielding's two anti-Pamelist novels, because of—their promise to do the opposite, words evince qualities that oppose the ambitions of their users. These words do not, as Jürgen Habermas argues, cloak limiting personal particularity in the representative abstraction of the man of letters.[57] On the contrary, they render printed characters the scene of literary mobility's failure. Fielding's novel *is* the venue where printed enactments of greatness are refused by print itself.

Slipslop's game stymies Joseph. When he fails to discern that she desires him, Slipslop rages: "'you must treat me with *Ironing*? Barbarous Monster! how have I deserved that my Passion should be *resulted* and treated with *Ironing*?' 'Madam,' answered *Joseph*, 'I don't understand your hard Words'" (28). The humorous "*Ironing*" of this exchange proceeds from its overlay of lexical and sexual oblivion: Joseph, who is in "want of a liberal Education" (20), cannot supply meaning for words he does not apprehend as homophones; furthermore, he fails to register Slipslop's corporeal hints of longing. But Joseph's obtuseness does more than affirm the redundancy of Slipslop's lack of amorous "Success" (27), because his failure to understand shows the sincerity of his avowal, made after Adams asks whether he would, in better circumstances, have wished to "have indulged his Talents and Desire of Knowledge": "[Joseph] was perfectly content with the State to which he was called . . . he should endeavour to improve his Talent . . . but not repine at his own Lot, nor envy those of his Betters" (20). Crucially, Joseph's solitary exercise of homophony thus attests to different motives than do Slipslop's: "when I have waited behind my Lady in a Room hung with fine Pictures . . . when it hath been asked whose Picture that was, it was never once answered, the Master's of the House, but *Ammyconni*, *Paul Varnish*, *Hannibal Scratchi*, or *Hogarthi*, which I suppose were the Names of the Painters" (203). One time in *Joseph Andrews*, Joseph's transcribed speech assumes the density of Slipslop's. Audible to him as pure sound, proper names become homophonic echoes and vernacular slang at once. But if these sobriquets are no less garbled than Slipslop's, Joseph is not an affecter of art historical competence bestowed by the liberal education he lacks. Whereas Slipslop's nonsensical but still leading cues undo her claim to understanding, Joseph does not recombine overheard syllables to mask irreparable ignorance. Instead, because his recitation is unwittingly acted, it vindicates his avowal of perfect contentment with the "State to which he was called." Rather than the inevitably absurd outcome of a servant's aspiration to understanding, the thingness of Joseph's speech proves his disinterest in the elevating promise of letters. Unlike "*Ironing*," "*Ammyconni*" does not fuse rivalry with his superiors to that rivalry's manifest failure. As evidence of unpretending simplicity, not assumed learning, Joseph's errors define his refusal to aspire *as* his virtue. More precisely, Joseph's virtue is his refusal of *literary* ambition, as his feminine complement Fanny makes clear. Fielding's starkest rejoinder to the literary affectations of the servant class, "poor *Fanny* could neither write nor read" (42).

Slipslop's italicized locutions define her as a gross double of Cibber, whose laureateship crowns Fielding's career-long complaint, voiced in *Joseph Andrews* when the playwright Wilson bemoans tastelessness that exhausts the literary marketplace: "some Persons finding it not so easy a Task to discern good from bad Authors, or to know what Genius was worthy Encouragement, and what was not" (187). Slipslop concretizes this lapse in public discernment, summoning a linguistic correlate for the difference between good and bad. It is not Fanny's embodiment of illiteracy, or Joseph's preemptive removal from the literary field, but the logic of Slipslop's game that adjudicates crises of social value in this novel.

An instance at *Joseph Andrews*'s end solicits the reader's corrective acumen in the arena of criminal justice. Convicted of "a kind of felonious larcenous thing" at the behest of the spurned Lady Booby, Joseph is rescued by Squire Booby from imminent delivery to Bridewell. Upon Booby's appeal to the Justice of the Peace, the latter produces "a Deposition, which in the Absence of his Clerk he had writ himself, of which we [Fielding's narrator] have with great difficulty procured an authentick Copy" (252) (Fig. 9). The justice's misspellings and inadvertent cognomens lend this text a distracting density. As its writer in "the Absence of his Clerk," the justice's compromised literacy is as perceptible Shamela's and Slipslop's. But because the justice stands as an impersonal administrator of the social contract, Fielding aggravates the thingness of a "Copy" whose assurance of equity should be underwritten by the abstracting idiom of the law. Most tellingly lent typographical objecthood by "*Justasses*," Fielding's target is legal corruption whose textual evidence repudiates whatever vestigial pretensions to fairness would be secured by the medium of discourse. Typography concretizes—and, in the eyes of the reader whose literacy enlists her in Fielding's game, rectifies—an abuse of justice also enacted in words.

"Cumfarting" obviously invalidates the juridical text's claim to propriety. In this capacity, language in *Joseph Andrews* acts: by turning words into things, Fielding lends them qualities that override the prejudicial enactment of the law they would enable. With "a Kind of Chymical Operation" that does not dissolve corrupted language but makes it sensibly meretricious, Fielding transforms the deposition that would prove Joseph's guilt into an artifact that palpably sustains his innocence. The egregiousness of "*Justasses*" effects its own justice, both in the medium of words and, on Locke's terms, outside it, in the world of perceptible things. My chapter's final section suggests that Fielding's *Jonathan Wild* takes the form of an alchemical object.

Jonathan Wild and Alchemical Form

> But that a Man in Chains, in Prisons, nay, in the vilest Dungeon, should with persevering Pride and obstinate Dignity, discover that vast Superiority in his own Nature over the rest of Mankind, who, to a vulgar Eye, seem much happier than himself; nay, that he should discover Heaven and Providence (whose peculiar Care, it seems, he is) at that very Time at work for him; this is among the *Arcana* of GREATNESS, to be perfectly understood only by an Adept in that Science.
>
> —Henry Fielding, *The Life of Mr. Jonathan Wild the Great* (1743)[58]

> Gold has particles which are mutually in contact: their sums are to be called sums of the first composition and their sums of sums, of the second composition, and so on. Mercury can pass, and so can Aqua Regia, through the pores that lie between particles of last order, but not others. If a menstruum could pass through those others or if parts of gold of the first and second composition could be separated, it would be liquid gold. If gold could ferment, it could be transformed into any other substance.
>
> —Isaac Newton, *De natura Acidorum* (1692)[59]

In the epigraph from *Jonathan Wild* cited above, the novel's eponymous protagonist has been set adrift at sea in a longboat with six biscuits. Fielding's mock biography of the real Wild, an infamous dealer in stolen goods who was executed in 1725, *Jonathan Wild* reflects on the imperviousness of its hero's "GREATNESS" to narrative circumstance at a conspicuously desperate juncture. This passage satirizes Defoe's *Robinson Crusoe* (1719), whose bereaved castaway redeems himself by elaborating the novel-length revelation that he is providence's "peculiar Care." (Self-professed testimony to its imperiled narrator's favor with "Heaven and Providence," *Crusoe* is a signal antecedent to *Pamela*.[60]) But in *Jonathan Wild*, puritanism is not the impetus for lowness's reconversion into evidence of "vast Superiority," because spiritual pretension is only one instance of aggrandizement that Fielding subsumes under the

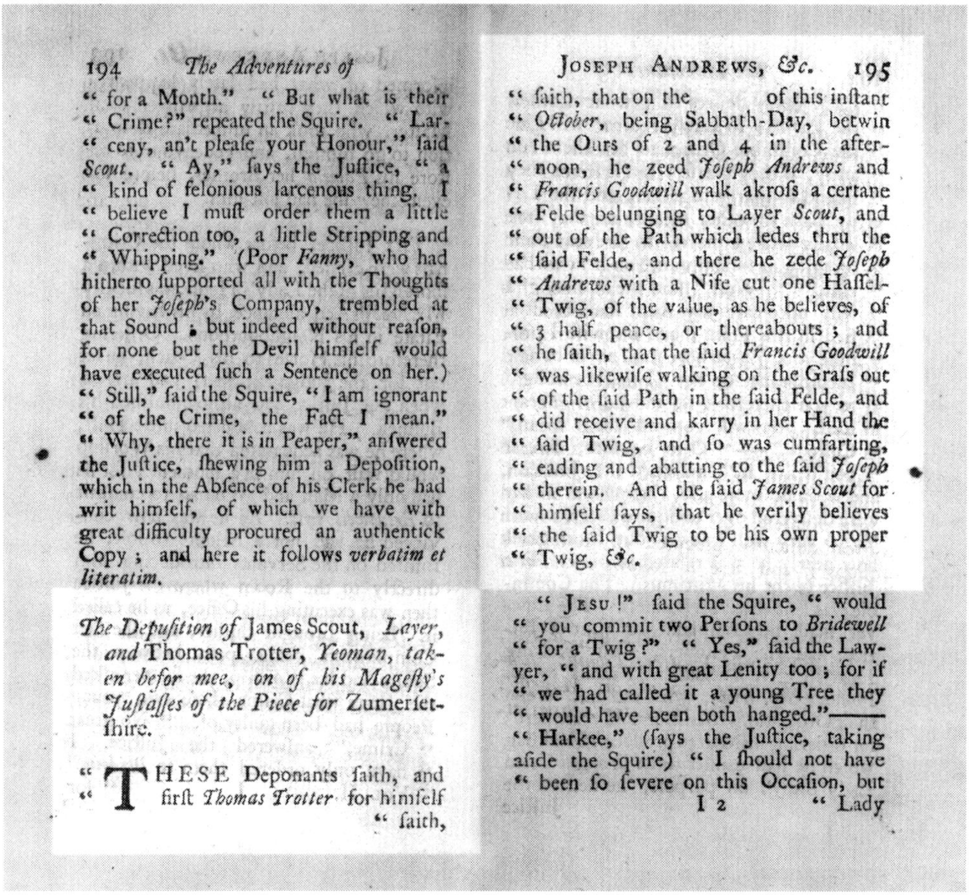
194 *The Adventures of*

"for a Month." "But what is their "Crime?" repeated the Squire. "Lar-"ceny, an't pleaſe your Honour," ſaid *Scout*. "Ay," ſays the Juſtice, "a "kind of felonious larcenous thing. I "believe I muſt order them a little "Correction too, a little Stripping and "Whipping." (Poor *Fanny*, who had hitherto ſupported all with the Thoughts of her *Joſeph*'s Company, trembled at that Sound; but indeed without reaſon, for none but the Devil himſelf would have executed ſuch a Sentence on her.) "Still," ſaid the Squire, "I am ignorant "of the Crime, the Fact I mean." "Why, there it is in Peaper," anſwered the Juſtice, ſhewing him a Depoſition, which in the Abſence of his Clerk he had writ himſelf, of which we have with great difficulty procured an authentick Copy; and here it follows *verbatim et literatim.*

The Depuſition of James Scout, *Layer, and* Thomas Trotter, *Yeoman, taken befor mee, on of his Mageſty's Juſtaſſes of the Piece for* Zumerſetſhire.

"THESE Deponants ſaith, and "firſt *Thomas Trotter* for himſelf "ſaith,

JOSEPH ANDREWS, &c. 195

"ſaith, that on the of this inſtant "*October*, being Sabbath-Day, betwin "the Ours of 2 and 4 in the after-"noon, he zeed *Joſeph Andrews* and "*Francis Goodwill* walk akroſs a certane "Felde belunging to Layer *Scout*, and "out of the Path which ledes thru the "ſaid Felde, and there he zede *Joſeph* "*Andrews* with a Nife cut one Haſſel-"Twig, of the value, as he believes, of "3 half pence, or thereabouts; and "he ſaith, that the ſaid *Francis Goodwill* "was likewiſe walking on the Graſs out "of the ſaid Path in the ſaid Felde, and "did receive and karry in her Hand the "ſaid Twig, and ſo was cumfarting, "eading and abatting to the ſaid *Joſeph* "therein. And the ſaid *James Scout* for "himſelf ſays, that he verily believes "the ſaid Twig to be his own proper "Twig, &c."

"JESU!" ſaid the Squire, "would "you commit two Perſons to *Bridewell* "for a Twig?" "Yes," ſaid the Law-yer, "and with great Lenity too; for if "we had called it a young Tree they "would have been both hanged."—— "Harkee," (ſays the Juſtice, taking aſide the Squire) "I ſhould not have "been ſo ſevere on this Occaſion, but

I 2 "Lady

Figure 9. "*The Depusition of* James Scout, *Layer, and* Thomas Trotter, *Yeoman*," Henry Fielding, *The History of the Adventures of Joseph Andrews* (1742). Photo courtesy of the Newberry Library, Chicago. Vault Case Ruggles 120 vol. 2.

transformative "*Arcana* of GREATNESS, to be perfectly understood only by an Adept in that Science." By referring to arcana kept secret by adepts, Fielding aligns "GREATNESS" not visible to "vulgar" eyes with a science that acts on metals to elicit the ennobling qualities of gold.

By invoking alchemy, Fielding seems to flag his protagonist's grandiosity as a case of linguistic abuse so extreme it lacks even the empirical substrate afforded by Vinegar's louse. But historians of science from Richard S. Westfall to Newman affirm that the conversion of corruptible metals into gold is not precluded by a particulate episteme in which qualities proceed from corpuscular texture rather than elemental essence.[61] Matter that does not fix qualitative identity at the level of the corpuscle may contribute to a new texture whose secondary powers make it perceptible as something else. *Jonathan Wild* adopts the form of the alchemical object to structure the reader's experience of an hagiographic history whose pretender to greatness may represent the prime minister Walpole. A limit case of improvement or "perfection," Wild's assumption of superiority crowns the improbable ascent of a base man.

The corpuscular logic of alchemy—exemplified by Newton's "Mercury [that] can pass . . . through the pores" of gold—implicates texture that harbors qualities within opposing qualities. Metals composed of such texture manifest antagonistic attributes not once but multiple times as they undergo sublimation, distillation, slagging fusion, or calcination, defining alchemical science as the ultimately perfecting movement between one property and its opposite. Paul of Taranto's medieval Latin appropriation of Jābir ibn Ḥayyān's or Geber's "*70 Books*,"[62] the influential *Summa Perfectionis*, dictates: "Then let the fixed be made volatile and the volatile again fixed, and let the volatile be made fixed, the fixed volatile, and the volatile fixed so often that easy fusion upon firing appear. The very precious secret is completed in this order."[63] Geber's veiled transmission of his "precious secret" signals another register in which *Jonathan Wild* recurs to alchemy, Fielding's generic and typographical play on the obscurity with which the science is relayed. Before its translation by seventeenth-century adepts like Starkey's pseudonymous alter ego Eirenaeus Philalethes, alchemical knowledge was mediated by romance, textual dispersion, planetary cognomens, and typographical tropes that demand decoding.[64] As hagiography that solicits demystification, *Jonathan Wild* is *Pamela* to Fielding's *Shamela*, reversing Fielding's previous strategy. Wild's perfected history compels the reader to behave like Fielding—to disclose the corrupt material from which this text's greatness has been conjured. *Jonathan Wild* prompts the reader to intuit a vile original on the basis of cant, idiom, illiteracy, and other debasing clues.

Newton claims that because gold is composed of regressive assemblages of particles, it can "be transformed into any other substance" if a sufficiently subtle solvent is found. By passing through "the pores that lie between particles," subtler and subtilizing mercury enters other metals to make them denser and thereby incorruptible. Newton illuminates the alchemical corollary of corpuscular science, whose anti-elemental doctrine sustains the historical revision glossed by Betty Jo Teeter Dobbs: "it is apparent that mechanical philosophies did nothing to strike at the root of alchemy. Indeed, if anything, they encouraged the concept of transmutability and gave it a new rationale."[65] Of alchemy's role in the corpuscular philosophy of Boyle, Locke, and Newton, Newman affirms of Eirenaeus Philalethes's (Starkey's) alchemical treatise *Introitus Apertus ad occlusum Regis Palatium* (1667) (*Secrets Reveal'd: or, An Open Entrance to the Shut-Palace of the King* [English, 1669]): "the *Introitus apertus* was still being printed in 1742 and 1749, and the transmutational theory upon which it was based was not utterly discredited until [Antoine-Laurent] Lavoisier published his discovery of chemical elements at the end of the eighteenth century."[66] Newton owned a copy of *Secrets Reveal'd*, Dobbs recounts, "heavily annotated in Newton's own hand."[67] Locke's copies, in both Latin and English, are listed in his library catalog.[68] While alchemy was intermittently censured by the church, the state, and some chemists, Boyle's, Locke's, and Newton's chrysopoeiac enterprises testify to its congruity with the corpuscular doctrine of qualities.[69] As Lawrence Principe affirms, the presumed antagonism of "mechanical corpuscularianism" and alchemy marks a "false dichotomy": "Many chrysopoeians from pseudo-Geber to Philalethes use an explicitly corpuscularian system to explain metallic transmutation."[70]

Geber's *Summa* defines alchemy as a corpuscular science while elaborating corpuscularianism as an alchemical theory of qualitative identity. To proceed to transmutation's corpuscular rationale: Geber identifies mercury (which cannot be equated with today's element but is rather what Dobbs calls a variable "mercurial principle,"[71] a caveat that applies to all chymical names of present-day elements) and sulfur as components of the six known metals tin, lead, copper, iron, silver, and gold. While sulfur lends metals their flammability and color, the "viscous water"[72] mercury meets Geber's criteria for the alchemical "medicine" able to perfect metals as well as itself: "the medicine itself ought to be of very subtle, pure substance, adhering of its own nature to the quicksilver, of very easy, subtle liquefaction like water, and fixed during the attack of fire. For such a medicine will coagulate it [the base metal] and convert it into a solar [golden] or lunar [silvery] nature."[73] Newman stresses the "consistently corpuscular"[74] operation of Geber's self-perfecting medicine, which lends qualities to baser metals not by transfusing some essential quiddity but because its "subtle" parts "adhere[]" to those of the object with which it mixes. Geber refines the corpuscular justification of alchemy's "goal": "The deficient in them [baser metals] is therefore the paucity of quicksilver in them, and the improper compaction of the same. Hence the goal in them will be the good multiplication and compaction, and permanent fixation of quicksilver . . . Since the quicksilver prepared by our craft in a medicine is cleansed and led back into a very pure, bright substance, if projected onto bodies removed from perfection it will brighten them and perfect them with its fixation."[75] "Compaction" and "fixation" define metallic perfection as a corpuscular standard. Quicksilver perfects because it penetrates, adheres, compacts, and fixes the parts of other metals. Density, nonreactivity, ductility, and incorruptibility are not inimitable essence but qualities produced by the changeable corpuscular structure Boyle later names texture.

Yet to invoke texture is to obscure a nuance evident when the *Summa* designates sulfur and quicksilver "the matter of the metals."[76] Geber's specification of this "matter" differs from Boyle's restriction of particles' physical attributes to size, shape, motion, and cumulative structure. To employ Locke's terms, the difference turns on particulate matter's primary qualities: for Geber, these are not size, shape, and so on, but Aristotle's "four elemental qualities, hot, cold, wet and dry."[77] Geber "violated the Aristotelian principle of [the] homogeneity"[78] of unmixed things to assign elemental qualities to the particulate units that comprise his six metals. Aristotelian qualities endow infinitesimal parts, as when the *Summa* discusses how sublimation—heating in a furnace to induce change from a solid to a vapor or so-called volatile "spirit,"[79] which ascends into a glass vessel or aludel—purifies alchemical medicines that otherwise "burn, blacken, and defile" the metals they are meant to perfect: "[They] do this from a double cause. The first is that the burning oiliness of the sulfureity whose property is to be easily ignited . . . But the second cause is earthiness, which also had not been removed from them . . . Therefore we have devised a way to cleanse all these from burning unctuosity and earthy superfluity . . . For when fire rises, it always raises the smaller particles with it; hence it leaves behind the larger. And this appears through sublimation—that the spirits are cleansed of earthiness, which also impeded ingression and gave a dirty color."[80] "Sulfureity" refers to a principle that dictates the disposition to flammability or fieriness. "Earthiness" or "earthy superfluity" engenders "a dirty color" and hampers the subtlety of medicine that must enter the pores of another body. But it is the mechanical difference between "smaller" or "larger" corpuscular dimension that cleanses still adulterated spirits of fiery and earthly excess. Failure of perfective ingress and inflammability are elemental qualities fortified by opposing excesses of particulate size. Geber thereby recombines Aristotelian and corpuscular attributes: chemical reactions like sublimation are driven by the mechanical affection of size, but reacting particles are composed *of* the peripatetic qualities humidity, earthiness, inflammability, and dryness. Metals are perfectible because they are made of alienable units of essential principles. The *Summa* marshals elemental substances whose embodiment as particles renders them agents of both mechanical and qualitative change.

Boyle distinguishes the imperceptible but real determinants of body—size, shape, motion, and texture—from the ideas they stimulate in persons. But in the *Summa*, the elemental qualities dry, wet, fiery, and earthy undergird other, more immediately manifest properties, because, according to Geber's Aristotelian deployment of primary endowments, sensible qualities disclose other, deeper sensible qualities. For example, a process like calcination or open heating over flame—"The calcination of a thing by fire is, therefore, its conversion into powder, due to the removal of the humidity consolidating its particles"[81]—extirpates "humidity" sealed inside the object. As Geber writes, because "jupiter [tin] has a fugitive substance of quicksilver closed up in the depth of its nature," calcination leaves the metallic body "deprived of humidity." By calcining mercury, Geber aims to undo "the concealment of the sulfureity in the continuity of quicksilver's substance."[82] Whether it is mercury ("humidity") or sulfur ("sulfureity") that a differently qualified metal conceals in its "depth," Geber posits parts composed of qualitatively antipathetic parts.[83] As Newman explains, Geber defines what is concealed "in the depth" and what is manifest as mutually qualitative and spatial difference: "The Jābir school assumed that every material substance contains its opposite, but in a 'hidden' fashion. Thus every substance has a *bāṭin* and a *ẓāhir* (*occultum* and *manifestum* in Latin), an 'occult' and a 'manifest,' an 'interior' and 'exterior.' Jabir employs this terminology of 'occult' and 'manifest' interchangeably with 'center' and 'circumference.' Hence silver, which is cold and dry 'externally,' contains the opposite qualities, hot and wet, 'internally.' "[84] As the *Summa* stipulates, metals must be "known both in the depth and manifest of their nature."[85] "Depth" and "manifest" likewise structure the corrective—in Fielding's case, debasing—impetus of *Jonathan Wild* as the reader encounters the printed book.

The influence of the "Jābir school" from Geber to eighteenth-century chymistry is reflected in Newton's *Opticks*, whose third edition (1730) theorizes the particulate stickiness Newton names attraction. Beginning with the premise that "Salts are dry Earth and watry Acid united by Attraction,"[86] Newton offers a planetary analogy for a base or salt's corpuscular construction:

> As Gravity makes the Sea flow round the denser and weightier Parts of the Globe of the Earth, so the Attraction may make the watry Acid flow round the denser and compacter Particles of Earth for composing the Particles of Salt. For otherwise the Acid would not do the Office of a Medium between the Earth and common Water, for making Salts dissolvable in the Water . . . Now, as in the great Globe of the Earth and Sea, the densest Bodies by their Gravity sink down in Water, and always endeavour to go towards the Center of the Globe; so in Particles of Salt, the densest Matter may always endeavour to approach the Center of the Particle: So that a Particle of Salt may be compared to a Chaos; being dense, hard, dry, and earthy in the Center; and rare, soft, moist, and watry in the Circumference.[87]

Placed in "common Water," the humid or "watry" part of Newton's salt performs "the Office of a Medium" between the environing fluid and salt's differently qualified inner part, thereby facilitating the dissolution of the salt's "dry Earth." (Just as "mercury" can signify more or less perfected "mercury," "common Water" is not identical to the watery "Acid" that comprises the moist part of Newton's salt.) Newman cites this passage to affirm that its characterization "of the salt corpuscle is strikingly resonant with the Jābirian theory of the occult and the manifest, according to which every mineral contained its opposite hidden within itself."[88] Dobbs remarks its "synthesis of the new corpuscularianism of particles and the old 'chymistry' of substances."[89] Indeed, Newton enlists "Gravity" to vindicate the alchemical organization of his salt's constituent qualities: the corpuscle's earthy "Center" is held in place by the same force that, on his chaos or "primordial earth,"[90] makes "the densest Bodies . . . sink down in Water." It is no analogy, but the extra- and intracorpuscular influence of the same gravitational force, that justifies the oceans' tides and the salt's submerged dryness. Gravity layers salt's opposing parts, although it cannot account for the anterior, Aristotelian integrity of the elemental primary qualities wateriness and earthiness.

Newman tracks the articulation of "spatially distinct layers of hidden and manifest corpuscles or components of corpuscles"[91] from Jābir, to Geber, to Paracelsus, to Helmont, to Starkey, and thence to Newton.[92] My foreshortened juxtaposition of Geber and Newton illuminates the aspect of alchemy germane to Fielding's novel: an *alchemical* articulation of primary-secondary difference whereby one sensible quality—in Fielding's novel, a sensible attribute of print—hides its secreted opposing trait. Writing to Boyle, Newton

Figure 10. "Saline particles still encompassing ye metallick ones," Isaac Newton to Robert Boyle, *Correspondence*, ed. H. W. Turnbull (1960). Image adapted by author.

provides an annexed "figure" to portray the layering of "saline particles still encompassing ye metallick ones as a coat or shell does a kernell" (Fig. 10).[93] Taking Newton's particulate "coat" and its qualitatively opposed "kernell" as a form of primary-secondary difference, we can return to Fielding. Fielding's library catalog includes no works by Starkey or Philalethes; however, he owned the abridged *Philosophical Transactions of the Royal Society* (1716–35) to 1733, as well as Boyle's *Works, with his Life* (1744) and Henry Pemberton's *View of Newton's Philosophy* (1728).[94] These sources affirm Fielding's familiarity with Lockean empiricism and its corpuscular underpinnings as well as Newton's elaboration of the role of gravity and attraction in chymistry. Indeed, *Joseph Andrews* identifies the alchemical or spatial organization of opposing qualities as Fielding's "Province in the present Work" (6).

Joseph Andrews's narrator declares that he targets the practice, appearance, and form of human "Affectation" (6). Two motives drive that practice: "Vanity," whose end is to "purchase Applause"; and "Hypocrisy," motivated by the effort "to avoid Censure by concealing our Vices under an Appearance of their opposite Virtues" (6). Because they sustain the discontinuity of a performer's surface from the qualities "under" it, vanity and hypocrisy share one form. Fielding discriminates between them by virtue of the extremity of the break between visible affectation and what it cloaks: "when [affectation] proceeds from Hypocrisy, it be nearly allied to Deceit; yet when it comes from Vanity only, it partakes of the Nature of Ostentation: for instance, the Affectation of Liberality in a vain Man, differs visibly from the same Affectation in the Avaricious; for tho' the vain Man is not what he would appear, or hath not the Virtue he affects, to the degree he would be thought to have it; yet it sits less aukwardly on him than on the avaricious Man, who *is* the very Reverse of what he would *seem* to be" (6–7). "Vanity only" is less vicious than hypocrisy, because while the vain man affects a characteristic he does not harbor to the same "degree," the hypocritical man manifests "the very Reverse" of the attribute he conceals. A human avatar of Newton's shell and kernel whose structure is motivated by the endowment he hides, Fielding's hypocrite incarnates the alchemical form of primary-secondary difference.

Published in Fielding's *Miscellanies* (1743), "An Essay on the Knowledge of the Characters of Men" refines his indictment of hypocrisy.[95] Offered to assist the reader's understanding of persons who cannot be sounded by looks alone, *Characters* cites "censorious Sanctity" as a distillation of hypocrisy whose exposure requires a kind of dismemberment: "I shall take some Pains in the ripping it up, and exposing the Horrors of its Inside . . . and at the same Time will endeavour so plainly to describe its Outside."[96] But Fielding does not plumb "the Horrors" of further depth, for the evils of "sanctified Hypocrisy" reside in spatial organization whose extroversion of goodness is itself vicious: "It confines all Merit to those external Forms which are fully particularized in Scripture."[97] Scriptural sanction of the "Merit" of external observance hardens Fielding's prior appraisal of vain showiness.[98] While in *Joseph Andrews* vanity is no worse than ostentatious display, *Characters of Men* declares that "Nothing can, in Fact, be more foreign to the Nature of Virtue, than Ostentation."[99] This is an explicitly formal precept: "Ostentation" articulates a declension of moral personhood that hides qualitative dissonance in the depths of the alchemical object.

Joseph Andrews extracts a warrant for humor from a sequence of discoveries that reveal "any one to be the exact Reverse of what he affects":

> [N]or do I believe any Man living who meets a dirty Fellow riding through the Streets in a Cart, is struck with an Idea of the Ridicu-

> lous from it; but if he should see the same Figure descend from his Coach and Six, or bolt from his Chair with his Hat under his Arm, he would then begin to laugh, and with justice. In the same manner, were we to enter a poor House, and behold a wretched Family shivering with Cold and languishing with Hunger, it would not incline us to Laughter . . . but should we discover there a Grate, instead of Coals, adorned with Flowers, empty Plate or China Dishes on the Side-board, or any other affectation of Riches and Finery . . . we might then indeed be excused, for ridiculing so fantastical an Appearance. (7)

A "dirty Fellow" is initially "in" an unprepossessing vehicle; his dirtiness is aggravated when he instead "descend[s] from" a "Coach and Six." A starving family merits "Laughter" only when "we . . . enter" a lavishly appointed "House." Fielding predicates comedy on the containment of one quality inside its other, although he selects states and accoutrements whose capacity to characterize persons is surprisingly weak: would-be elemental primary attributes like dirtiness or hunger are housed within their qualitative antagonists luxuriousness, warmth, and satiety. Absurdity ensues not because either set of endowments claims ontological privilege but because, as Fielding writes of vanity, the outermost property "sits . . . aukwardly on" the one below. The dirty fellow who exits a coach evinces dissonance akin to that manifested by mercury or "dry water."[100] But in anticipation of *Jonathan Wild*, he animates properties that cannot be seamlessly blended.

Jonathan Wild, like *Shamela* and *Joseph Andrews*, confronts the crisis in public discernment evidenced by Cibber's and Walpole's ascent. Disclaiming in Fielding's mock-empiricist narrative mode, *Jonathan Wild*'s narrator compares comets, or "Occurrences of the Phænomenous Kind which have lately appeared in this our Hemisphere," to the spectacle of useless men whose "Preferment" likewise vindicates nature's "unlimited Power" (81). This narrator ponders:

> [H]ow Flattery should denote a Judge, or Impiety and Atheism, a Bishop, he [mankind] is not capable of comprehending. And indeed, we ourselves, wise as we are, are forced to reason *ab effectu*, and if we were asked what Nature had intended such Men for, before she herself had by the Event demonstrated her Purpose, it is possible we might be sometimes puzzled to declare; for . . . great

> Fortitude of Mind with a vast Capacity and Knowledge, might induce a Belief in the Beholder, that such Endowments were by Nature designed for Power and Honour rather than the reverse; whereas daily Experience convinces us of the contrary. (82)

Comets are no less striking than the mismatch of merit and preferment modeled by an impious bishop. Yet Fielding's narrator refuses to interrogate his "daily Experience" of the misplaced "Endowments" that McKeon designates "status inconsistency."[101] Fielding's mock surrender to the empirical inscrutability of nature's "Purpose" defines status inconsistency as a union of opposed qualifiers whose testimony to unearned reward is realized as alchemical form. This form reflects corruption so pervasive that honorifics published in print infallibly signal their opposite.[102] *Jonathan Wild* incarnates a limit case.

Like the coach holding a dirty fellow, *Jonathan Wild*'s manifest content sits awkwardly over antipathetic qualities portended by the novel itself. Sporadic eruptions of corrupt typography, bad spelling, and cant oppose the narrator's historicizing ostentation, compelling the reader to recognize Wild's deeper depravity. In *Jonathan Wild*, the nested qualities that render preferment empirically puzzling do not presume a referent detached from the language that makes him. *Jonathan Wild*'s puffing sustains status inconsistency as the opposition of words to what they should really signify, as when Fielding denominates the murderous thief Wild a "GREAT MAN" (82) or "the GREAT WILD" (90). With the modifier "GREAT," other cues invite the reader to transmute Fielding's history into its typographic, idiomatic, and qualitative opposite. One such prompt is Fielding's footnotes: inserted to define italicized criminal cant like "**Priggism*," asterisked definitions like "*Thievery" (14) occupy the page's bottom margin to intimate the visual encroachment of the novel's canting double. (This variant of "prig" and "priggers," listed in neither the *New Canting Dictionary* nor Nathaniel Bailey's *Collection of Canting Words and Terms*, is Fielding's own neologism.)

Fielding recapitulates the moral posture of *Jonathan Wild*'s great man to insinuate some hint of baseness:

> The Character which he most valued himself upon . . . was that of Hypocrisy. His Opinion was, that no one could carry *Priggism* very far without it; for which Reason, he said, there was little GREATNESS to be expected in a Man who acknowledged his

> Vices; but always much to be hoped from him, who professed great Virtues; wherefore, though he would always shun the Person whom he discovered guilty of a good Action, yet he was never deterred by a good Character, which was more commonly the Effect of Profession than of Action: For which Reason, he himself was always very liberal of honest Professions, and had as much Virtue and Goodness in his Mouth as a Saint . . . nay, tho' he held Good-nature and Modesty in the highest Contempt, he constantly practised the Affectation of both. (177)

This distillation of Wild's real "Opinion" does not proceed straight from the character's own "Mouth," an organ dedicated to the external "Profession" of sanctified hypocrisy. Instead, it comes close to the narrative approximation of interiority known as free indirect speech. Although the "he's" punctuating this third-person portrait accentuate Fielding's historicizing omniscience, its accretion of clausal caveats seems to mimic the internal logic of Wild's own valuation of self-declared "good Character" over empirically proven "good Action." As in most of the novel, Fielding's polite narrator supplies the default diction in which Wild's history and his opinion transpire. Even "*Prigism*," a neologized elevation of cant, does not disrupt the fluid wit with which Fielding expresses his protagonist's aversion to persons "discovered guilty of a good Action."

The novel's solitary presentation of Wild's unassisted exercise of literacy, a missive written to his future wife Lætitia or Tishy Snap, thus delivers a genuine shock:

> Most Deivine and adwhorable Creture,
>
> I dout not but those IIs, briter than the Son, which have kindled such a Flam in my Hart, have likewise the Faculty of seeing it. It would be the hiest Preassumption to imagin you eggnorant of my Loav. No, Maddam, I sollemly purtest, that, of all the Butys in the unaversal Glob, there is none kapable of hateracting my IIs like you. Corts and Pallaces would be to me Deserts without your Kumpany, and with it a Wilderness would have more Charms than Haven itself . . . I am konvinced you must be sinsibel of my violent Passion for you, which, if I endevored to hid it, would be as impossible as for you, or the Son to hide your Butys. I . . . therefore

> hop you will, out of Kumpassion, let me have the Honour of seeing you this Afternoon; for I am, with the greatest Adwhoration,
>
> Most Deivine Creeture,
>
> *Iour most pessionate Amirer,*
>
> *Adwhorer, and Slave,*
>
> JOHANATAN WYLD. (100)

More than Shamela's ruined orthography or Slipslop's homophones, this letter stages illiteracy for maximum phenomenological impact. In something of a kitchen sink strategy of textual debasement, Fielding lards Wild's letter with inadvertent cognomens ("adwhorable"), potential cant ("Flam"), obtrusive misspelling ("Glob"), and even a near ideogram ("IIs"). But as evidence of epistolary ineptitude extending to the rendition of his own name, Wild's letter nonetheless remains recognizable as an iteration of sentiment dictated by precedent akin to Vinegar's FRMS. For while Wild's illiteracy vitiates individual words, it does not detract from the grammatical integrity of this letter's highly sophisticated punctuation and sentence structure. Wild's letter represents illiteracy that is still, paradoxically, legible: the palpability of his ruined words is not impeded by grammar that should be correspondingly wrecked. By leaving intact its syntactical scaffolding, Fielding amplifies the empirical effect of this unlettered letter's sensible qualities. Laid over an unblemished syntactical foundation, Wild's spelling intensifies the perceptibility of his lack of literary value. Fielding exploits the form of primary-secondary difference: because of its grammatical propriety, the letter's primary infrastructure claims the secondary power to aggravate Wild's orthographic failings. His viciousness is rendered an object of empirical sensation by dint of the impeccable texture of Fielding's syntax.

More than any other artifact in Fielding's novelistic canon, Wild's letter shows Fielding's investment in the phenomenal power of printed words to expose status inconsistency. To muster the letter's shock, Fielding assumes that his reader will hold Wild to a literary standard—indeed, that this *is* the standard by which status inconsistency can be rectified. That assumption is evident in the letter's conservatism. Not only does the missive's address to Tishy presume a reader's generic fluency in the tropes of romance; Fielding sublimates that presumption as the ground for the palpability of Wild's superadded incompetence. The power of typographically qualified text to expose baseless preferment turns on an anterior standard of literacy or, more precisely, literariness. What propels the empirical impression of Wild's baseness is a standard whose integrity Fielding conserves in the very act of its formal repression.

The reader cannot fail to be struck by this singular evidence of Wild's epistolary ineptitude. Fielding immediately accounts for its difference from the rest of his text: "if it should be observed that the Style of this Letter doth not exactly correspond with that of our Hero's Speeches . . . we answer, it is sufficient if in these the Historian adheres faithfully to the Matter, though he embellishes the Diction with some Flourishes of his own Eloquence" (100). With a defense of substance that should be impervious to embellishments of "Diction," Fielding satirically contravenes the evidence that cues his reader to take unassisted illiteracy as the truth of Wild's character. Rather than "Flourishes" extrinsic to merit, "Style" and "Diction" *are* the medium which palpably deflate Wild's greatness. By gesturing toward the remainder of his novel's capacity, *Pamela*-like, to elevate its hero, Fielding indicts historians and "Epistolary" (99) novelists who reconvert viciousness into puffery. Fielding mobilizes print to goad his reader the other way: the failure of the bulk of *Jonathan Wild* to "exactly correspond" with the baseline set by Wild's letter compels the game of downward rectification. This game is often propelled by text that departs from third-person narrative, as when the recently married Wild and Tishy pursue a dialogue "taken down," Fielding avers, "*verbatim*" (103):

> *Jonathan*
>
> My Dear, I wish you would lie a little longer in Bed this Morning.
>
> *Lætitia*
>
> Indeed I cannot: I am engaged to breakfast with Sir *John*.
>
> *Jonathan*
>
> I don't know what Sir *John* doth so often at my House. I assure you I am uneasy as it; for though I have no Suspicion of your Virtue, yet it may injure your Reputation in the Opinion of my Neighbours.
>
> *Lætitia*
>
> I don't trouble my Head about my Neighbours; and they shall no more tell me what Company I am to keep than my Husband shall.
>
> *Jonathan*
>
> A good Wife would keep no Company which made her Husband uneasy.
>
> *Lætitia*
>
> You might have found one of those good Wives, Sir, if you had pleased, I had no Objection to it.
>
> *Jonathan*
>
> I thought I had found one in you. (103–4)

To add one extradialogic detail, Fielding makes clear that Tishy is cuckolding Wild with the equally ignoble Sir John. As with Shamela and Slipslop, one referent of preposterous greatness is the unsavory embodiment of an "adwhorable" woman whose metonymic proximity supplements the literary debasement of Fielding's just-married great man. In the register of diction, this dialogue is striking for its lack of idiomatic and typographical inflection: "Kumpany" is restored to "Company," Tishy's proclivities are transposed into Wild's delicate demurral of "Suspicion of your Virtue," and she rebukes to her husband with the genteel salutation "Sir." But aligned with Wild's sensationally erroneous letter, the stilted politeness of this dialogue renders Fielding's refusal to render the idiomatic stuff of Tishy and Wild's speech an equally perceptible reversal. Following Wild's epistle, the absence of particularizing attributes makes the propriety of this fight likewise striking. With all historians who obfuscate status inconsistency, Fielding corrects the "Diction" of Tishy and Wild's argument. Juxtaposed with Wild's missive, however, the act of textual elevation becomes phenomenally egregious.

By dint of perusing Fielding's plot in order, even the least alchemically inclined reader recognizes that the novel manifests reversals of greatness and anterior depravity. Certainly, the intimation that Fielding perfects unlettered speech extends to all remaining stretches of polite—or literate—text. Fielding's reader thereby experiences status inconsistency as a particular mode of empirical attunement to the sensible qualities of print. Books may assist the preferment of persons without merit, but Fielding's novel prompts the alchemical corrective that is reading.

Awaiting execution at Newgate after England passes a law criminalizing the traffic or fencing of stolen goods, Wild inspires one final revelation of the alchemical form of high and low. While incarcerated, he is preached at by an "*Ordinary*" whose references to such theological topics as "THE DEVIL AND HIS ANGELS" are interrupted by the countervailing thingness of words in a footnote to the word "whole*": "*He pronounced this Word HULL, and perhaps would have spelt it so" (166). Fielding's hint at orthographic infelicities insinuated by ear devolves into a field of asterisks interspersed with typographical fragments (Fig. 11). As *Jonathan Wild* nears its ignoble end with the intimation of total typographical opacity, Fielding's asterisks are legible as indices to villainy whose resistance to puffing has, for once, overtaken the novel. While most of *Jonathan Wild* solicits the lowering exchange of "HULL" for "whole," asterisks refuse the switch of cant for history. As print that refuses to elevate,

390 *The LIFE of* Book IV.

Ordinary.

* are * Atheiſt. * * Deiſt * * * Ari * * * cinian * * * hanged * * burnt * roiled * oaſted. * * * * * * * * Dev * * his An * * * * ell Fire * * ternal Da * * * tion.

Figure 11. Ordinary's instructions, Henry Fielding, *Miscellanies* (1743). Photo courtesy of the Newberry Library, Chicago. Case Y12. F456 vol. 3.

asterisks mark both the zero degree of typographical thingness and hypocrisy no historian can redeem.

Fielding inserts a footnote before the asterisks to assert: "This Part was so blotted that it was illegible" (167). Signaling his novel's surprise recurrence to a found manuscript, Fielding echoes Pamela's inaugural inscription of the empirically nonsensical feint that defines epistolary form: "O how my Eyes run!—Don't wonder to see the Paper so blotted!" (11). But if *Pamela* asserts that her letter is unreadable in clearly intelligible printed words, Fielding's "blotted" "Part" honors its typographical declaration of illegibility. For Fielding, *Pamela*'s conceit forfeits the novel's honesty in the very medium of type. But *Jonathan Wild*'s printed signifiers stay true to their word. *Shamela*, *Joseph Andrews*, and *Jonathan Wild* mobilize the palpable experience of print as the truth of character's literary value. Virtue—and its opposite—become empirically manifest as typography whose sensible qualities engage readers in the corrective mechanism of Fielding's game: reading, and seeing, characters.

Status may reside in blood, but it is proven in Fielding's novels by print. To vindicate social value, Fielding deploys the very medium that historically elevated the persons whose pretensions he spoofs. With this gesture to the paradox sustained by the secondary qualities of typography in Fielding's novel, I turn to Richardson's *Clarissa* and the substance of sex.

CHAPTER 6

Fixing Sex

Richardson's *Clarissa*

And God made
The Firmament, Expanse of liquid, Pure,
Transparent, Elemental *Air.* MILTON.
—Stephen Hales, *A Description of Ventilators* (1743)

The Sex is turn'd all Whore.
—John Dryden, *The Satires of Decimus Junius Juvenalis* (1693)

In *Hints of Prefaces for Clarissa,* Samuel Richardson describes writing to the moment, the conceit that first-person epistles flow from the writer's pen in the midst of the events she relates: "Those Characters sink deeper into the Mind of the Reader, and stamp there a perfect Idea of the very Turn of Thought, by which the Originals were actuated."[1] Richardson's defense of the "perfect Idea" communicated by epistolary form encompasses rigorously Lockean concerns. Effusions continuous with the scenes they relay, letters written to the moment preclude both abstraction and dissimulating distance. They thus promise to transform another person's "Thought" into a sensible "Idea" for the person who peruses it: "Originals" in another mind become "Characters" that impress a "stamp" on "the Mind of the Reader." The provenance of Richardson's perfect idea—not of things but of a fictional letter-writer's thought—is Lockean. Through Locke, Richardson innovates a novel that makes thought belonging to one mind into sensory ideas perceived by another.

A professional printer, Richardson did not publish Boyle's work. Although it is therefore unlikely he read Boyle,[2] Richardson printed the *Philosophical Transactions* of the Royal Society from 1753 until his death in 1761 and, as William Sale suggests, his business relations with multiple Royal Society fellows position him as the Royal Society's "incumbent" printer before then.[3] Richardson's publication of the *Transactions*, contemporary texts on medicine and chemistry, and Swift's *Gulliver's Travels* assures his familiarity with corpuscular treatments of matter and its human apprehension. While his biographers T. C. Duncan Eaves and Ben D. Kimpel hazard that he "must have read a good many of the works that came from his press,"[4] Richardson's engagement with corpuscular philosophy is most resonantly evidenced, this chapter argues, by his novelistic engagement with Locke.

Richardson's second novel *Clarissa; or, the History of a Young Lady* (1747–48) also refers to Jonathan Swift. My chapter begins by exploring Richardson's wary but totalizing recourse to Swift's representation of human corporeality. *Hints of Prefaces for Clarissa* anticipates Swift's importance when Richardson justifies Clarissa's stubborn refusal to wed her rapist Lovelace: "The just Detestation that injured Lady had of Lovelace's vile Attempt to corrupt her Mind as well as Person, was surely a sufficient Argument against uniting her untainted Purity (surely we may say so, since the Violation reached not her Soul) in Marriage with so gross a Violator; and must for ever continue in Force, till the eternal Differences of Vice and Virtue shall coalesce, and make one putrid Mass, a Chaos in the Moral and Intellectual World."[5] In Richardson's most alarmist appraisal of the impropriety of Clarissa's marriage, her union with Lovelace would reverse biblical time. The world's declension into primordial "Chaos" catalyzes a metaphysical disaster, auguring consummation whose connotative resonance is less figural than substantive: the couple fuses into "one putrid Mass." This chapter's first two sections assess Clarissa's untainted purity—a refusal to marry that is at once a refusal to "coalesce"—in light of eighteenth-century respiratory theory. Trapped inside the bawd Sinclair's house, Clarissa shares air with women whose contrasting vileness Richardson amplifies by reference to Swift. But Clarissa's lack of taint is challenged by *Clarissa*'s reliance on a corpuscular account of breath that undermines her metaphysical difference.

Needless to say, Richardson's prediction of postnuptial chaos checks his faith in the power of marriage to convert illicit force into conjugal right. This withdrawal of confidence in the domestication of rakes justifies Richardson's damning reappraisal of his first novel *Pamela* (1740–41): writing of Pamela's putative reward, marriage to her master Mr. B, Richardson evokes her acquiescence to her husband's will as "implicit Obedience, and slavish Submission."[6] As we will see, slavishness represents the outcome Richardson abjures for Clarissa, an outcome that ruins the attribute left undamaged by her assault—"since," as Richardson stipulates, "the Violation reached not her Soul." Ruin not incarnated by sex affirms *Clarissa*'s overall refusal to locate Clarissa's femininity in an anatomical schematic of rape. Neither marriage nor "Lovelace's vile Attempt" operates in *Clarissa*, I argue in my chapter's last section, as a primary determinant of Clarissa's sex. *Clarissa*'s failure to isolate the source of Clarissa's sexed virtue reflects the novel's engagement with a metaphysics and an ontology engendered by corpuscles.

Air's Grossness: *Clarissa* and the Metaphysics of Purity

After Clarissa dies, Richardson's surrogate metaphysician Belford names her "the angel."[7] Become an angel, Clarissa finally sheds the corporeality whose tenacious appeal had made her both "a lovely skeleton" (1231) and "the lovely corpse" (1367). But at the same time that *Clarissa* plots the confluence of Clarissa's virtue and her achievement of the medium of spirit, the novel stages a parallel, reverse process: the disclosure of feminine corruption as the dirty physicality of the novel's prostitutes. Contained in one letter, the alignment of feminine viciousness and palpable revulsion is also enabled by Belford. Summoned Sunday morning as he prepares to forward his reformation by attending church, he visits the bedside of Clarissa's former tormenter, the brothel-keeper Sinclair, an experience that unleashes his concretizing depiction of the women who surround their ailing mistress:

> The other seven seemed to have been but just up, risen perhaps from their customers in the fore-house, and their nocturnal orgies, with faces, three or four of them, that had run, the paint lying in streaky seams not half blowzed off, discovering coarse wrinkled skins: the hair of some of them of divers colours; obliged to the blacklead comb where black was affected; the artificial jet, however, yielding apace to the natural brindle: that of others plaistered with oil and powder; the oil predominating: but every one's hanging about her ears and neck in broken curls, or ragged ends; and each at my entrance taken with one motion, stroking their matted locks with both hands under their coifs, mobs, or

> pinners, every one of which was awry. They were all slipshod; stockingless some; only under-petticoated all; their gowns, made to cover straddling hoops, hanging trollopy, and tangling about their heels; but hastily wrapped round them as soon as I came upstairs. And half of them (unpadded, shoulder-bent, pallid-lipped, feeble-jointed wretches) appearing from a blooming nineteen or twenty perhaps overnight, haggard well-worn strumpets of thirty-eight or forty. (1387–88)

The shock of Belford's discovery that women "paint" is belied by its long literary provenance, which extends from Ovid's *Metamorphoses* and Juvenal's Sixth Satire through Swift's poem "The Lady's Dressing Room" (1732).[8] What distinguishes the novelty of Richardson's version, along with Swift's, is its delineation of the formal parameters of the male viewer's acquisition of this knowledge. Indeed, the knowledge Belford gains is relayed as his experience *of* a form: he apprehends that "streaky seams," "artificial jet," "oil and powder," and deflated "gowns" lie on top of a body with which they are not coextensive. Broken into the deluded john's perception of "a blooming nineteen or twenty" and Belford's privileged intimation of underlying "haggard well-worn strumpets of thirty-eight or forty," the prostitute's articulation as primary-secondary difference is neatly replicated by Richardson's syntax. His parenthetical encasement of the whore's real person represents core materiality which few male experimentalists have the occasion to intuit: "(unpadded, shoulder-bent, pallid-lipped, feeble-jointed wretches)."

Composed of fucus laid over "well-worn" scaffolding, the cohort of whores is "each at my entrance taken with one motion." Their collective response to Belford's unprecedented access to their formal articulation is an attempt to dissimulate it, an effort whose futility Belford proceeds to underscore:

> I am the more particular in describing to thee the appearance these creatures made in my eyes when I came into the room, because I believe thou never sawest any of them, much less a group of them, thus unprepared for being seen[a] . . . If thou *hadst*, I believe thou wouldst hate a profligate woman as one of Swift's Yahoos, or Virgil's obscene Harpies squirting their ordure upon the Trojan trenchers; since the persons of such in their retirements are as filthy as their minds—Hate them as much as I do; and as much as I admire and next to adore a truly virtuous and elegant woman: for

> to me it is evident that as a neat and clean woman must be an angel of a creature, so a sluttish one is the impurest animal in nature. (1388)

In justifying his epistolary publication of "particular" details, Belford assumes the posture of the natural historian or travel writer who simulates "the appearance these creatures made in my eyes" to conjoin descriptive precision and his own sensation of the thing's empirical impact. The mandate for this double recapitulation is the fact that Lovelace, despite his formidable experience, has "never sawest" a prostitute "unprepared for being seen." As the observer whose eyes dictate the normative content of Lovelace's projected response, Belford deploys his disclosure of primary-secondary structure to leverage a more holistic indictment. His incursion into the prostitutes' room galvanizes a metaphysical collapse: "the persons of such in their retirements are as filthy as their minds." Shifting from a break rendered irreparable by the difference between twenty-year-old nymphs and forty-year-old strumpets—the break between ideas entertained by customers like Lovelace and the primary reality of trollopy wretches—Richardson asserts the equivocation of body and mind effected by nastiness that extends to both media. The wretchedness beneath a nymph's false bloom marks a temporarily formalizing expedient, whose clinching reference to the vileness of the prostitute's mind *Clarissa* cannot bring itself to plumb. It is not perceptually recessive particulate texture but mental filth to which the john's titillating ideas finally refer. As disclosed by Letter 499, both the whore's person and her mind should, Belford insists, stimulate disgust assured by their antipathy to Clarissa's angelic virtue.

Numerous commentators stress the metaphysics of feminine virtue polarized by Clarissa and the prostitutes, noting two seemingly discrepant outcomes: first, the bifurcation of the category woman; and second, the likeness augured by the near-anagrammatic identity of the typographical constituents of "Sinclair" and "Clarissa."[9] This paradox invites a return to the prospective site of Clarissa's virtue, the metaphysical ideality Richardson opposes to filth. "Ordure" signals *Clarissa*'s most redolently repellant correlate for the substance of feminine profligacy. But its capacity to encroach on Clarissa's purity is evidenced by Richardson's effort to regulate the instrumentality of Swift to *Clarissa*. This effort is anticipated by the most metatextual of all the novel's footnotes, appended to Belford's claim for the uniqueness of his vantage on whores "thus unprepared for being seen[a]": "*a* Whoever has seen Dean Swift's *Lady's Dressing Room* will think this description of Mr Belford not only more natural but more decent painting, as well as better justified by the design, and by the use that may be made of it" (1388). An indelible violation of the letter's pretense to epistolary immediacy, Richardson's footnote registers the literary debt encoded in Belford's surprise encounter as well as the difficulties of a source whose "design" and "use" *Clarissa* must respecify. But what makes Swift's poem an ambivalent resource for *Clarissa* is the same thing that invites Richardson to deploy it: Swift's insistence that matter never recedes into sensory ineffability. In *Gulliver's Travels*, Swift projects what can be called the reactionary counter-analogization of body, which spoofs the consistency of macroscopic and submicroscopic bodies so crucial to corpuscular chemistry and empiricism. Minuscule Gulliver does not perceive blown-up matter as particles stripped down to the primary qualities or, as Boyle calls them, mechanical affections—size, shape, motion, and cumulative texture—that produce sensory ideas in the minds of persons. Flouting Boyle's refusal to elaborate a category of primary ideas, as well as Locke's vision of the incomprehensible things microscopical eyes would reveal, Brobdingnagian bodies are simply grosser versions of what Gulliver perceives at normal scale. The likeness of body to body that guarantees the familiar physicality of primary matter operates, in Swift's satire, to amplify the sensory impact of infinitesimal things. Bodies just like bigger bodies, Swift's particles are directly—and disgustingly—experienced. For Swift, air is no transparent vehicle of untainted freshness, but the repository of the revolting odor that accompanies the visually revelatory climax of "The Lady's Dressing Room": "Oh! *Celia*, *Celia*, *Celia* shits!"[10]

Locke anticipates Swift's counter-analogy. Writing of "Duration and Expansion, considered together" (the second term is cognate with the Cartesian designator "extension"), Locke concedes their indefinite divisibility. But he dismisses the relevance of this concession to his operational unit of empirical understanding, the uncompounded sensory idea: "But the least Portions of either of them [duration or expansion], whereof we have clear and distinct *Ideas*, may perhaps be fittest to be considered by us, as the simple *Ideas* of that kind, out of which our complex modes of Space, Extension, and Duration, are made up."[11] Boyle disparages debates over infinite divisibility because fragments too tiny to react fall below the experimental threshold of chemical intelligibility. Locke likewise cuts the measure of empirical understanding to the standard of "clear and distinct *Ideas*" whose referent is not indefinitely minuscule parts but human perceptual capacity: "I know not whether I may be allowed to call [it] *a sensible Point*, meaning thereby the least Particle of Matter or Space we can discern, which is ordinarily about a Minute, and to the sharpest eyes seldom less than thirty Seconds of a Circle, whereof the Eye is the centre."[12]

In the case of Locke's sensible particle, "simple *Ideas* of that kind" do not claim the extradiscursive self-evidence of the color red or the taste sweet, because Locke specifies the lowest physiological and geometrical limit at which they provoke "clear and distinct *Ideas*" in the sensoria of persons.[13] The neologism "*a sensible Point*" constitutes a hybrid of ordinary perceptual experience and its physical referent: rather than an object that stimulates an idea, the circumference of Locke's point maps the smallest perceptible impression persons clearly register. Unlike Locke's own equivocal recurrence to imperceptible microstructure, the sensible point instantiates perceptual understanding ordained by the natural aptitude of the human eye.

Locke's sensible point is appropriated by the most vociferous eighteenth-century critic of primary-secondary difference, David Hume. Hume's *Treatise of Human Nature* (1739–40) offers an appraisal of "*the modern philosophy*" that pivots on rudiments of human understanding whose lower limit comprehends Locke's least particle. For Hume, extension is a complex idea which, broken down, consists of smaller units that must retain some *other* still perceptible quality: " 'tis impossible to conceive extension, but as compos'd of parts, endow'd with colour or solidity . . . These simple and indivisible parts . . . must be non-entities, unless conceiv'd as colour'd or solid."[14] Hume radicalizes Locke, whom he footnotes as the "great philosopher" credited with specifying empirical understanding's still-perceptible "bounds,"[15] to ascribe the perdurability of particulate matter to its existence as a perceived idea:

> That compound impression, which represents extension, consists of several lesser impressions, that are indivisible to the eye or feeling, and may be call'd impressions of atoms or corpuscles endow'd with colour and solidity. But this is not all. 'Tis not only requisite, that these atoms shou'd be colour'd or tangible, in order to discover themselves to our senses; 'tis also necessary we shou'd preserve the idea of their colour or tangibility in order to comprehend them by our imagination . . . Upon the removal of the ideas of these sensible qualities, they are utterly annihilated to the thought or imagination.[16]

Crucially for his refusal of what he calls the "hypothesis . . . of the double existence of perceptions and objects,"[17] Hume derives the signification of the words "atoms or corpuscles" not from a speculative standard of "indivisible" matter but from a phenomenological standard of "indivisible . . . feeling." Humean atoms and corpuscles are atoms and corpuscles *of feeling*, whose status as "impressions . . . reduc'd to a *minimum*"[18] constitutes the guarantor of their status as simultaneous "perceptions" and "objects," materiality merged with its apprehension by persons. Hume banishes the double existence of sensory ideas and perceptually elusive particles by claiming human perception as matter's simultaneously real and thinkable "*minimum*." By so doing, Hume argues, he conserves rather than erodes its physicality. While "a mathematical point is a non-entity," indivisible particles whose materiality is entrenched by color cannot occupy the same space at once: "A blue and a red point may surely lie contiguous without any penetration or annihilation. For if they cannot, what possibly can become of them? Whether will the red or the blue be annihilated?"[19] In this scenario, the perceptual ideas "red" and "blue" secure the durability of Humean matter. Red and blue entrench the indissolubility of minima which, unlike Descartes's "mathematical point," cannot be superimposed and reduced to mathematical unity. This expedient precludes the "double existence of perceptions and objects," a doubling which, for Hume, renders empirical philosophy liable to the charge of "the most extravagant skepticism."[20] By claiming the sensible quality color *as* the substance of matter's smallest unit, Hume invests particles with phenomenal vividness that prevents them from being either physically or conceptually "annihilated." The qualitative quiddity of red and blue secures the ineradicable difference of Hume's two points.[21]

As Hume's recourse to red and blue indicates, his defense of the perceptual constituents of corpuscular minima depends heavily on the capacities of the human eye: "Put a spot of ink upon paper, fix your eye upon that spot, and retire to such a distance, that at last you lose sight of it; 'tis plain, that the moment before it vanish'd the image or impression was perfectly indivisible."[22] Whether thirty minutes of a circle or the perfect unity of a vanishing dot of ink, sensory indivisibility recurs for both Locke and Hume to the powers of unassisted sight.[23] But for Swift, sensible points and particles of feeling recruit another sense. Central to his debunking rendition of modern experimental knowledge as well as his repudiation of the sham ideality to which ladies aspire, Swift's assertion of the inextinguishable sensory potency of particles attaches a different quality to matter. Sensible points Hume incarnates as minimal units of sight are, for Swift, equally imperishable vehicles of smell.

Most germane to *Clarissa*'s attempted use of Swift is the chymical genealogy found in *A Tale of A Tub* (1704)—a text to which *Clarissa* glancingly refers[24]—when Swift ridicules the "*light*" textual "productions"[25] of postclassical thinkers. Swift's central target is the medical reformer Paracelsus

(1493–1541), as evidenced in *Tub*'s mock recipe for the ongoing creation of modern knowledge:

> *You take fair correct copies . . . of all modern bodies of arts and sciences whatsoever, and in what language you please. These you distil* in balneo Mariæ, *infusing* quintessence of poppy Q.S., *together with three pints of* Lethe, *to be had from the apothecaries. You cleanse away carefully the* sordes *and* caput mortuum, *letting all that is volatile evaporate. You preserve only the first running, which is again to be distilled seventeen times, till what remains will amount to about two drams. This you keep in a glass vial,* hermetically *sealed, for one and twenty days. Then you begin your* Catholic *treatise, taking every morning fasting . . . three drops of this* elixir, *snuffing it strongly up your nose. It will dilate itself about the brain (where there is any) in fourteen minutes, and you immediately perceive in your head an infinite number of* abstracts, summaries, compendiums, extracts, collections, medullas, excerpta, quædams, florilegias, *and the like, all disposed into great order, and reducible upon paper.*[26]

As the key antecedent to Fielding's dismissal of vernacular claims to medical competence, Swift's fake alchemical receipt foregrounds its appeal to any operator who affects spagyrical knowledge. In this capacity, it implicates both Paracelsus's departure from Galenical medical precedent and the English-language publication of alchemical arcana first authored by the adept Starkey under the pseudonym Eireneus Philaletha Cosmopolita in *Secrets Reveal'd: or, An Open Entrance to the Shut-Palace of the King* (1669). Starkey's preface stresses the democratizing and, in his opinion, divine mandate for his disclosure of Paracelsian and Geberian practice: "*I profess there is none that ever writ in this Art so clearly; and that many a time in writing I laid aside my pen . . . but God compelled me to write . . . many will become Blessed in this last Age of the World with this* Arcanum, *by reason I have written faithfully; nor have I willingly left any thing doubtful for a young Beginner.*"[27] Swift's references to "*what language you please*" and "*your* Catholic *treatise*" indict the shoddiness of chymical knowledge as a function of the vernacular leveling that extends expertise to the "*young Beginner.*" Claimed by Starkey as a spur to the diffusion of alchemical knowledge, the unprecedented inclusivity of sense-based understanding canonized by Locke finds an antecedent whose leveling promise is also clear. It is not gratuitous snobbery that propels Swift's and Fielding's contempt for practitioners whose formation is restricted to English, but the promise at the heart of Starkey's defense of Englished arcana that will make "*many . . . Blessed*": "Gold and Silver would at last be of as mean in esteem as Dirt, which hath been hitherto the great Idol adored by the whole World; then we who know these things should not need so studiously to hide our selves."[28] Modern accomplishments whose perpetuation requires Swift's empty-headed operators to drink "poppy" and "Lethe" are welded to a chymical horizon that renders gold as common as "Dirt"—the latter a signifier whose excremental connotations were experimentally well tested.

As laboratory procedure, Englished alchemy prescribes changes of state from volatile to solid and back again mimicked by Swift's account of knowledge that passes from rising gas into voluminous excrescences of printed "collections" and "medullas" like Starkey's *The Marrow of Alchemy* (1654–55). Having distilled, cleansed, and redistilled "*all modern bodies of arts and sciences*," Swift's author-operator (like Starkey, who professes that "*I am one who have for many Years been one of* Gebers *Cooks*"[29]) performs the gesture that affirms the signal capacity of Swift's particulate matter to bear qualities: "*snuffing it strongly up your nose*." The once indispensable labor of learning abridged by breathable effluvia soon to be reembodied as yet more "abstracts, compendiums, extracts," modern arts and sciences claim their Swiftian correlate in the entity that also characterizes Swift's ladies: "*wind*."[30] As the bridge between would-be ideality—for Swift's *Tub*, the suspect ideality of empirically acquired ideas—and indissoluble embodiment whose signature qualitative taint is odor, Swift's suggestion that moderns "affirm the gift of BELCHING to be the noblest act of a rational creature" takes a grosser turn with *Tub*'s obsessive reprisal of the equivalence spelled out in Swift's footnoted reference "to that renowned *cabalist Bumbastus**": "*This is one of the names of Paracelsus; he was called Christophorus, Theophrastus, Paracelsus, Bumbastus."[31] Swift's cognomen for the name Walter Pagel identifies as "the best documented designation of Paracelsus,"[32] Bombastus of Hohenheim, "Bumbastus" locates the source of *Tub*'s surest de-idealization of spirit into wind. Indeed, in a metaphysical collision going back centuries, so-called spirit of urine was a vital chymical reagent (and source of the discovery of phosphorus) prepared by, among others, Boyle's long-suffering laboratory assistants. As Lawrence Principe reports, "there remained a focus on excrement—both urine and feces—which at least had the advantage of being starting materials easy and cheap enough to acquire in large quantities in the streets of early modern Europe."[33]

Swift's mock-experimental portrayals of scatological changes of state protract the experimentalist's exposure to sensory input well after his release from the visual burden imposed by the object. In a footnote, Swift credits his ultimate figure for modern knowledge production: "Paracelsus, who was so famous for chemistry, tried an experiment upon human excrement to make a perfume of it."[34] Alchemical change, to whose effacement of value Swift superadds the sublimation of dirt into "perfume," marks a trope for epistemological modernity further refined when *Tub*'s narrator proposes the occupants of "Bedlam" as candidates for "the several offices in a state":[35] "Accost the hole of another kennel, first stopping your nose, you will behold a surly, gloomy, nasty, slovenly mortal, raking in his own dung and dabbling in his urine. The best part of his diet is the reversion of his own ordure which, expiring into steams, whirls perpetually about and at last reinfunds. His complexion is of a dirty yellow with a thin scattered beard, exactly agreeable to that of his diet upon its first declination; like other insects, who having their birth and education in excrement, from thence borrow their colour and their smell."[36]

Although Richardson credits his rendition of *Clarissa*'s prostitutes to "The Lady's Dressing Room," the holistic sweep of Belford's repugnance finds a closer antecedent in the "slovenly mortal" entirely composed of "dung." Dung through and through, this person is not comprised of imperceptible texture that effects the narrator's sense of smell. As Swift's most explicit disavowal of Boyle's doctrine of qualities, this man's "colour" and "smell" are occasioned by nothing other than sensible points of dung. But the undifferentiated substance from which he is made does admit changes of state. If qualities like stench and "dirty yellow" emanate directly—not unlike Lucretius's visible films of objects—from his nasty body, this matter subsists as solid, "steams," and "reinfund[ed]" gas once more. A person whose empirical "education" funds his holistic filthiness, Swift's slovenly mortal enacts the equivocation of steam and solid practiced by Paracelsian cooks. The trope is recycled when Gulliver visits Laputa's Academy of Projectors and is "almost overcome with a horrible stink . . . [The projector's] Face and Beard were of a pale Yellow; his Hands and Clothes daubed over with Filth. When I was presented to him, he gave me a very close Embrace (a Compliment I could well have excused.) His Employment . . . was an Operation to reduce human Excrement to its original Food."[37] Swift's scatological changes of state do not deny modern thought the metaphysical status of the idea. Rather, Swift aligns an excremental spectrum of metaphysical difference with the solid and gaseous states of dung and wind.

The variant ideality of the latter may remain invisible to the eye, but it is horribly qualified by human feeling registered through the "nose."

The referent to which Richardson's claim for the filthiness of the prostitutes' bodies and minds most urgently gestures is Swift's holistically nasty man. Turning from the excremental constituents of Swiftian ideas to the threat of "a very close Embrace" with persons comprised of dirty matter, we reach Swift's representation of ladies who pretend to an ethereality stripped of the qualitative marker of stink. Yet the lady whose debris Strephon, the hapless narrator of "The Lady's Dressing Room," inspects does ultimately claim a resemblance to Swift's holistic shit-man. A sample of Strephon's discoveries defines Celia's dressing room as a repository of waste preemptively removed from her body:

> Hard by a filthy Bason stands,
> Fowl'd with the Scouring of her Hands;
> The Bason takes whatever comes
> The Scrapings of her Teeth and Gums,
> A nasty Compound of all Hues,
> For here she spits, and here she spues.
> But oh! it turn'd poor *Strephon*'s Bowels,
> When he beheld and smelt the Towels,
> Begumm'd, bematter'd, and beslim'd,
> With Dirt, and Sweat, and Ear-Wax grim'd. (ll. 37–46)

Although Swift also lists "Gallypots and Vials . . . fill'd . . . with Pomatum, Paints and Slops" (ll. 33–35), the overwhelming impetus of the poem is driven not by Celia's recourse to cosmetic overlay but by her evacuation of "nasty" excess. A person who "spits," "spues," "bematter[s]," "beslim[es]," and finally "shits," Celia emits filth which extends in the end to the "excremental Smell" (l. 111) whose gassiness portends its capacity to escape confinement by her dressing room. As the legacy of his foray into Celia's chamber pot, Strephon's still idealized visual apprehension of ladylike femininity is subject to uncontrollable qualitative taint by his more stubbornly retentive sense of smell: "His foul Imagination links / Each Dame he sees with all her Stinks" (ll. 121–22). In a momentous twist of invisibility's reflexive equation with imperceptibility tout court, Swift de-idealizes the medium of sight by recourse to the medium of smell. Celia is consigned to the metaphysics modeled by Swift's dung-man not as a brake on her claim to epistemological modernity but as the failure of her effort to rid herself of filth before she makes an appearance. Swift's refusal of etherealized femininity is rendered durable not by the visual impact of exposed waste but by the variant ideality of waste's stink.

Celia's attempt to evacuate, rather than her application of makeup, defines her aspiration to ladylike femininity. Swift intimates that her dressing room contains debris whose removal may define femininity as such:

> A Pair of Tweezers next he found
> To pluck her Brows in Arches round,
> Or Hairs that sink the Forehead low,
> Or on her Chin like Bristles grow. (ll. 55–58)

By revealing that Celia possesses "Bristles" on "her Chin," Swift does not simply intimate that the excrescences she strives to hide undermine a sanitized construction of femininity. As extrusions plucked from her chin, bristly "Hairs" must be extirpated to *make* femininity refined at this juncture in the poem as the deferral of a generically masculine—or, at least, hirsute—corporeal state. If Swift anticipates the impossibility of depilatory travails which aspire to realize femininity on these terms, then at the level of *form*, Celia's effort to be ladylike does not match the schematics of sex and gender Judith Butler motivates as "[t]he performance of drag": "If the anatomy of the performer is already distinct from . . . the gender of the performance, then the performance suggests a dissonance not only between sex and performance, but sex and gender."[38] In Butler's account, the difference between genital "sex" and behavioral "gender" is leveraged by cosmetics and clothing (gender) laid over anatomy (sex) that does not normatively solicit them. Performance straddles dissonant anatomical sex to disrupt the supposedly natural or inevitable consonance of sex with a cultural standard. But while "The Lady's Dressing Room" also divorces Celia's body—or, at least, her smell—from the idealized criterion she strives to instantiate, this divorce is not driven by femininity's split into anatomical and performative strata. If Swift can be said to depict femininity as gender—or, preparation that takes Celia "Five Hours" (l. 1)—this proceeds less from sartorial artifice than from scraping, plucking, sweating, spewing, and purging. Celia's *metaphysical* artifice does not demand supplemental layering, whether this outer layer ultimately shores up or denies the naturalness of an underlying sex.[39] Celia becomes a lady less by dressing than by defecating, less by applying paint to her surface than by leaving behind her a closet full of filth.

By translating femininity's idealizing pretenses into Celia's preparatory attempt to purge, "The Lady's Dressing Room" invites alignment with Michael Stolberg's important refutation of Thomas Laqueur. Whereas Laqueur argues for the radical consistency of male and female anatomy in Western theories of sexual difference up to the eighteenth century—according to Laqueur's "one-sex" model, women's genitalia are homologous, internalized versions of men's—Stolberg examines sixteenth-century anatomies of the human skeleton to show their investment in "explicit sexual dimorphism."[40] Such dimorphism entails, for example, women's "wider" "iliac bones," "more curved" "lower lumbar spine" and "more curved" "pubic bones and . . . coccyx."[41] The sixteenth-century entrenchment of dimorphism in skeletal anatomy reveals sexual difference that denies the mutability ascribed it by Laqueur. Incarnated by bones, sex persists in the morphology of Clarissa's remains, whose designation as "a lovely skeleton" accrues feminizing specificity. Evoked by Stolberg as "findings that modern anatomy does not accept,"[42] curved pelvic bones conform to an extrinsic norm whose ultimate discrepancy from the body cannot be catalyzed by clothes. That the skeleton obeys an anterior specification of womanhood is instead revealed, for Stolberg, by history (or "modern anatomy"). Strephon's discovery that femininity adheres to an external standard likewise does not recur to a harbinger of dissonance whose blueprint is drag. As an imposition of ideality that implicates the particulate constituents of the whole person, Celia's femininity solicits debunking that must reciprocally extend to particles. The critical force of Swift's undifferentiated dung-man becomes evident in this capacity: for Swift, the revelation of the impossible ideal to which Celia aspires coincides with Strephon's discovery that invisible particles of air will not support it. Femininity cannot be embodied by persons whose smallest parts are sensibly smelly points.

A poem that plots this failure as the climax of a wedding night, Swift's "Strephon and Chloe" (1734) stipulates the metaphysical ideal from which, after Celia, Chloe will also depart:

Her graceful Mein, her Shape, and Face,
Confest her of no mortal Race:
And then, so nice, and so genteel;
Such Cleanliness from Head to Heel:
No Humours gross, or frowzy Steams,

No noisom Whiffs, or sweaty Streams,
Before, behind, above, below,
Could from her taintless Body flow . . .
You'd swear, that so divine a Creature
Felt no Necessities of Nature.[43]

By specifying femininity as "Cleanliness" that leaves "no" sensibly apprehended trace "Before, behind, above, below," Swift sets the poem up for a dénouement realized, of course, by Strephon's nose. The de-idealization of formerly "taintless" Chloe occurs when, having drank "Twelve Cups of Tea" before the wedding night, she is driven to urinate in a "Vessel" while the couple lies in bed:

STREPHON who heard the fuming Rill
As from a mossy Cliff distill;
Cry'd out, ye Gods, what Sound is this?
Can *Chloe*, heav'nly *Chloe—?*
But, when he smelt a noysom Steam
Which oft attends that luke-warm Stream . . .
And, though contrived, we may suppose
To slip his Ears, yet struck his Nose:
He found her, while the Scent increas'd,
As *mortal* as himself at least. (ll. 175–86)

As companion satire to "The Lady's Dressing Room," the latter involves Strephon's discovery of "*Shit*" while the former relays his impression by "[*piss*]." These complementary epiphanies rely on the pervasiveness of "Scent" whose obtrusiveness discloses waste that might otherwise elude the eye and "Ear." A consummation of "Strephon and Chloe" forced by the enclosure of bad air in the couple's bedroom, the reclassification of Chloe as "*mortal*" transforms the newlyweds' imminent celebration of sexual difference into their enactment of contaminating sameness: while Strephon "Let fly a Rouzer in her Face" (l. 192), Chloe is no longer "asham'd . . . when a-bed, to let out Wind" (ll. 215–18).

Chloe's achievement of equivalence with her husband seals an erosion of metaphysical polarity powered by farting. The rendition of air as wind taints the medium of idealized femininity, a restoration of sensible qualities to body that threatens, for some readers, to assign grossness to one sex only.[44] But particulate things like "frowzy Steams" place Swift's ladies—and *Clarissa*'s prostitutes—not in a metaphysically retaliatory margin but squarely in the mainstream of an hydrostatic science that weighs air and quantifies breathing's grossness.

Amphibious Air: Attraction, Respiration, and Frowzy Corpuscles

Swift's dung-eating experimentalists represent modern knowledge—in both its solid and gaseous states—as waste. Yet Swift is indebted to the chymistry he spoofs, whether Paracelsus's spagyrical method or Boyle's air-pump. In another episode at Laputa's Academy, disgust may obscure the directness of Swift's reference to contemporary laboratory practice. Debilitated by "Colic," Gulliver visits a "great Physician . . . who was famous for curing that Disease by contrary Operations from the same Instrument": "He had a large pair of Bellows with a long slender Muzzle of Ivory. This he conveyed eight Inches up the Anus, and drawing in the Wind, he affirmed he could make the Guts as lank as a dried Bladder. But when the Disease was more stubborn and violent, he let in the Muzzle while the Bellows were full of Wind . . . I saw him try both Experiments upon a Dog . . . After the latter, the Animal was ready to burst, and made so violent a Discharge, as was very offensive to me and my Companions. The Dog died on the Spot."[45] Swift's satire touches on Boyle's vivisectionist trials of arterial, venous, and respiratory pressure; air-pump trials that gauge the inflation of gases as pressure descends in the receiver; and the fields of hæmastatics and hydrostatics advanced by two Newtonian iatromechanists, the anatomist and physician James Keill (1673–1719) and the curate, plant and animal physiologist, pneumatic chemist, and trustee of the Colony of Georgia Stephen Hales (1677–1761). Flatulence reinsinuated by means of "a large pair of bellows" mimics and inverts hydrostatical assays on the medium of breath pioneered in Keill's *Account of Animal Secretion, The Quantity of Blood In the Human Body, And Muscular Motion* (1708): "Now that I might know by what force the Air is thrust out of the Lungs in Expiration, I took a thin Hogs-bladder, which I could easily blow up with the Breath of one Expiration; and . . . fix'd a small Tube . . . to the Neck of the Bladder; then filling the Bladder with Air, I put a Weight . . . on the top of it."[46] Swift's wind reinforces the quantitative impact of once-insensible processes of bodily ingress and egress confirmed by Keill's "Hogs-bladder" experiment: "100 *lib*: which is the force by which the Air is thrust out of the Lungs every Expiration."[47] With Alexander Pope and Thomas Twining,

Gulliver's Travels indicts hydraulic trials of animal systems as "barbarities"[48] like those retailed by Keill's *Animal Secretion*: "I took a piece of the Intestine of a Dog . . . I blew it up, and hung it up to dry."[49]

To calculate the amount of liquid, Keill theorizes after Newton, that glands strain in the manner of particulate filters, Keill cites "a full bodied Jovial Taylor [who] lost Ten pounds of *Blood* by the *Hæmorrhoides*," "a Nun of a thin Habit of *Body*, who by bleeding at the Nose, spitting of *Blood*, and with Urine, voided eighteen pound of *Blood*," and "a Lady of a bleeding at the Nose; the *Blood* . . . was eighteen Pound."[50] Cases whose import anticipates Swift's leveling treatment of feminine corporeality, Keill's bleeding "Nun" and "Lady" void volumes in excess of what is shed by the tailor (even though, Keill repeats, "it is particularly said of the Nun . . . that she was a spare and thin Woman").[51] Needless to say, pounds of extruded blood admit the hydrostatical idealization of neither divinized beauties nor spare nuns. In *Vegetable Staticks* (1727), Hales credits Keill for quantifying the volume of far less empirically impressive fluids: "The quantity perspired by a man in twenty four hours is 31 ounces, as Dr. Keill found, *vide Medicina Statica Britannica* [1718]."[52] Bracketing respiration, Hales calculates the amount of sweat exuded by a representative man to be "1/50 part of a cubic inch perspired off a square inch in twenty four hours"[53]; he later reports that "nineteen Ounces and a half of Matter perspires from a Man here in *England* in twelve-Hours,"[54] while exhaled breath adds up to "1.39 Pound"[55] per day. Hydraulic extrapolations of expired sweat and breath whose totals can be estimated in pounds and ounces add up to something like Locke's sensible point or Hume's corpuscle of feeling in the register of experimentally palpable magnitudes of effluvia. As calculated by Keill and Hales, once imperceptible exudations of "Matter" are as ineffaceable as Swift's imperishably smelly wind. Proof of susceptibility to measurement that extends to the sweat of ladies, Keill's and Hales's numbers affirm the congruence of Swift's treatment of apparently rarefied bodies with the modern science he satirizes.

Relating *Clarissa* to this science entails attention to the qualities of exhaled corpuscles defined in Query 31 of Newton's *Opticks* (appended to Latin ed., 1706; appended to English ed., 1718). Citing "an attractive Power in Nature" distinct from "that other attractive Principle, which . . . was first discovered by the incomparable Sir *Isaac Newton*,"[56] Keill claims Newton's query as the source for his explication of an "Animal Oeconomy" driven by "Corpuscles of various Figures and Magnitudes, and endued with various Degrees of an attractive Power."[57] Keill and Hales theorize secretion and respiration to break from the rigorously empirical method of a practitioner like Sydenham, because they accommodate the "attractive Power" that augments the complement of Boyle's mechanical affections.[58] While Boyle does not explicitly include intracorpuscular stickiness among the primary endowments figure, magnitude, motion, and texture, for Newton, simple mechanical "Contact" is inadequate—while imaginatively figuring "hooked Atoms . . . is begging the Question"[59]—to justify the intensity of chemical reactions where "Parts in mixing run towards one another with an accelerated Motion, and clash with the greatest Force."[60] Newton thus posits "Agents in Nature able to make the Particles of Bodies stick together by very strong Attractions,"[61] a force that justifies motion's acceleration because it increases with proximity: "a very potent Principle, which acts upon them only when they approach one another."[62] Offering as an example the immersion of iron filings in "*Aqua fortis*, or Spirit of Vitriol" (nitric acid or sulfuric acid, respectively), Newton identifies the differential chemical affinities whereby "the acid Parts of the Liquor rush towards the Parts of the Metal with violence, and run forcibly into its Pores."[63] While aqua fortis dissolves iron and silver, it fails to work on gold, a discrepancy Newton ascribes not to acid parts whose figure is insufficiently "subtil" but to the "want of an attractive virtue"[64] specific to the latter reagents. Even without isolating differential affinities in the lab, Newton asserts "the assistance of something which causes them to be attracted or press'd towards one another" as the answer to the structural conundrum posed by any hard porous body whose constituent parts "can scarce touch one another in more than a few Points."[65] Newton's "something" makes a porous body like water capable of "freezing"[66] into a solid, answering Locke's complaint that this verb masks a blind spot mechanical causation cannot solve. "Something" also anticipates a physiological rationale for the incorporation of formerly discrete parts: "And therefore where the distance is exceeding small, the Attraction must be exceeding great."[67]

Keill's estimate of the force with which he inflates the hog's bladder operates to power a human physiology in which Newtonian attraction propels the red corpuscles of blood Keill observes "with a Microscope": "the Globules, attracting one another, unite like Spheres of Quicksilver."[68] Indrawn breath whose weight Keill calculates as "100 *lib*" serves not to imbibe any given substance but rather to impart opposing pressure: "The Particles of the Blood returning by the Veins mutually attract one another, and cohering form Globules too big for any Secretion; and therefore there was an absolute necessity, that they should be broken and divided in the Lungs by the force of Respiration."[69] Breathing impairs the union of attracting globules whose cohesion intensifies as they return to the heart through the narrow circumference of the capillary veins. For Keill, blood's attraction to itself warrants the quantitative obtrusiveness of indrawn breath: its weight is dictated not by anti-idealist metaphysics but by Newton's claim for the inverse relation of attraction and distance.

Hales's career as a curate and Royal Society fellow spans his insight into the respiratory cycle of plants and his design of ventilators for use in the holds of slave ships. He specifies the necessity of human breathing by citing the debt of subsequent "chymio-statical Experiments" to "the first great Author of this important discovery, Sir *Isaac Newton*." But in a departure from Keill, Hales stresses the corollary that as a core attribute of parts of air, attraction may become its opposite: "particles . . . are capable of being thrown off from dense bodies by heat or fermentation into a vigorously elastick and permanently repelling state: And also of returning by fermentation, and sometimes without it, into dense bodies: It is by this amphibious property of the air, that the . . . principle operations of Nature are carried on."[70] To elaborate the physiological and environmental significance of "amphibious" particles, Hales cites Newton, whose gloss of the orders of magnitude of difference between repellant and attractive forces cannot be justified by air's Boylean spring: heated or fermenting aerial parts "sometimes . . . take up above a million of times more space than they did before in the form of a dense body, which vast contraction and expansion seems unintelligible, by feigning the particles of Air to be springy and ramous, or rolled up like hoops, or by any other means than by a repulsive power."[71] Made of particles which may be either attracting or "a million of times" more repelling, air's quantitative difference from itself demands Newton's attribution of reversible power rather than Boyle's "feigning" of springy figure.

Elastic air fills "space" while attracting air enters "the form of a dense body." A crucial proof that air can be elastic or dense is its incorporation by the blood, demonstrated when blood emits bubbles as atmospheric pressure descends inside the air-pump.[72] Hales measures the hydrostatic force of the volumes of air released at low pressure by other things, like "pounded Apples," to argue that in the absence of its parts' attracting power, "so great an expansive force in an Apple would certainly rend the substance of it with a strong explosion."[73] But the embodiment of air with which *Clarissa* is concerned marks the outcome of an experiment Hales tries on himself:

> Having blown up the bladder, I put the small end of the fosset into my mouth; and at the same time pinched my nostrils close that no air might pass that way . . . In less than half a minute I found a considerable difficulty in breathing . . . and at the end of the minute, the suffocating uneasiness was so great that I was forced to take the bladder from my mouth. Towards the end of the minute the bladder was become so flaccid, that I could not blow it above half full with the greatest expiration that I could make: And at the same time I could plainly perceive that my lungs were much fallen . . . Whence it is plain that a considerable quantity of the elasticity of the air contained in my lungs, and in the bladder was destroyed.[74]

By exhaling breath that renders containing vessels "flaccid" and "fallen," Hales deduces a mandate for respiration which departs from Keill's: the lung's "vesicles would in a little time subside and fall flat, if they were not frequently replenished with fresh elastick air at every inspiration."[75] Or in his variant synopsis of this experimental proof: "the elasticity of the air is greatly destroyed by the *respiration of human lungs*."[76]

Hales's foray into self-asphyxiation may fall among the empirical practices Swift mocks with the patient who reimbibes his own gas. More broadly, Swift's scatological changes of state mimic the amphibious quality of "chymio-statical" air that fills space or enters into a dense body. Most topically, amphibious gas powers an imminently excremental account of digestion whereby, Hales explains, "the variety of mixtures in the stomach appear sometimes to generate, and sometimes to absorb air . . . In a true kindly digestion, the generating power exceeds the absorbing power but a little: But whenever the digestion deviates . . . to generate a greater portion of elastick air, then we are troubled more or less with distending *Flatus's*."[77] In a collapse the exactness of whose transcription by Swift need not be belabored, the hydrostatic etiology of unkind digestion mobilizes Newton's "elastick air" *as* wind. Indeed, Boyle gestures to a variant title of his own celebrated apparatus: "the *Wind-Pump* (as some body not improperly calls it)."[78] But it is not the explosive connotations of distending air that most decisively enter the referential frame of *Clarissa*. Rather, Richardson recurs to the tendency of expired particles to accrete, as Hales explains of confined environments whose insalubrity may be "attributed to the loss of a considerable part of the air's elasticity, and the grossness and density of the vapours, which the air is charged with; for mutually attracting particles . . . will readily coalesce into grosser combinations."[79] If

"*Flatus's*" propel the leveling consummation of Chloe's wedding night, particles that "coalesce" charge closed rooms with imminently gross corpuscles.

As Belford withdraws from the bedridden Sinclair's conference with her surgeons, *Clarissa* cites the "thick noxious air" in which, Hales states, "unelastick, sulphureous, saline, and other floating particles will most easily coalesce, whereby they are rendred too gross to enter the minute vesicles."[80] Of his experience in the sickroom, Belford narrates: "I retired to the further end of the room, and threw up a window for a little air, being half poisoned by the effluvia arising from so many contaminated carcasses; which gave me no imperfect idea of the stench of gaols, which corrupting the ambient air give what is called the prison distemper" (1390). When he cites "prison distemper," Belford reprises an experimental extrapolation reported in Hales's *Hæmastaticks; or, An Account of some Hydraulick and Hydrostatical Experiments made on the Blood and Blood-Vessels of Animals* (1733). After breathing for "two Minutes and half" into a receiver whose other opening is immersed in water, Hales measures the resulting elevation of the water line inside the chamber to observe: "I found that eighteen cubic Inches, or 1/29th Part of the whole Air, was in that time reduced from an elastic to a fixed State." He concludes: "We see in this Experiment that near two Gallons of Air . . . being breathed to and fro for two Minutes and half, became thereby unfit for Respiration: Whence no wonder that the Air should be infected, and apt to breed Distempers in close Prisons, where not only the Breath, but also the plentiful Perspiration of many confined together stench the Air, and make it apt to breed what are called Goal Distempers."[81] In a scientifically current assessment of the badness of Sinclair's air, Richardson refers to the grossness of exhaled breath's inelastic particles. By likening Belford's empirical "idea" of the brothel's chamber to "the stench of gaols, which corrupting the ambient air give what is called the prison distemper," *Clarissa* adopts the Newtonian causality that justifies the noxiousness of corpuscles expired by confined persons. Belford's experimental reenactment of choking uneasiness recruits not a metaphysics of flesh but air itself rendered "too gross."

The amphibious tendencies of air cannot, in the end, be confined to Sinclair's bedroom. As the most critical signal of the historical urgency of corpuscular stench, prison distemper spurs the freshening apparatus Hales promotes in *A Description of Ventilators: Whereby Great Quantities of Fresh Air May with Ease be conveyed into Mines, Goals, Hospitals, Work-Houses and Ships*

(1743). *Ventilators* underscores the overwhelming grossness of chambers invested with both breath and sweat:

> There is so great a Quantity of Vapours carried off by Respiration or Breathing . . . that more than a Pound-weight of Moisture goes off by the Breath in twenty-four Hours . . . [A] close confined Air, in which there are many Persons, is filled not only with the Vapours arising from their Breath, but also what perspires off their Bodies; which Respiration and Perspiration both together, are equal to the Quantity of half the Meat and Drink which we take in daily; which is estimated to be about thirty-nine Ounces in *England*, and is much greater in hot Climates. And if the Quantity of Vapours which arise from one Man in twenty-four Hours is thirty-nine Ounces, then in an Hundred Men it will amount to Two Hundred Forty-three Pounds.[82]

In an estimate that bypasses scatology, Hales posits a physiology that entails food's sublimation into effluvial suspensions of "Meat and Drink." The heft of airborne victuals is further stressed by the geometric increase of the number of contained men. A scene of confinement in which air is also "filled," Hales's projected containment of an hundred persons scales up Newtonian attraction to depict the crowding of life-size bodies. The hundredfold aggravation of inelasticity sustained by particles that add up to "Two Hundred Forty-three Pounds" solicits Hales's endorsement of his machine: " 'tis no wonder that when we breathe an Air, thus loaded with Vapours, it should be apt to cause, what are called Goal-Distempers; which Inconvenience might in a great measure be prevented, if such close Places were ventilated with fresh Air."[83] "Goal-Distempers" locate Richardson's prostitutes in a textual relay that ranges from the lady's dressing room to Sinclair's house to jails to ships. It is not the misogynist commitment of Richardson's metaphysics but, as we also saw in Defoe's *Journal*, the enclosure of exhaled breath in receivers that forces minuscule particles to accrete.

As he justifies the utility of ventilators to "close Places" like mines, jails, ships, hospitals, workhouses, and granaries, Hales deploys a keyword repeated by Swift: "frowzy." For example, ventilators are advised "in hot Climates, *viz.* because thereby the frowzy Vapours, which exhaled from those Persons, being carried off, a freer Perspiration is thereby promoted, which refreshes and exhilarates."[84] Contrary to Chloe's "frowzy Steams" and Celia's "Petty-coats in frowzy Heaps" (l. 48), Hales's frowziness qualifies vapors exuded by men. An attribute as indiscriminately sourced as Swift's stink, frowziness presumes the soon-to-be palpable release of effluvia from all human persons. Yet *Ventilators* does once project the anterior susceptibility of this quality to feminization. Writing of shipboard ventilators worked by sailors, Hales dismisses their trivial expenditure of labor: "Shall it be said of the brave and undaunted *British* Sailor; that rather than pull his Hand out of his Bosom, and work for half an Hour, once in ten Days, he will chuse to lie down and suffer that brave manly Spirit to be suffocated in a frowzy Stench, a Stench that has destroyed the Lives of Millions of the stoutest and bravest."[85] The effects wreaked by frowziness are already portended in the refusal of the "undaunted *British* Sailor" to operate the ventilator: he will not "pull his Hand out of his Bosom." His lassitude anticipates the influence of deflating vapors on any representative human person: he "suffer[s] that brave manly Spirit to be suffocated." Frowzy stench communicates the emasculating powers of bad breath. As Newtonian science's most decisive normalization of Swift's metaphysics, grossness overtakes men who "chuse to lie down and suffer" the containment of their own inelastic particles. The counter-analogization of body enacted when Gulliver confronts a gigantic breast is redundant in an atmosphere whose palpably unmanning frowziness is sustained by breath itself.

Given the destination of Hales's ventilator, his sailor demands a more precise referent than manliness soon to be ruined by frowzy stench. Hales projects the sailor's stint at the ventilator as "half an Hour, once in ten Days" based on the anterior estimate: "And suppose there be in a Transport, or *Guinea* Slave-Ship, two Hundred Men."[86] As cargo that eludes a tally restricted to British "Men," the ultimate historical referent for the asphyxiating proximity of incarcerated persons is not prisoners in a jail but Africans suffering the Middle Passage. Located at one end of the referential chain leading from Newton to the "*Guinea* Slave-Ship," *Clarissa* cites the noxiousness of air endured by bodies that sustain British history's most suffocating infliction of human entrapment. Hales does not evoke these persons' experience, although he once asks: "why do they suffer the Air in Transport-Ships, and especially in *Guinea* Slave-Ships, to be so intolerably nauseous?"[87] As we saw in Chapter 4, the refusal of corpuscles to entrench racializing difference does not operate to elevate the personhood of bodies who expire the same particulate matter. "Intolerably nauseous" vapors do not compel Hales to recognize the competing sufferance of air whose inelasticity recurs to a human physiology indifferently driven by particulate attraction. But if the chamber described in *Clarissa* claims an unbroken referential line from Hales's Guinea slave ship, then the prostitutes enclosed in Sinclair's room bear some resemblance to that adjectivally specified human cargo. To argue that Newtonian particles refuse to sustain Belford's distinction of "contaminated carcasses" from the angelic Clarissa is, therefore, to extend a likeness powered by attraction to all the persons implicated by Richardson's text, at whatever distance air may bind them. This likeness blurs the antagonism Ian Baucom stresses when he opposes novelistic idealism to Britain's intractably real past, because a metaphysics of breath that incorporates would-be angelhood as well as suffocating filth entangles both history and literary history in the intolerable grossness of trapped air.[88] *Clarissa* cannot distinguish angels from prostitutes and slaves because—in British science, empire, and the novel—the medium of ideality cannot be detached from its nauseating obverse.

As we have seen, Belford enjoins Lovelace to embrace the difference between the prostitutes' newly disclosed filth and the hygienic appeal of a woman like Clarissa: "If thou *hadst* [seen the unprepared whores], I believe thou wouldst hate a profligate woman as one of Swift's Yahoos, or Virgil's obscene Harpies squirting their ordure upon the Trojan trenchers; since the persons of such in their retirements are as filthy as their minds—Hate them as much as I do; and as much as I admire and next to adore a truly virtuous and elegant woman: for to me it is evident that as a neat and clean woman must be an angel of a creature, so a sluttish one is the impurest animal in nature." But because the discovery that a blooming nineteen recurs to holistic dirt proceeds from the Sunday-morning revelation of whores not ready to be seen, the opposition of "elegant" and "sluttish" women is destabilized by the scene Richardson stages. The contingency of sluttishness assigned prostitutes when they are "unprepared" is amplified by the criterion that qualifies Clarissa as an angel: she must reciprocally keep herself "neat and clean." Swift's lady poems anticipate the failure of Belford's synopsis to provide any better safety for the durability of feminine metaphysical virtue. As Richardson's footnote to "The Lady's Dressing Room" concedes, filthiness and cleanliness are contingently produced. Clarissa is an angel only as the amelioration of early-morning sluttishness. The metaphysics enacted by slipshod as opposed to neat women fails to secure an ontology of intrafeminine difference in *Clarissa*. This failure is perpetuated by the grossness of the air that even angelic Clarissa lets out.

In *Ventilators*, Hales suggests that good housekeeping assists his machine's removal of "grosser combinations" of inelastic particles: "But whatever Method is used to ventilate Ships, Goals, Hospitals or Work-Houses with fresh Air, in order to make it more effectual, it is absolutely necessary to use all Methods of Cleanliness by frequent washings, &c. And notwithstanding all these Means, there must needs be some degree of Frowziness, where many Persons are inclosed in a small compass."[89] In anticipation of Belford's exhortation to Lovelace, Hales underscores the flimsiness of the distinction between clean and dirty. Just as neat women exude the same particles as sluttish ones, so Hales's ventilator cannot—to repeat Hume's verb—annihilate inelastic air's grossness. Clarissa cannot be extricated from the indissoluble grossness of air where, as long as she breathes, there remains "some degree of Frowziness." The ventilator facilitates not total purity but less frowziness, not the elimination of taint but the freer circulation of more exuded matter. In a gesture that assimilates the moralizing valences of Belford's indictment of sluts, Hales "observed the Air to be very disagreeable in the Churches of some populous Parishes."[90] Sanctified and unhallowed spaces alike hold exhaled breath to threaten even consecrated receptacles of spirit—as Swift's *Tub* also insists—with encroaching stench.

One of *Clarissa*'s answers to the subspecification of more from less frowzy women is to designate its angel "*a Clarissa*" (1054), to hive her off from the sex, or the species, that mandates her emission of grossness. As *Clarissa*'s most infamous and deathly expedient, she internalizes the metaphysical polarity previously shared between herself and the whores. Richardson dramatizes Clarissa's assumption of this polarity when, succumbing to her lingering demise, she attends church: "the shortness of her breath, her extreme weakness, and the fervour of her devotions when at church, were contraries, which pulling different ways (the soul aspiring, the body sinking) tore her tender frame in pieces" (1308).[91] Clarissa introjects the antipathy of soul and her own body to herald the former's ascension: notably, Richardson lists "the shortness of her breath" among the indicators of corporeal frailty soon to be extinguished by her fervent devotional ambition.

But *Clarissa*'s final severance of breath from soul is facilitated by the grosser denial evident in her appearance as a lovely skeleton, a denomination she earns over a hundred pages before she dies. Belford relays the content of that denial to offer the novel's solitary physiological justification for Clarissa's "extreme weakness": "though emaciated . . . her features are so regular and exact, her proportion so fine, and her manner so inimitably graceful, that were she only skin and bone, she must be a beauty" (1351). As the novel's sole medical rationale for her sinking, "emaciated" Clarissa inhabits the skeletal state of "skin and bone." What vindicates angelhood is not the single moment when death severs soul from body but a much more protracted refusal to take matter in—whether that refusal is willed or whether, as the novel intimates, it proceeds from the peculiarly deviant agency of digestion that fails to achieve sustenance. Clarissa's inability to incorporate nutritive particles—a process of microrenewal enacted by eighteenth-century vegetables and animals alike—reactivates the referential pressure of the novel's footnote to Swift. Whereas Celia's pretense to divinized femininity fails because she excretes, Clarissa literalizes the mandate incumbent on any woman who would avoid the disclosure of her body's Swiftian metaphysics. As only skin and bone, Clarissa circumvents Celia's attempted removal of waste by exempting her person from its production. *She* may not do this, but emaciation signals less a radical foreclosure of will than physiological mechanism that Richardson finally exempts from the imperatives of corpuscular science.[92] In this extreme coalescence of air into body, Clarissa realizes the primary attribute of "proportion" by becoming skin and bone. As in Swift's poem, the medium of air makes living angelhood an untenable aspiration. The grossness of air's sensible points compels Clarissa to cede not food but breath itself to other women.

Aping Virtue: Feminine Desire and Lovelace's Natural History

Richardson dramatizes the antagonism of angels and sluts in an attempt to fix Clarissa's difference, an effort that marks a crisis in his specification of the qualities that delimit femininity as such. This is not *Clarissa*'s only instance of the same crisis. In registers other than metaphysics, the novel fails to stabilize femininity's core determinant: first, in the register of conjugal power or marriage; and second, in the register of anatomy or, more precisely, penetrability. These crises solicit a subversive reading of *Clarissa*'s adjudication of the ontology of Clarissa's sexed virtue: specified not as bone but, Belford writes, "manner," this virtue is a secondary effect that other persons might also possess the power to produce.

As the legal guarantor of what Carole Pateman designates masculine sex-right, or sexual access to a wife's person granted her husband by her consent to marry, the doctrine of coverture proceeds from the man's claim to naturalized or customary—Locke vacillates on this point—superiority over his spouse.[93] But in *Clarissa*, it is Lovelace's professions of inferiority to Clarissa which occupy the manifest content of the text. As Lovelace puts it, he is reluctant to marry Clarissa because she claims "So *visible* a superiority, to so proud a spirit as mine!" (658). Lovelace thus anticipates a demoralizing postmarital dilemma: "My wife (as I have so often said, because it so often recurs to my thoughts) to be so *much* my superior! . . . —Will not every act of her duty (as I cannot deserve it) be a condescension . . . ?" (734). For some stretches of Richardson's novel, Lovelace assumes Locke's perspective on marriage, which locates a husband's right to govern his wife in masculine superiority buttressed by nature, custom, and divine fiat: as a result, sex is sufficient to compel wifely duty.[94] But most of the time, Lovelace shares the perspective of the feminist critic of conjugal power Mary Astell, because when he makes the parenthetical aside "(as I cannot deserve it)," Lovelace concedes that his sex cannot justify Clarissa's obedience. Libertine, liar, and seducer he may be, but Lovelace refuses to fall back on Locke's rationale for masculine dominance: when it comes to the vindication of his right to command Clarissa, he admits that his manhood fails him. One need not be a psychoanalytic reader of *Clarissa* to remark the typographic and thematic stress Richardson lays on feminine superiority that is "So *visible*" and "so *much*"; when he relates how "sublimely great" Clarissa appears to him, Lovelace writes: "Her whole person was informed by her sentiments. She seemed to be taller than before" (853). The excessive visibility of Clarissa's superiority seems sensibly to attach to her—or, in Lovelace's variant experience of her greatness, to make him shrink—which is to say that by divesting Lovelace of the right to make her obey, Clarissa takes something from him. Locke, not Lacan, enables the suggestion that if conjugal authority is predicated on sex, and if Clarissa robs Lovelace of his claim upon it, then marrying her threatens to unman him. Indeed, Clarissa refuses to accept Lovelace's tardy offer and retroactively efface her own rape because—in a dizzyingly symptomatic mise en abyme—by failing to merit the power he would wield as a husband, Lovelace cannot reciprocally stimulate the eagerness to submit that Clarissa insists a virtuous wife must feel.[95]

In *Clarissa*, marriage neither fixes nor ratifies masculine superiority as the linchpin of sexual difference. Of the physical index of sex that is penetrability, *Clarissa* has little to say about it as a determinant of corporeal dominance or pleasure. Lovelace eschews the topic, declaring: "More truly delightful to me the seduction progress than the crowning act—for that's a vapour, a bubble!" (616). Notoriously, *Clarissa* will offer no further elaboration of this "act," figuring it instead over the novel's first third by the suitor Solmes's ravishment of an unsexed appendage, Clarissa's "hand" (319; 322).[96] But whereas "a vapour, a bubble" seem to characterize even Lovelace's person in terms of the ethereal density of the female hymen, *Clarissa* is saturated with representations of mind that lend it figural weight and shape. When Clarissa explains why she writes, the novel offers as the zero degree of epistolary subjectivity receptiveness that seems to map on to her sex: "You have often heard me own the advantages I have found from writing down everything of moment that befalls me . . . this helps to form one to a style, and opens and expands the ductile mind" (483). Clarissa's mind is ductile, expansive, and can be opened; it is thus tempting to superimpose her claim, after she is raped, that "My will too [is] unviolated" (1162) upon the crude schematics sketched by Lovelace when he refers to her before the rape as "corporally inviolate" (879). By citing her inviolate will, Clarissa seems to affirm the particularizing tendencies of *Clarissa*'s representation of mind even as she denies Lovelace's capacity to impinge upon her own. And yet the novel's claims for her capacity to be opened do not always line up with Clarissa's sex. As she wastes away, Lovelace sends Belford the following remedial fantasy: "thou, Belford, art in a plot with the dear creature . . . in order to work up my soul to the deepest remorse and penitence; and . . . , when she is convinced of the sincerity of both, and when my mind is made such wax as to be fit to take what impression she pleases to give it, she will then raise me up with the joyful tidings of her returning health and acceptance of me?" (1335). Here Lovelace fantasizes that Clarissa will penetrate him. This fantasy replicates the developmental narrative found in Locke's *Essay*, which offers "wax" as the metaphor for a baby's mind to stress both its lack of innate content and its susceptibility to external sensory influence. When Lovelace anticipates remorse that will render his mind amenable to Clarissa's "impressions," he imagines regressing to the state of infantile plasticity that marks empirical understanding's starting point. In this scenario, Clarissa would replace Lovelace's mother, whose inadequacy as a Lockean parent he often bemoans. For Richardson, the mutually figurative and epistemological workings of mind are delimited by Locke, and these cannot be distributed according to sex.[97] As Clarissa writes Anna Howe at a moment which promises to induce Lovelace's salutary receptivity: "(rejoice with me, my dear, since I seem devoted to him, that the man is not absolutely impenetrable!)" (599).

Marking the novel's effort to adjudicate between mental penetrability and corporeal entrance, *Clarissa* is obsessed with the epistemological supercompetence of women who have accumulated at least as many impressions as Lovelace himself. As he remarks, "A fallen woman, Jack, is a worse devil than even a profligate man" (535), and the taxonomic difficulties caused by women who, according to Lovelace's empiricist rubric, rank "worse" than rakes exacerbate *Clarissa*'s difficult specification of this sex's anatomical correlate. Most notably, *Clarissa* does nothing to disturb Lovelace's claim that the prostitutes on "Dover Street," and not Lovelace himself, enact Clarissa's violation. In a provocative response to the novel's refusal decisively to isolate anatomical liability, Judith Wilt suggests that "the women of the house [are] . . . the ones perhaps responsible for the deed itself." Yet Wilt's claim for "the secondary position of Lovelace in the real affair"[98] does not essay corporeal refinements left untested by her referential deference to the "real affair." Some interrogation of the primary constituents of sexual difference is requisite to justify Lovelace's distribution of blame: "All, all, but what is owing to her relations is the fault of that woman, and of her hellborn nymphs" (1430). To query the bodily articulation of the final event in a sequence initiated by "relations" is not to reinstate genital reality but rather to isolate the question *Clarissa* compels: How does the novel figure rape in the absence of ontological determinants for Clarissa's sex that rape itself fails to supply?[99]

Before the rape, Lovelace tropes on the amorous fiction of Eliza Haywood to dismiss the chance that Clarissa might, in response to seduction, mimic the legitimizing enactment of volition that renders sexual force redundant inside marriage: "As to the means, thou dost not imagine that I expect a *direct* consent—My main hope is but in a yielding reluctance, without which I will be sworn, whatever rapes have been attempted, none ever were committed one person to one person" (719). A variant of willingness that is not "*direct*" but nonetheless counts as "consent," Lovelace's hope for Clarissa's "yielding reluctance" refers to the gradually manifested sexual reciprocity consistently evinced by Haywood's female protagonists. As feminine responsiveness that defers masculine liability for rape, the literary-historical efficaciousness of yielding reluctance leaves little resistance left to specify, except when the inconceivable obduracy of the victim's refusal drives Lovelace to overpopulate the scene: "whatever rapes have been attempted, none ever were committed one person to one person." The ascription of rape to a superfluity of force illuminates what service the prostitutes provide, as Clarissa's sole first-person report of her assault testifies: "I was so senseless that I dare not aver that the horrid creatures of the house were personally aiding and abetting: but some visionary remembrances I have of female figures flitting, as I may say, before my sight" (1011). What defines this scene as a rape is not the prostitutes' sex but the fact that, with Lovelace, they add up to more than one. Insofar as they distinguish slowly evidenced submission from superhuman resistance subdued only by agential excess, the most salient attribute of these "female figures flitting" is not their femaleness but their number. On Lovelace's own terms, the designation of this scene as rape disqualifies his masculinity as its deciding criterion. While the anatomical referent of yielding and unilateral force may remain the same, it is the prostitutes who make a rape occur.

Before the episode that triggers Clarissa's visionary remembrances, Belford begs Lovelace to protect her from Sinclair: "let it not be one of thy devices . . . to give her, no not for one hour . . . into the power of that villainous woman, who has, if possible, less remorse than thyself; and whose *trade* it is to break the resisting spirit, and utterly to ruin the heart unpractised in evil.—Oh Lovelace, Lovelace, how many dreadful stories could this horrid woman tell the sex!" (714). Motivated by "*trade*," Sinclair divests Clarissa's rape of desire on both sides.[100] (In this light, episodes like the fire scene, which evoke Clarissa's pulchritude through Lovelace's eyes, ratify the mechanical irrepressibility of *his* desire.) Instead, the ultimate ruin Sinclair portends is perpetrated in the medium of "many dreadful stories": she marshals "power" whose vehicle is not sexed anatomy but empirical ideas.

With Sinclair, the prostitutes pose a threat whose difference from corporeal penetration is intimated by their wish to prosecute it after the rape, as Lovelace writes on the eve of his departure from London later the same month: "[the prostitutes] wish for nothing so ardently as that I will leave the perverse lady to their management while I am gone to Berkshire; undertaking absolutely for her humility and passiveness on my return; and continually boasting of the many perverse creatures whom they have obliged to draw in their traces" (940). Clarissa's rape goes unrepresented, but it is nonetheless visited, in some fashion, upon her body, for as Clarissa argues, it involves only her body: because she did not consciously yield, "My will is unviolated" (1254). But the prostitutes' proposed management claims Clarissa's will as its privileged object. Whatever may become of her person in the process, Clarissa's "humility and passiveness" mark the irremediable outcome of the whores' effort "to break the resisting spirit." The conversion of resistant volition into passivity, breaking escapes figural or causal closure, as Locke cautions in his treatise on liberal childhood pedagogy *Some Thoughts Concerning Education* (1693): "if the *mind* be curbed and *humbled* too much in children, if their *spirits* be abased and *broken* much by too strict a hand over them . . . *dejected minds*, timorous and tame, and *low spirits* are hardly ever to be raised."[101] Rather than a discrete application of violence, it is the ominous interminability of "too strict a hand over them" to which Locke attributes dejected minds. Such minds are henceforth unable to summon the child's liberalizing enactment of consent: "Such a sort of *slavish discipline* makes a *slavish temper*."[102]

Traced by the whores, the still "perverse" Clarissa would not become depraved. By suffering ineradicable mental traces—a process inflicted, this time, while her mind is conscious to take things in—she would manifest the incorrigible passivity of Locke's broken child. Because breaking, not penetration, would ruin Clarissa, the schematics of her projected humility do not anchor sexual difference in genital anatomy. Far from it: Clarissa shares her vulnerability to the prostitutes' management with the object of slavish pedagogy specified by Locke—usually, a young boy.[103] Recalling the physiological consistency mobilized by jail distemper, we can place Clarissa, with Locke's boy, in a referential circuit extending to other spirits that endure the breaking imposition of "passiveness."

Neither metaphysics, masculine superiority, nor anatomical penetrability supply the qualities that determine Clarissa's sex. As a novel about rape, and not in spite of it, *Clarissa* fails to leverage an ontological guarantor for Clarissa's femininity from any of these attributes.[104] However, Richardson grapples with another candidate for the endowment that might, to use Karen Barad's phrase, offer an "ontologically primitive"[105] constituent of feminine difference. Posited by what I will call Lovelace's natural history, this attribute is presumed by the hypothesis that defines his attempted seduction as a taxonomic enterprise: "yet is she not a *woman*? Cannot I find one but half-yielding moment, if she do not absolutely hate me?" (657). Lovelace's appeal to the infallibility of "half-yielding" may read as his adherence to the precedent established by novelists like Haywood. Proof that would denominate Clarissa a "*woman*," its inevitability is vindicated by accumulated assays whose truth is transmitted in the words of Sinclair's minions: "If I would treat her as flesh and blood, I should find her such" (535).

But the "flesh and blood" Lovelace hopes to find is a woman fails to act half-willingly. Awed by her virginal fear as he "ardently" clasps her knees, Lovelace speculates on the proceedings of Clarissa's childhood: "Dear creature!—Did she never romp! Did she never from girlhood to now, hoyden? The *innocent* kinds of freedom taken and allowed on these occasions would have familiarized her to greater. Sacrilege but to touch the hem of her garment!—Excess of delicacy!—Oh, the consecrated beauty!—How can she think to be a wife!" (646). The etymology of "hoyden," the variant into which the verb "romp" collapses, dates the advent of the former word's connotation of sexual difference only to the Restoration.[106] The most striking aspect of the performance on which Lovelace pins the feminizing attribute of half-yieldingness is thus its historically indiscriminate enactment by all children. Lovelace's natural history diverts the "*innocent* kinds of freedom" enjoyed by youth of any sex into the particularized trajectory observed by girls already "familiarized . . . to greater." By reclaiming unsexed play as feminized tolerance of the encroachment of men, Lovelace poses the standard of immobile girls as safety for women's resistance to liberties to which their own liberty has prepared them to succumb. Become readiness to be familiar, innocent freedom converts a still extrinsic stipulation of womanhood into the deeper predicate of femininity that is flesh and blood.

A responsiveness to seduction furthered by childhood games might fuel sensitivity to less venal motives for the prophylactic effort to idealize spoofed by Swift. Indeed, Clarissa's emaciation is legible as a denial—even a feminist denial—of the half-willingness that Lovelace's natural history knits into her very person.[107] But insofar as Lovelace invokes it as the aptitude that would make Clarissa "a wife," feminine sexual reciprocity is not gainsaid as a virtue by *Clarissa*. The novel projects the virtue of Clarissa's willingness in the negative: well before Lovelace wonders whether she has hoydened, Richardson gestures toward the variant of Lockean breaking portended by Clarissa's forced marriage to the morally and perceptually repulsive suitor Solmes. This entails conjugal subjection to an unwanted man, as Clarissa remarks: "What a dreadful thing must even the love of such a husband be!" (129). Her confidante Anna Howe imagines the compensatory fakery requisite, after marriage, to counteract Clarissa's "past declarations of aversion": "In short, my dear, you must then blandish him over with a confession that all your past behaviour was maidenly reserve only: and it will be *your* part to convince him of the truth of his impudent sarcasm, *that the coyest maids make the fondest wives*" (239). Whether propelled by wifely duty or "that *fear* he proposes to govern by" (239), Clarissa would be driven to palliate her brutal husband by simulating yearning for "the hideous fellow" (238). But because she is barred from hypocritical blandishment by the overwhelming force of her disgust, Clarissa illuminates the role wifely desire plays in the stipulation of sex-right granted by coverture. As a predilection that ratifies this right rather than rendering it dreadful, Clarissa's capacity for wifely fondness is implicated in the force of her aversion. The former as well as the latter sustain the virtue of her refusal: "to marry a man one *cannot endure* . . . is enough to make a creature who wishes to be a good wife, a bad or indifferent one" (255). The reservoir of impossibility that contravenes a premarital creature's "wishes" is signaled by italics which elude volitional control. Sustained by "a man one *cannot endure*," irredeemably miserable duty draws its resources from flesh and blood whose pleasures cannot be mustered by Clarissa's will.

In a letter to Anna, Richardson permits Clarissa to refine her justification of her negative: "the marriage-intimacies (permit me to say to you, my friend, what the purest, although with apprehension, must think of) so *very* intimate: myself who never looked upon any duty, much less a voluntarily-vowed one, with indifference; could it have been honest in me to have given my hand to an odious hand . . . ?" (507). Stipulated as a "voluntarily-vowed" will to submit to her husband's sex-right, Clarissa's conjugal duty must, like all her other duties, solicit not "indifference" but eagerness. Instead of the odium Solmes provokes, what makes this compliance "honest" is desire. By contrast with relations who would "force my hand into his and hold it there till the service be read" (322), thus rendering parental tyranny continuous with an inversely figured event of non-consensual entrance, Clarissa's willingly "given . . . hand" promises "marriage-intimacies" that cannot be passive.

But while Clarissa's fitness for "*very* intimate" service is relayed, for the most part, by the inverse fierceness of her disgust, Richardson's vindication of the source of that disgust is ambivalent.[108] This is because *Clarissa* proves unable to secure *another* intrafeminine difference: not between angelhood and grossness, but between desire that legitimizes coverture and desire that sates rakes. If the former would make Clarissa a good wife, the latter turns Lovelace's cast-off lovers into the prostitutes Sally Martin and Polly Horton. Clarissa's reaction to the odious Solmes may be routed into her "purest" imagining of conjugal duty's extravolitional supplement, but Richardson does not further discriminate between the female person that ratifies masculine sex-right and the female person that yields to illicit familiarity.

The problem of feminine desire in Richardson's novels can be broached, on his terms, as the problem of good-looking men. Richardson's early vision of this dilemma appears in *Pamela* during a dream scene that occurs after Pamela, trapped at Mr. B's Lincolnshire estate, meets his grotesque Swiss lackey Colbrand: "when I went to-bed, I could think of nothing but his hideous Person, and my Master's more hideous Actions; and thought them too well pair'd; and when I dropt asleep, I dream'd they were both coming to my Bedside, with the worst Designs."[109] As what Sigmund Freud calls a "composite figure,"[110] the "pair'd" duo of Colbrand and Mr. B enacts the work of condensation not to overdetermine the dream's expression of content—since it is B alone who wishes to molest Pamela—but rather to redress the perceptual charm of handsome rakes. Because her dream harnesses Colbrand's "hideous Person" to her master's "more hideous Actions," Pamela immediately perceives the repulsiveness of the intentions B harbors. The composite Colbrand-B renders inner viciousness as striking as outer good looks whose appeal the dream thereby concedes. The same concession is intimated when Pamela insists that invisible depravity undoes B's manifest attractions: "it is impossible I should love him; for his Vices all *ugly him over*, as I may say."[111] An explicitly formal solution, uglying over offers an antidote to rakishness not always reviled at the level of Pamela's eye.[112]

Clarissa evokes Lovelace: "I never saw a man, whose *person* I could like, before this man" (507). Writing to Lady Bradshaigh, Richardson affirms that Lovelace "had some good Qualities given him in Compliment to the Eye and Ear of Clarissa."[113] Empirically sensed ideas, the "good Qualities" that ratify Clarissa's discernment are identical to the good qualities that stimulate Sally and Polly. To bestow a "Compliment" that implicates Lovelace's sensible attractions, Richardson cannot discriminate between "the Eye and Ear of Clarissa" and the eyes and ears of other percipients who also like the persons of rakes. Indeed, *Clarissa* acknowledges that Clarissa's organs of sense, and the impressions they entail, may operate like anybody else's when the novel broaches the empirical effects of performance. This turn is also presaged by *Pamela*, which tracks a crucial facet of *Clarissa*'s plot in reverse. In an unprecedented show of worldliness, Pamela denounces her guardian Jewkes as "a vile *London* prostitute" because Jewkes "talks . . . Quite filthily!"[114] Much later, Pamela registers surprise after she rises to become Jewkes's mistress: "Her Talk and Actions are intirely different from what they us'd to be, quite circumspect and decent; and I should have thought her virtuous, and even pious, had I never known her in another Light."[115] Gauged by a behavioral standard, Jewkes's enactment of propriety obviates suspicion that she is anything other than "virtuous, and even pious." Without the hard-won experiential understanding bestowed by her plot, Pamela's impression of Jewkes's piety would preclude her knowledge that it is assumed. *Pamela* thus explicitly withdraws an internal warrant for the perceptual impact of virtue produced by this novel's double of Sinclair. Clarissa's duping by Sinclair is justified in reverse as the experimental outcome of never having "known her in another Light," a rationale Lovelace sanctions when he describes a prostitute he has employed to impersonate his aunt: "She has always been admired for a grandeur in her air that few women of quality can come up to: and never was supposed to be other than what she passed for" (875). The divorce of this apparition of "quality" from any underlying specification is sustained at the level of novelistic cross-reference when *Clarissa* echoes *Pamela*: of the women disguised as his aunt and cousin, Lovelace remarks, "Both are accustomed to ape quality" (875). In Richardson's first novel, Pamela herself is accused of "aping thy Betters."[116] Between them, Richardson's two novels confuse Pamela's laboriously internalized vindication of status and depraved women's equally efficacious acts of passing.

In a key aping scene, *Clarissa* flouts the effort to distinguish flesh and blood that funds intimate duty from flesh and blood that half-yields. As Clarissa wastes away, Sinclair tempts the disconsolate Lovelace with a "new face," an unbroken woman who, Sinclair reports, "will not be at the word of command yet—is but just in the trammels." Upon meeting this enticing object, however, Lovelace recognizes "that little toad, Sally," the fallen ex-mistress who still adores him:

> Oh my dear, dear Mr Lovelace? cried she, I am glad anything will bring you to me! And so the little beast threw herself about my neck, and there clung like a cat. Come, said she, what will you give me, and I'll be virtuous for a quarter of an hour and mimic your Clarissa to the life.
>
> I was *Belforded* all over. I could not bear such an insult upon the dear creature . . . and cursed her most devoutly for taking her name in her mouth in such a way. But the little devil was not to be balked; but fell a crying, sobbing, praying, begging, exclaiming, fainting, so that I never saw my lovely girl so well aped; and I was almost taken in; for I could have fancied I had her before me once more. (1217)

This passage stages a crisis in the negotiation of intrafeminine difference so acute that Lovelace is driven to denominate its agent not "her" but rather a "toad," "beast," "cat," and "devil." Yet his denial of species likeness climaxes in the surprise acknowledgment that Lovelace, by far the most knowing arbiter of these women's difference, succumbs to Sally's performance. Although Richardson may incriminate the rake's blunted faculties, this scene animates a process of denomination akin to the one Locke posits when he decries the empirical and not spiritual constituents of the complex idea man. Empirical understanding excludes such things as soul from the sensible qualities to which words refer, justifying the humanizing moniker "man" as a product of extrinsic approximation. In the event of Clarissa's aping, even though Lovelace *knows* Sally produces the impression he experiences, his privileged causal insight fails to mitigate the sensation she effects.

This outcome deeply subverts the ontology of Clarissa's virtue, because it is Sally who claims the power, as she puts it, to "be virtuous"—that is, Sally is the primary body whose power to produce ideas renders angelic femininity a secondary perceptual effect. But if, in this scenario, virtue finally recurs to Sally's power to make Lovelace perceive it, Richardson does not thereby cement a misogynistic appeal to the base corporeality, also credited to Swift, into which all women devolve. Like the failure of air to conserve angelhood, Sally enacts feminine virtue innately resident in neither breath nor any other kind of matter antecedent to Lovelace's experience. Rather than a metaphysical essence or substance, Clarissa's virtue emanates from the same body as does Sally's simulation of it, whether the inelastic particle or the hoydening child that refuses to accede in advance to its marital or extramarital destiny.

To pursue the claim that feminine virtue in *Clarissa* is a disposition or power which does not preemptively distinguish body from body, I turn to Richardson's most expansive gloss of the economy that sustains Clarissa's pretense to angelhood. Begging Lovelace to spare her, Belford advances a famously Platonic defense: "I am ready to regret that such an angel of a lady should even marry. She is, in my eye, all mind" (555). But Belford's plea does more than exempt Clarissa from the labor of childbearing and "the vulgar offices of domestic life" (555), because he proceeds to suggest that as a result of Clarissa's ethereality, Lovelace would not enjoy her: "Thinkest thou, truly admirable as this lady is, that the end thou proposest to thyself, if obtained, is answerable to the means . . . ? In every real excellence she surpasses all her sex. But in the article thou seekest to subdue her for, a mere sensualist of her sex, a Partington, a Horton, a Martin, would make a sensualist a thousand times happier than she ever will or can" (555–56). "A Partington, a Horton, a Martin" refer to Sinclair's whores. In an extension of metaphysics which opens Swift's dressing room to expose the work of other, less ladylike women, Belford suggests that the novel's prostitutes act as accessories to Clarissa. For her to remain "an angel of a lady," fallen females must supply the flesh that will satisfy Lovelace "in the article." *Clarissa*'s attribution of ethereal excellence to Clarissa is sustained by an only contingent division of metaphysical labor whereby some women can be angels because others appease the needs of rakes. Anna's exasperated appraisal of Lovelace's appeal—"Well may our sex be the sport and ridicule of such libertines! Unthinking eye-governed creatures!" (1137)—enables an outcome useful to *Clarissa*, because "eye-governed creatures" who succumb to libertines, like Polly Horton and Sally Martin, stock London's brothels with the sensualists who sustain angelic ladies' status as all mind. Clarissa's eyes, as we have seen, admit the same ideas as do Polly's and Sally's. With this indication of the historically pervious border between ruined female sensualists and Platonic female minds, *Clarissa* hints at patriarchy's connivance in the metaphysical collapse effected at the level of the corpuscle by gross air.

Richardson's engagement with corpuscular science is marked at the sites of both respiratory hydrostatics and feminine desire. While *Clarissa*'s plot hints at the arbitrariness of Clarissa's structural distinction from the eye-governed Sally, *Clarissa*'s metaphysics affirm that prospective indifference as the gross particles every body, even a lovely skeleton, exhales. The anti-essential person Richardson inherits from Locke, with the inelastic corpuscles Richardson adopts from Newton and Hales, dictate the impossibility of fixing Clarissa's saving virtues in matter, even matter ethereal as air.

Epilogue

Denominating Oxygen

Since then air is found so manifestly to abound in almost all natural bodies . . . may we not with good reason adopt this now fixt, now volatile *Proteus* among the chymical principles . . . notwithstanding it has hitherto been overlooked and rejected by Chymists, as no way intitled to that denomination?

If those who unhappily spent their time and substance in search after an imaginary production, that was to reduce all things to gold, had, instead of that fruitless pursuit, bestowed their labour in searching after this much neglected volatile *Hermes*, who has so often escaped thro' their burst receivers, in the disguise of a subtile spirit, a mere flatulent explosive matter; they would then instead of reaping vanity, have found their researches rewarded with very considerable and useful discoveries.

—Stephen Hales, *Vegetable Staticks* (1727)

In the meetings we held we all tried to act in the same spirit. We forgot what had been done, as well as what each of us had accomplished, so we could concentrate on what needed to be done. Only after reviewing all aspects of chemistry several times and after reflecting deeply on the metaphysics of language and on the relationship between ideas and words did we dare form a plan.

—Louis-Bernard Guyton de Morveau, Antoine-Laurent Lavoisier, Claude Louis Berthollet, and Antoine François de Fourcroy, *Méthode de Nomenclature Chimique* (1787)

Two crucially modernizing moments, one in science and one in literature, mark the endpoint of this book. In the history of science, the chemist Antoine-Laurent Lavoisier departs from a corpuscular regime mobilized by anti-elemental texture. Most famously, Lavoisier identifies oxygen as an irreducible element, breaking with the non-essentialist deployment of corpuscles advanced by Boyle. In the history of literature, the novelist Jane Austen perfects a narrative technique known as free indirect discourse, which occupies the grammatical third person but nonetheless privileges the point of view of an individual character. Claiming Austen's accomplishment as a consummate stage in the development of the novel, Ian Watt argues: "She was able to combine into a harmonious unity the advantages both of realism of presentation and realism of assessment, of the internal and of the external approaches to character."[1] By reconciling "internal and . . . external approaches," Austen's novels, according to Watt, sit at the horizon of modern realism. This epilogue rethinks the place of Lavoisier's oxygen and Austen's novel—in particular, *Pride and Prejudice* (first written 1796–97, published 1813)—in the history of science and literature.[2] While Lavoisier denies any continuity with the chymistry he repudiates, the name he gives to the element oxygen betrays the debt of the new discipline of chemistry to its corpuscular predecessor. In Austen's case, *Pride and Prejudice* shares ontological and epistemological preoccupations with eighteenth-century novels in which, I have argued, qualitative knowledge is *produced*. No less than Defoe, Fielding, Richardson, Chetwood, Aubin, or Haywood, Austen refers to empirical knowledge that implicates perceptible ideas and insensible powers. Epitomes of realism in science and literature, Lavoisier's elemental system of nomenclature and Austen's marriage plot refer at once to experimentally felt qualities and productive dispositions that cannot be directly sensed. In modern chemistry and modern novels, empirical reality continues to implicate the figures and relations that effect it.

I conclude this book with Lavoisier, the "father of modern chemistry,"[3] not because Lavoisier discovered oxygen—that honor goes to the English experimenter Joseph Priestley—but because Lavoisier named it. First, I evoke Lavoisier's debt to Stephen Hales, whose *Vegetable Staticks* (1727), translated in 1735 by Georges-Louis Leclerc, Comte de Buffon, was "the predominant influence in interesting Lavoisier in the chemical role of air."[4] Hales's impact may be encapsulated by the "pneumatic collecting apparatus"[5] represented in *Vegetable Staticks* (Fig. 12). Peering in at the upper border of the page, Hales's self-asphyxiating avatar exhales used or inelastic air into a bladder. With the invention of this so-called pneumatic trough, which measures the gas released as

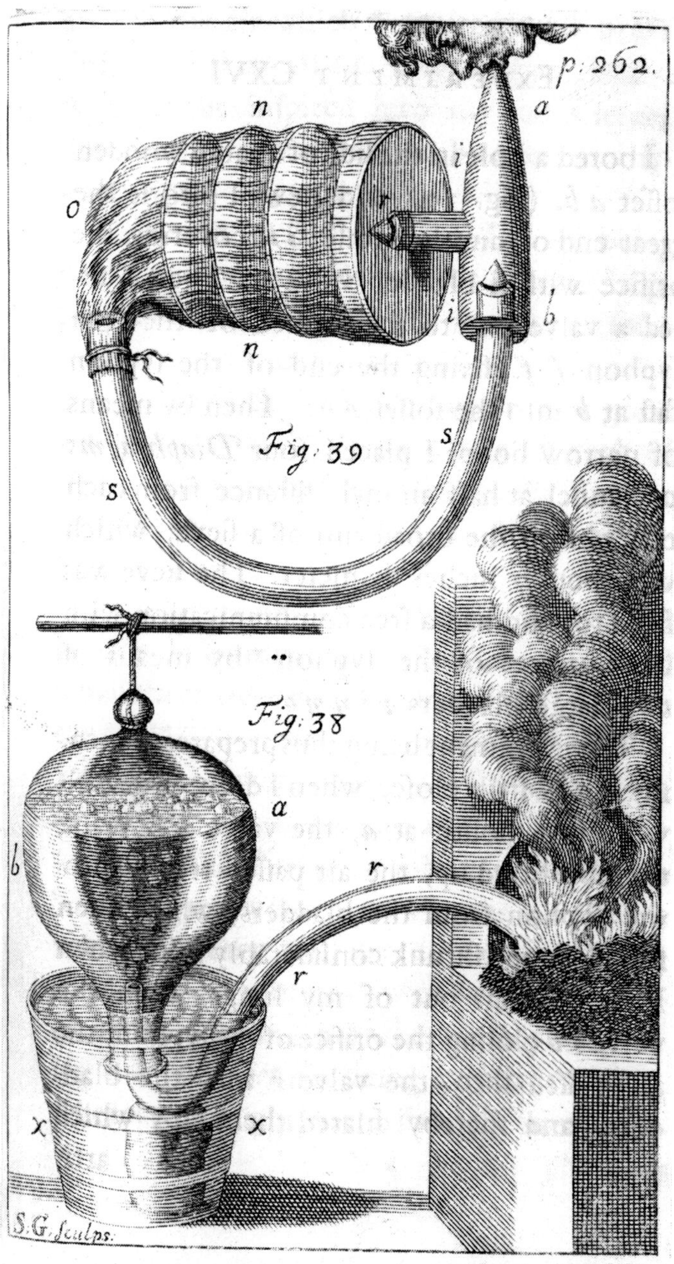

Figure 12. "Bladders" and "inverted chymical receiver," Stephen Hales, *Vegetable Staticks* (1727). Courtesy of the Lilly Library, Indiana University, Bloomington.

objects combust, Hales traps elastic air formerly fixed inside a multitude of bodies: "Horn, *calculus humanus*, Oystershell, Oak, Mustard seed, Indian-wheat, Pease . . . Sugar, Amber, Coal, Earth, *Walton* Mineral, sea Salt, Salt-petre, Tartar, *Sal Tartar*, Lead, Minium."[6] Hales's hydrostatic experiments on things like minium (red lead) are succeeded by Priestley's demonstration that a candle burns brightly in this combustible's unfixed air. But in one of the history of science's most famous misnomers, Priestley dubs the stuff "dephlogisticated air,"[7] reflecting the popular conviction that combustibles contain phlogiston, a fiery substance emitted as they burn. Because it is not overcharged by phlogiston—which enters an enclosed atmosphere to stifle fire or breath—dephlogisticated air, according to believers, best sustains the candle's flame. It is Lavoisier who places combustion's source not in burning bodies but in the air they consume. He thus explains why metals that are heated or calcined in an open vessel do not lose weight, as phlogiston theory would dictate, but gain it.[8] Lavoisier's revision compels him to encode the agency of the simple substance to which he refers as "eminently respirable air" ("air éminemment respirable")[9] in its new name.

Echoing Hales's disdain for alchemists who treat air as "mere flatulent explosive matter," Lavoisier testifies before the French Royal Academy of Sciences that he and his colleagues have attempted to empty their minds. Drawing upon the Lockean empiricist Étienne Bonnot de Condillac, Lavoisier avows his prefatory willingness to "forget what has been done" as his *linguistic* commitment to a chemistry from which redundancies like phlogiston have been purged. A revolutionized chemical language demands the one-to-one correspondence of word and idea ("rapport des idées avec les mots"), as Lavoisier elaborates: "The perfection of chemical nomenclature . . . consists in rendering ideas and facts in their exact truth, without suppressing and above all without adding anything to what they present: nomenclature must be nothing but a faithful mirror."[10] By declaring a mimetic standard for chemical names, Lavoisier underscores an ideal of referential fidelity whose guarantor is the chemist's assumption of Lockean childhood. In this crucial capacity, Lavoisier relies heavily on Condillac, whose sole critique of Locke—albeit a critique which preoccupies much of Condillac's philosophical career—is conveyed in his claim that Locke "passes too lightly over the origins of our knowledge."[11] Lavoisier affirms that while experimenters should acquire knowledge in the same way as empirical babies, they are instead steeped in chymical prejudice. What precludes the chemist's regression to a state of epistemological innocence modeled, Lavoisier writes, by a "child" ("l'enfant"[12]) is alchemical terminology that either secretively masks meaning or occludes it beneath deluding signifiers. Lavoisier stresses the absurdity of words that fail to reflect laboratory experience: "The names of *oil of tartar per deliquium*, *oil of vitriol*, *butter of arsenic and of antimony*, *flowers of zinc*, &c. were still more improper, because they suggested false ideas: For, in the whole mineral kingdom, and particularly in the metallic class, there exists no such thing as butters, oils, or flowers."[13] Just as Locke condemns the indelible associative superstition infused into children by their irrational nurses, Lavoisier bemoans "names" which, like "*oil of tartar per deliquium*," cement chemical unreason in the minds of scientists.

For Lavoisier, alchemy perpetrates the empirical outrage of inexistent referents like butters of metals. This science's bankruptcy is evidenced by words that recur to "chimerical or absurd ideas."[14] With these examples, Lavoisier underscores the figurative excess of butters, oils, or flowers whose pictorial vividness strays from perceptual reality. I turn now to his defense of the new name he coins for eminently breathable or, as it was also called, vital air. As what Lavoisier and his team classify a simple or experimentally indivisible substance, eminently breathable air solicits a denomination that accords with the rule: "With the new names, we have tried to express the most general or most characteristic property of the body which they designate."[15] A properly empirical stimulant invested with sensory memory and meaning (the double significance of the French *sens*), "oxygen" represents a recombination of Greek roots that evoke the following chemical attribute:

> When the name dephlogisticated air was changed to that of vital air, without doubt this choice conformed better to the rules by substituting for an expression based on rank hypothesis an expression derived from one of the most striking properties of this substance . . . but now it has been clearly shown that this body does not always exist in a gaseous or aerial state . . . we must consider this substance and designate it in its most simple state; furthermore, the logic of nomenclature demands that it be named first, so the word that recalls its idea becomes the standard for the names of its compounds; we have satisfied these conditions by adopting the expression "*oxygen*," in drawing, as Mr. Lavoisier has long proposed, from the Greek οξυς "*acid*" & γείνομαι "*I make*," because of the constant property of this principle, the base of vital air, of transforming into an acidic state a large number of substances with which it combines,

> or indeed because it seems to be a principle necessary to acidity. We will therefore say that vital air is the gas oxygen.[16]

Like oxygen, the new moniker for another air, so-called inflammable air—hydrogen—refers to what this simple substance reacts to make. (The British scientist Henry Cavendish first combined inflammable and vital airs to produce water.) The stem *hydro* thus denotes the properly empirical idea of water. In the case of oxygen, *oxy* cements Lavoisier's claim that vital air is, in Henry Guerlac's words, an "acidifying principle"[17] which combines with combusting substances to lend them that attribute. Guerlac notes that despite its claim to radical novelty, Lavoisier's nomenclature presumes an equivalence of qualitative endowments and denominational specificity that dates to earlier chemical practice: "In the older chemistry, from which Lavoisier is unable to free himself completely, the elements . . . were thought of as the bearers and the causes of the distinctive qualities of bodies into which they enter."[18] The name "oxygen" affirms qualitative understanding as the guiding principle of Lavoisier's new system: oxygen's most experimentally manifest tendency—its capacity to acidify burning things like "phosphorus," "sulphur," or "charcoal"[19]—vindicates the Greek neologism. But Guerlac's observation points to a more resonant link between Lavoisier's nomenclature and an older chymistry, the problem of empirical words that refer to insensible texture. For while both water and acids are humanly perceptible, the connotative resonance of the prefix *oxy* extends the descriptive pretenses of Lavoisier's name into the realm of particulate features that cannot be sensed.

The Greek prefix οξυς returns us to Chapter 1 and the wounded guts of Robert Boyle's nuns. Their innards are pierced by fragments of glass, we recall, that model the mechanical affections of insensible corpuscles. Blown up to amplify Boyle's claim that qualities proceed from relations between bodies rather than existing as such in any given thing, ground glass activates poisonousness as an encounter between straight-edged shards and the tenderer texture of human flesh. The Greek οξυς claims a range of meanings that communicate just as vivid a delineation of vital air's figure. As Henry George Liddell and Robert Scott's *Greek-English Lexicon* relays, these meanings include: "sharp, keen, whether of a point or an edge . . . of weapons or anything made of metal"; "sharpened" or "brought to a point"; "in reference to the senses . . . sharp, keen," "piercing," "severe"; "of sound, shrill, piercing"; and "of taste, sharp, pungent, acid."[20] In answer to Lavoisier's prefatory mandate, οξυς summons the sensible quality of taste to supply a sensory referent for oxygen's capacity to acidify the reactants with which it mixes. But the latter property cannot be directly apprehended, a fact aggravated by the meanings of οξυς which seem to limn the physicality of acidifying air: sharp, keen, brought to a point. Jean-Pierre Poirier confirms that Lavoisier is reticent on the topic of corpuscles: "Chemistry was now to be regarded as the science of chemical changes rather than the investigation of the composition of matter."[21] But if Lavoisier's nomenclature does not designate units of matter below the simple substances that resist analysis in the lab, the connotations of οξυς ineluctably posit particles. More precisely, οξυς refers to the particulate attribute of figure. The denomination oxygen does not mirror empirically derived ideas: it figures a particulate attribute that can never be perceived. Oxygen refers to both the relational power of vital air and its corpuscular antecedent.

Lavoisier's reliance on Condillac amplifies this paradox, because not even the infantile chemists promised by *Méthode*'s preface can perceive oxygen's sharp points. Boyle's doctrine of qualities persists in the name of Lavoisier's inaugural element, because the denomination "oxygen" summons powers evinced in relation to phosphorus, sulfur, charcoal, or human lungs, powers motivated by a moniker that conjures the shape of insensible bodies. Lavoisier scorns the figurative extravagance of chymical language, but he does not expunge imperceptible figure from the simple substance that solicits an empirical name. Oxygen embeds the figural elaboration of insensible matter's power at the heart of modern chemistry.

How might I frame the closing connection between Lavoisier's *Méthode* and Jane Austen's *Pride and Prejudice*? Both texts and both projects—chemistry's modern nomenclature and Austen's modern marriage plot—define qualities as empirical effects produced in relation. Turning to one of Austen's most well-documented debts to eighteenth-century literary history, her admiration for the novels of Samuel Richardson, we can affirm her interest in Clarissa's susceptibility to the sensory appeal of a rake.[22] *Pride and Prejudice*'s protagonist Elizabeth Bennet recapitulates this problem when she is repulsed by Fitzwilliam Darcy and charmed by George Wickham. Her reaction to the first man is evoked by Slavoj Žižek, whose dialectical gloss of Austen's plot suggests that Elizabeth and her future husband Darcy take the duration of a novel to change as their perceptions of the other change: "the failure of their first encounter, the double misrecognition concerning the real nature of the other, functions as a positive condition of the final outcome . . . In other words, Darcy's pride is not a simple, positive state of things existing independently of his relationship with Elizabeth, an immediate property of his nature; it appears, *only from the perspective of her prejudice*; vice versa, Elizabeth is a pretentious empty-minded girl *only in Darcy's arrogant view*."[23] Žižek cogently observes that qualities like Darcy's hauteur do not "exist[] independently" of Elizabeth's experience of them. I will return to his important point that the relational status of attributes like snobbery provides the "positive condition" of Austen's marriage plot. But first we can note that Žižek's Hegelian synopsis does not mention the obverse of Elizabeth's repugnance for Darcy, her attraction to Wickham. Austen underscores the strength, and even the venality, of this attraction during a long conversation undertaken while the Bennet sisters dine at their aunt's. Echoing Anna Howe's indictment of eye-governed creatures, Wickham is "the happy man towards whom almost every female eye was turned" (110) as soon as he enters the dining room. Elizabeth does not display unique perspicuity, but instead confirms Wickham's generic allure, when she finds herself—depicted in Austen's free indirect discourse, which tinges the appraisal with some discordant irony—"the happy woman by whom he finally seated himself; and the agreeable manner in which he immediately fell into conversation . . . made her feel that the commonest, dullest, most threadbare topic might be rendered interesting by the skill of the speaker" (110). Austen stresses, even overstresses, the appeal of charms whose superficiality is enforced by the fact that they work on Elizabeth "immediately." The homophony of Elizabeth's sobriquet Lizzy and her coquettish sister's name Lydia gathers hermeneutic force in this scene: Lydia *is* the person who will be ruined (at least, for a time) by Wickham's appeal to both her and her sister's "eye."

As Wickham slanders Darcy while avowing his reluctance to expose him, Elizabeth "honoured him for such feelings, and thought him handsomer than ever as he expressed them" (113). This occlusion of Lizzy's judgment condenses Austen's genealogical ties to both Richardson and Fielding: his display of hypocrisy renders Wickham "handsomer" in Elizabeth's view.[24] Austen collapses Wickham's heightened appeal and his assumption of hypocrisy's form, putting words straight in Lizzy's mouth to preclude the mitigation of his charms by free indirect discourse: "A young man too, like *you*, whose very countenance may vouch for your being amiable" (114). Located in Wickham's "countenance," the empirical guarantor of his honesty is summed by Elizabeth in a precipitous elision of Clarissa's effort to muster the alignment—or, in Clarissa's redeployment of a keyword in Helmontian medicine, the "*consentaneousness*"[25]—of rakish body and soul: "there was truth in his looks" (119). Wickham's good "looks" repeat Lovelace's. The actor who plays Lizzy's attraction to Wickham out until almost the end, Lydia barely avoids the fate of Lovelace's discarded paramour Sally Martin: as Austen writes with an explicitness that cannot be gainsaid, "it would have been more for the advantage of conversation, had Miss Lydia Bennet come upon the town" (316).

But Austen's plot departs from Richardson's at this point. Lydia embarks on a patched-up marriage that Austen represents not as the declension into moral chaos predicted by *Clarissa* but as the banal unhappiness sustained by spouses who deserve each other. *Pride and Prejudice*'s more momentous divergence concerns Lizzy. What crucially enables this break is the fact that Elizabeth's father leaves her free to choose: as his complete unwillingness to force Lizzy's marriage to the Solmes-ish Collins confirms, Mr. Bennet's paternal laxity licenses *Pride and Prejudice*'s better outcome. Elizabeth's liberty facilitates a different novel, one in which masculine value is neither—in a departure from the novels of Austen's equally revered predecessor Frances Burney—preemptively perceptually evident nor, as in Clarissa's case, fatally delusive.[26] In Lizzy's plot, characters render each other marriageable as the result of how they react.

In *Clarissa*, the novel asserts the fatality of perceptual misprision that leads to Clarissa's demise. But in *Pride and Prejudice*, the novel corrects for too hasty denomination by plotting the changes to oneself involved in eliciting changes from another. In this capacity, Wickham's appeal is not incidental, but the concrete reinforcement of Lizzy's initial likeness to Lydia. That likeness is decisively altered when Elizabeth confronts the history relayed in a letter Darcy pens to clear his name from Wickham's smears. In the letter, Darcy absolves Elizabeth of her eagerness to believe "what falsehood he has imposed on you; but his success is not perhaps to be wondered at, ignorant as you previously were of every thing concerning either" man (222). The wondrousness of Elizabeth's endorsement of masculine calumny, like the wondrousness of Clarissa's failure to see through the faked virtue of prostitutes, is "perhaps" assuaged by her lack of experience. Wickham's duplicitous appeal must ultimately be undone by the epistolary revelation of family secrets too sordid—and, in retrospect, too convincing—to contravene. Darcy's letter fixes history. Elizabeth's subsequent visit to his estate Pemberley, where she encounters him in the flesh, corrects sensation. Free indirect testimony to Elizabeth's feeling underscores the shock of the phenomenal difference: "Never in her life had she seen his manners so little dignified, never had he spoken with such gentleness as on this unexpected meeting" (266). The palpable excess of Elizabeth's superlative—since "Never in her life" entails under a year of infrequent contact—magnifies her sense of a change that *is* pure experience. For Elizabeth, the revelation of Pemberley is her encounter with the quality she has been instrumental in eliciting, "gentleness." Of course, Darcy's gentleness reflects his power in, at least, the Lockean sense of his disposition to manifest this attribute. But as a quality stimulated by the person he can now marry, his gentleness is bound to Elizabeth's perception of it.

Elizabeth's aunt Mrs. Gardiner remarks a change to Darcy's character manifest in the same medium that conceals Wickham's viciousness: "And there is something of dignity in his countenance, that would not give one an unfavourable idea of his heart" (271). This reference to Darcy's heart does not seal Lizzy's triumph at excavating a brooding man's hidden virtues, as Joe Wright's film adaptation of *Pride and Prejudice* (2005) suggests by accommodating the novel to the ideological requirements of formula romance.[27] The latter schema is imposed on Austen from the vantage of a psychological depth model wedded to the still-patriarchal fiction that women might conjure ungentle men's redemptive interiority. Yet the critical repetition in *Pride and Prejudice*'s plot, the fact that Elizabeth rudely rejects Darcy's first proposal of marriage, might seem to insinuate another patriarchal demand. Darcy's forgiveness of her embrace of Wickham's slurs, evident in the solicitousness that precedes his second request, elicits Lizzy's "gratitude" (277). The conservatism of this response is augured by John Gregory's popular conduct book *A Father's Legacy to His Daughters* (1761), which prescribes gratitude as the gateway to feminine amorous preference forbidden before the man advances a legitimate offer.[28] But gratitude marks Lizzy's felt reaction not because Darcy proposes, but because he overlooks the alacrity with which she approved Wickham's lies. As such, gratitude is nowhere near as reactionary a mandate as the requirement that wives summon virtues from husbands who cannot manifest them.[29] If Elizabeth solicits Darcy's gentleness, he provokes first impressions whose revision sustains Lizzy's unprecedented freedom to react twice.[30]

To conclude: because his new nomenclature heralds the rise of elemental chemistry, Lavoisier's system marks the end of the corpuscular episteme in which, I have argued, particulate matter does not entrench ontology. But with historians like William Newman, I have stressed the persistence of corpuscular explanation into modern chemical science. This is no vestige of reductionist atomistic mechanism or alchemical magic, but a mode of access to qualitative reality that entails particles endowed with powers. We can thus ask: Are words like "oxygen" and "gentleness" realistic in the sense of Watt's stipulation that "the writer's exclusive aim is to make the words bring his object home to us in all its concrete particularity?"[31] I have argued throughout my book that they are not. For Watt, the metaphysics installed by "Descartes, the founder of modern philosophical realism," dictates a novel bifurcated "by the narrative equivalent of dualism—the problematic nature of the relation between the individual and his environment." Narrative "dualism" is sustained, Watt claims, by realism that places "the subjective and the external orientations of the novelist" on parallel referential tracks.[32] But the chymical antecedents to British corpuscular science and empiricism through Isaac Newton and John Locke do not advance reified Cartesian dualism. This science does not impose ideality on passive objects but, from Jābir or Paracelsus forward, propels an alternative genealogy of perceptual understanding that implicates insensible matter's powers, corpuscular endowments whose capacity to engender qualitative experience solicit figuration in and as language.

As inheritors of *this* metaphysics, neither Lavoisier's oxygen nor Austen's gentleness insulates "subjective" from "external orientations," whether the scientist's or the novelist's. We can recall Elizabeth's assessment of the domestic felicity promised by her match: "It was an union that must have been to the advantage of both; by her ease and liveliness, his mind might have been softened, his manners improved, and from his judgment, information, and knowledge of the world, she must have received benefit of greater importance" (318). Notably, this couple's prospects are expressed as a reaction. Rather than inviting transparently mimetic reference, their happy "connubial" (318) outcome is contained in a formula that conveys Lizzy's and Darcy's conjoined power to realize it.[33]

Austen's formula, which implicates insensible dispositions exerted over time and in relation, offers no static referent to ground the "scientific objectivity"[34] Watt claims as his standard of novelistic realism. The shibboleth of "scientific objectivity," I have argued, cannot accommodate the role of corpuscular power in eighteenth-century chemistry and empiricism. In the critical case of the father of modern chemistry, Lavoisier's oxygen recombines insensible causes and sensed effects, insinuating corpuscular shape into vital air not to mirror empirical experience but to figure the insensible particles that produce it. The word "oxygen" figures vital air's productive and relational power, the disposition to acidify exhibited in reaction with bodies like sulfur, the disposition to stimulate empirical sense experienced in reaction with bodies like lungs.

Does Darcy's gentleness affirm, in John Bender's phrasing, "realism's power to give an impression of the real 'thing,'" conveyed by a novel in which "[t]he illusion is of objects per se"?[35] Lavoisier represses his affinity with a chymistry—and an empiricism—in which words figure matter's power to effect ideas. Austen, on the other hand, foregrounds the always contingent state of empirical understanding as Elizabeth and Darcy's accession to qualitative knowledge that must be *produced.* Narratives like Austen's do not offer the illusion of an experiential scene inhabited by passive objects and distant witnesses. Along with chemistry through Lavoisier, eighteenth-century novels plumb the forms and relations that produce sensory knowledge, including the reader's. In this, Austen affirms the influence of corpuscular science on empirical understanding in the novel, and beyond.

NOTES

INTRODUCTION

Epigraphs: Robert Boyle, *The Origin of Forms and Qualities*, in *The Works of Robert Boyle*, ed. Michael Hunter and Edward B. Davis (London: Pickering and Chatto, 1999), 5: 418. Bruno Latour, *We Have Never Been Modern*, trans. Catherine Porter (Cambridge, Mass.: Harvard University Press, 1993), 86.

1. Boyle, *Forms and Qualities*, 381.

2. Michael McKeon, *The Origins of the English Novel, 1600–1740* (Baltimore: Johns Hopkins University Press, 1987), 80.

3. Cathy Cobb, Monty L. Fetterolf, and Harold Goldwhite, *The Chemistry of Alchemy from Dragon's Blood to Donkey Dung: How Chemistry Was Forged* (Amherst, N.Y.: Prometheus Books, 2014), 52.

4. Boyle, *Forms and Qualities*, 395.

5. Robert Boyle, *New Experiments Physico-Mechanicall, Touching The Spring of the Air, and its Effects*, in *Works of Robert Boyle*, ed. Hunter and Davis, 1: 245.

6. Boyle, *Spring of the Air*, all citations, 165.

7. Ibid., 257.

8. Ibid., 172.

9. Robert Boyle, *Of the Excellency and Grounds of the Corpuscular or Mechanical Philosophy*, in *Works of Robert Boyle*, ed. Hunter and Davis, vol. 8. Emphasis on penultimate word mine.

10. Boyle, *Spring of the Air*, 210.

11. Ibid., 212–13.

12. Boyle, *Forms and Qualities*, 353.

13. Robert Boyle, *Of the Mechanical Origine of Heat and Cold*, in *The Mechanical Origine or Production of divers particular Qualities* (1675–76), in *Works of Robert Boyle*, ed. Hunter and Davis, 8: 341.

14. Quentin Meillassoux, *After Finitude: An Essay on the Necessity of Contingency*, trans. Ray Brassier (London: Bloomsbury Academic, 2012), both citations, 124.

15. *After Finitude*, 129 n. 2, appends the acknowledgement: "We should add that, from a strictly Lockean point of view, secondary qualities, like primary qualities, *are intrinsic to perceptible bodies*, since they correspond to the latter's capacity to engender in the mind sensible qualities . . . But we are here following a common usage according to which secondary qualities are identified with sensible qualities" (emphasis mine). Meillassoux chooses a "common usage" which contravenes Locke's insistence that secondary qualities inhere in objects and minds at once. To divorce mathematical or Cartesian primary qualities from sensory ideas, Meillassoux departs from Locke's definition of secondary qualities.

16. Elizabeth Grosz, "Irigaray, Ethics, and the Arts: Ontologies, Topologies, Temporalities," talk delivered May 5, 2015, Northwestern University.

17. Boyle, *Mechanical Origine*, 341.

18. Boyle, *Spring of the Air*, 213.

19. Ibid., 220.

20. Boyle echoes Lucretius's "lucky concourse" of atoms, the materialist *clinamen* that obviates the creative agency of a god in Lucretius's cosmogony *On the Nature of Things*. This echo seems deliberate insofar as Boyle refers to Lucretius and other atomists in *Spring of the Air* to affirm the physicality of the primary attributes of the corpuscle. I discuss Boyle's selective appropriation of materialist atomism in Chapter 1.

21. Boyle, *Spring of the Air*, 288.

22. Cobb, Fetterolf, and Goldwhite, *Chemistry of Alchemy*, 86.

23. On the medieval alchemist John of Rupescissa, author of *On the Consideration of the Fifth Essence of All Things*, and on the integration of medicine and alchemy forwarded by John's use of alcohol as a therapeutic agent, see Lawrence M. Principe, *The Secrets of Alchemy* (Chicago: University of Chicago Press, 2013), 69–71.

24. Antonio Clericuzio, *Elements, Principles and Corpuscles: A Study of Atomism and Chemistry in the Seventeenth Century* (Dordrecht: Kluwer, 2000), 7.

25. Clericuzio, *Elements, Principles*, 7.

26. William R. Newman, *Atoms and Alchemy: Chymistry and the Experimental Origins of the Scientific Revolution* (Chicago: University of Chicago Press, 2006), xi. See also Newman and Lawrence M. Principe, "Alchemy vs. Chemistry: The Etymological Origins of a Historiographic Mistake," *Early Science and Medicine* 3: 1 (February 1998): 32–65.

27. William R. Newman and Lawrence M. Principe, *Alchemy Tried in the Fire: Starkey, Boyle, and the Fate of Helmontian Chymistry* (Chicago: University of Chicago Press, 2002), 319. They proceed: "[Chymist Joan Baptista] Van Helmont used no less mathematics than most modern-day chemists. Synthetic organic chemists, the majority within the most populous community of modern scientists, do very well with very little mathematics at all. They weigh starting materials and final products, calculate yields and compositions—predominantly the same sorts of things that Van Helmont . . . and eventually Lavoisier did" (319).

28. Clericuzio, *Elements, Principles*, 4. He proceeds: "[Boyle] explained chemical processes in terms of corpuscles endowed with chemical, not just mechanical properties" (5).

29. Steven Shapin and Simon Schaffer, *Leviathan and the Air-Pump: Hobbes, Boyle, and the Experimental Life* (Princeton: Princeton University Press, 1985), 156.

30. For refutation of Shapin and Schaffer's claim that the activity of witnessing marks a fundamental break with the alchemical tradition, see Lawrence M. Principe, *The Aspiring Adept: Robert Boyle and His Alchemical Quest* (Princeton: Princeton University Press, 1998), 24–25, 107–11.

31. Shapin and Schaffer, *Leviathan and the Air-Pump*, 205.

32. Newman, *Atoms and Alchemy*, 9.

33. Principe, *Aspiring Adept*, 111.

34. Boyle, *Spring of the Air*, 165.

35. On the term "spectator theory of knowledge," see Ian Hacking, *Representing and Intervening: Introductory Topics in the Philosophy of Natural Science* (Cambridge: Cambridge University Press, 1983), 130–46.

36. Shapin and Schaffer, *Leviathan and the Air-Pump*, 36.

37. Ibid., 67.

38. Ibid., 65.

39. Ibid., 140.

40. Ibid., 147.

41. Boyle, *Spring of the Air*, 214.

42. Ibid., 253.

43. Shapin and Schaffer, *Leviathan and the Air-Pump*, 337.

44. Boyle, *Spring of the Air*, 165.

45. See Hacking, *Representing and Intervening*, Chapter 1, "What Is Scientific Realism?"

46. Latour, *We Have Never Been Modern*, 29, 23, 29.

47. In *Forms and Qualities*, Boyle refers to "me, whose sense of Smelling is none of the Dullest" (395). On Boyle's sight, whose degradation seems to date from a fall from a horse in Ireland on May Day 1654, see Michael Hunter, *Boyle: Between God and Science* (New Haven: Yale University Press, 2009), 91–92.

48. Latour, *We Have Never Been Modern*, 62.

49. See the following key texts: Karen Barad, *Meeting the Universe Halfway: Quantum Physics and the Entanglement of Matter and Meaning* (Durham: Duke University Press, 2007); Ian Hacking, *Representing and Intervening*; Donna J. Haraway, *Simians, Cyborgs, and Women: The Reinvention of Nature* (London: Routledge, 1991); and Latour, *We Have Never Been Modern*. On the situated knower, see Haraway, "Situated Knowledges: The Science Question in Feminism and the Privilege of Partial Perspective," in *Simians, Cyborgs, and Women*, 183–202.

50. Hacking, *Representing and Intervening*, 141–42.

51. Ibid., 140, repeated 141.

52. Ibid., 274.

53. Barad, *Meeting the Universe Halfway*, 107, 19.

54. Ibid., 39.

55. Ibid., 354, 356.

56. Ibid., 337.

57. Ibid., both citations, 197.

58. Jane Bennett, *Vibrant Matter: A Political Ecology of Things* (Durham: Duke University Press, 2010), 30.

59. Bennett, *Vibrant Matter*, both citations, 5. Bennett cites Stephen Jay Gould. See Tom Sparrow, *The End of Phenomenology: Metaphysics and the New Realism* (Edinburgh: Edinburgh University Press, 2014), whose gloss of "[Graham] Harman's method" echoes Locke: Harman "proposes to deduce the primary qualities of objects from the phenomenal" (115).

60. Bennett, *Vibrant Matter*, 15.

61. Ibid., 58, 59.

62. On evidence from hair samples showing that Newton suffered from "mercury poisoning," see Principe, *Aspiring Adept*, 179.

63. Hacking, *Representing and Intervening*, 146. Hacking writes: "you learn to see through a microscope by doing, not just by looking" (*Representing and Intervening*, 189). Barad notes: "Clearly, we do not see merely with our eyes" (*Meeting the Universe Halfway*, 157). On these pages, Hacking cites the eighteenth-century philosopher George Berkeley on the spatial constituents of the visual perception of depth, while Barad cites visually impaired persons whose sight is restored late in life, an example formulated as the Molyneux problem in Locke's *Essay* to assert the interrelation of visual and spatial sensory data. Hacking and Barad rely on eighteenth-century epistemology to defend knowledge practices they deny eighteenth-century science.

64. Ian Watt, *The Rise of the Novel: Studies in Defoe, Richardson and Fielding* (Berkeley: University of California Press, 1957), 12.

65. For brilliant motivations of objecthood in the novel, which nonetheless do not query the reality of submicroscopic bodies, see Cynthia Sundberg Wall, *The Prose of Things: Transformations of Description in the Eighteenth Century* (Chicago: University of Chicago Press, 2006) and Sandra Macpherson, *Harm's Way: Tragic Responsibility and the Novel Form* (Baltimore: Johns Hopkins University Press, 2010).

66. Watt, *Rise of the Novel*, 294, 294, 295.

67. Ibid., 295.

68. Ibid., 292.

69. John Bender, *Ends of Enlightenment* (Stanford: Stanford University Press, 2012), 82.

70. Bender, *Ends of Enlightenment*, both citations, 40.

71. Ibid., 52.

72. Ibid., 83.

73. Ibid., 46.

74. See Betty Jo Teeter Dobbs's pioneering study, *The Foundations of Newton's Alchemy or "The Hunting of the Greene Lyon"* (Cambridge: Cambridge University Press, 1975).

75. Bender, *Ends of Enlightenment*, 66.

76. Ibid., 43.

77. Ibid., 42, 36, 42, 38, 46.

78. Ibid., both citations, 55.

79. Ibid., 71.

80. Ibid., 73.

81. Michael Fried, *Absorption and Theatricality: Painting and Beholder in the Age of Diderot* (Chicago: University of Chicago Press, 1980), 93, 103.

82. Fried, *Absorption and Theatricality*, 96, 104.

83. For the argument that Hogarth "seems surprisingly unconcerned with enhancing pictorial illusion," see Abigail Zitin, "Thinking Like an Artist: Hogarth, Diderot, and the Aesthetics of Technique," *Eighteenth-Century Studies* 46: 4 (Summer 2013): 555–70. Citation, 561.

84. Jayne Elizabeth Lewis, *Air's Appearance: Literary Atmosphere in British Fiction, 1660–1794* (Chicago: University of Chicago Press, 2012), 59. Lewis cogently queries Shapin and Schaffer's deployment of Boyle's *Spring* when she notes that their "Hobbesian interpretation of Boyle's experimental method holds its materialist sleights of hand responsible for the descent of the ideological mist that is modern science" (43).

85. Jonathan Brody Kramnick, *Actions and Objects from Hobbes to Richardson* (Stanford: Stanford University Press, 2010), 221.

86. The atheistic mandate of Epicurean materialism is clear: as Lucretius announces, "I teach of weighty things, and work / man's heart free of religion's garrotte-knot." *The Nature of Things*, trans. Frank O. Copley (New York: Norton, 1977), Book 4, ll. 6–7. But whether atheistic materialism excludes the complexity of embedded processes or ends is less clearly affirmed by Lucretius's text. See, for example, his treatment of species conservation in book 2, where he cites "sheep," "cows," and horses which "yet keep identity, kind by kind preserving": "Take all things else; survey them with like logic: / you'll find they hide within their mass the seeds / of many things, arrayed in varied patterns." Book 2, ll. 662, 663, 665, and 677–79. For an account of the epistemological influence of Epicureanism on Restoration culture that stresses the cognitive and cultural force of embodied objects of knowledge, see Richard W. F. Kroll, *The Material Word: Literate Culture in the Restoration and Early Eighteenth Century* (Baltimore: Johns Hopkins University Press, 1991).

87. William R. Newman, introduction to Isaac Newton, "Dibner MS. 101 B SCDIRB." *The Chymistry of Isaac Newton*, ed. William R. Newman (2006). Retrieved June 28, 2014 from: http://purl.dlib.indiana.edu/iudl/newton/ALCH00081. Newman dates the manuscript to "the first half of the 1670's" (introduction). I am grateful to him for this reference.

88. Newton, Dibner MS. 101 B SCDIRB, 5v.

89. See Walter Pagel, *Joan Baptista Van Helmont: Reformer of Science and Medicine* (Cambridge: Cambridge University Press, 1982). For discussion of "[t]he alchemical regime of *putrefactio*," see William R. Newman, *Gehennical Fire: The Lives of George Starkey, an American Alchemist in the Scientific Revolution* (Cambridge, Mass.: Harvard University Press, 1994). Citation, 140. See also Newman and Principe, *Alchemy Tried in the Fire*, 88–89.

90. On Boyle and Kenelm Digby, who "himself undoubtedly encouraged Boyle's chymical studies," see Newman and Principe, *Alchemy Tried in the Fire*, 261–62. Citation, 261.

91. Newton, Dibner MS. 101 B SCDIRB, 1r.

92. See 3v, where Newton writes of "minerall fumes" and which rise, are compressed by ether in the upper atmosphere, and then "tis gradually condensed & interwoven wth bodys it meets there . . . but in its descent it endeavours to beare along w^{t} bodys it passeth through, that is makes them heavy & this action is promoted by the tenacious elastick constitution whereby it takes ye greater hold on things in its way." We can note Newton's reliance on the agency of "elastick" air.

93. Principe discusses Boyle's quest for the alchemical reagent "animated" or "Philosophical Mercury," which "is capable of transmuting base metals into gold. Common gold is dead, but the animated Mercury animates the gold in turn and makes it grow." *Aspiring Adept*, 154, 155.

94. Boyle, *Forms and Qualities*, 467–68.

95. Hacking, *Representing and Intervening*, 24.

96. Barad, *Meeting the Universe Halfway*, 207.

97. Newman, *Atoms and Alchemy*, 224.

98. Ibid., 195.

CHAPTER 1

Epigraph: Robert Boyle, "The History of Fluidity and Firmness," in *The Works of Robert Boyle*, ed. Michael Hunter and Edward B. Davis (London: Pickering and Chatto, 1999), 2: 185.

1. Robert Boyle, *The Sceptical Chymist*, in *Works of Robert Boyle*, ed. Hunter and Davis, 2: 337.

2. Karen Barad, *Meeting the Universe Halfway: The Entanglement of Matter and Meaning* (Durham: Duke University Press, 2007), 353.

3. Ian Watt, *The Rise of the Novel: Studies in Defoe, Richardson, and Fielding* (Berkeley: University of California Press, 1957), 11.

4. Watt, *Rise of the Novel*, 13.

5. Jonathan Kramnick, *Actions and Objects from Hobbes to Richardson* (Stanford: Stanford University Press, 2010); Sandra Macpherson, *Harm's Way: Tragic Responsibility and the Novel Form* (Baltimore: Johns Hopkins University Press, 2010); Wolfram Schmidgen, *Exquisite Mixture: The Virtues of Impurity in Early Modern England* (Philadelphia: University of Pennsylvania Press, 2013).

6. Robert Boyle, *The Origin of Forms and Qualities*, in *Works of Robert Boyle*, ed. Hunter and Davis, 5: 309.

7. See Bacon's *The New Organon* (1620) on the relational status of qualities: "To the senses and to human touch heat is a variable and relative thing; so that tepid water feels hot to a hand in the grip of cold, but if the hand warms up, it feels cold." Francis Bacon, *The New Organon*, ed. Lisa Jardine and Michael Silverthorne (Cambridge: Cambridge University Press, 2000), 126.

8. Boyle, *Forms and Qualities*, 310.

9. Robert Boyle, *New Experiments Physico-Mechanicall, Touching the Spring of the Air, and its Effects*, in *Works of Robert Boyle*, ed. Hunter and Davis, 1: 219.

10. *Sceptical Chymist* makes the disclaimer: "because I seem not satisfi'd with the Vulgar Doctrines, either of the Peripatetick or Paracelsian Schools, many . . . have thought me wedded to the Epicurean *Hypotheses* . . . yet if you knew how little Conversant I have been with *Epicurean* Authors, and how great a part of *Lucretius* himself I never yet had the Curiosity to read, you would perchance be of another mind" (354).

11. Boyle, *Forms and Qualities*, 302.

12. My references to Aristotelian and scholastic "essence" reflect the distortion wreaked by antischolastic argument, including Boyle's and Locke's. See Monte Ransome Johnson, *Aristotle on Teleology* (Oxford: Clarendon Press, 2005): "the term 'essence' . . . is heavily laden with customs, baggage, and ambiguities . . . that do not really apply to Aristotle. It is common to attack Aristotelian 'essentialism' . . . but what is attacked . . . is more often than not a caricature of the position that Aristotle actually countenances" (47).

13. Boyle, *Forms and Qualities*, 479.

14. Ibid., 310.

15. William R. Newman, *Atoms and Alchemy: Chymistry and the Experimental Origins of the Scientific Revolution* (Chicago: University of Chicago Press, 2006), 98.

16. Boyle, *Forms and Qualities*, 302. In parentheses in original.

17. Ibid.

18. Barad, *Meeting the Universe Halfway*, 172.

19. On the challenge to natural order at the heart of the alchemical endeavor, see William R. Newman, *Promethean Ambitions: Alchemy and the Quest to Perfect Nature* (Chicago: University of Chicago Press, 2004).

20. Boyle, *Forms and Qualities*, 311–12.

21. Ibid., 321.

22. Boyle, *Sceptical Chymist*, 360.

23. Ibid., 242.

24. Boyle, *Forms and Qualities*, 316–17.

25. Ibid., 334.

26. Boyle, "History of Fluidity and Firmnesse," 137.

27. Boyle, *Forms and Qualities*, 472.

28. Charles Webster citing puritan preacher John Webster, in *The Great Instauration: Science, Medicine and Reform, 1626–1660* (London: Duckworth, 1975), 187. Webster cites John Milton's "finer expression of puritan anti-scholastic feelings": "scholastical trash" (188). Webster defends a puritan context for Boyle and shows the connection of puritan aspirations with reforms projected by the interregnum "State Agent for Universal Learning" (71) Samuel Hartlib.

29. Boyle, *Sceptical Chymist*, 292. On Boyle's contact with the Hartlib Circle, see Michael Hunter, *Boyle: Between God and Science* (New Haven: Yale University Press, 2009). Hunter deems it "likely" that Boyle and the puritan reformer Samuel Hartlib's "contact initially occurred through [Boyle's sister] Lady Ranelagh, who was close to influential parliamentarians" (65).

30. Barad, *Meeting the Universe Halfway*, 97.

31. Boyle, *Sceptical Chymist*, 324.

32. Ibid., 307.

33. Lawrence M. Principe shows that *Sceptical Chymist* must be read with Boyle's distinction of chymists from "vulgar" chymists in mind, thereby distinguishing his critique of charlatans from his chary expressions of approval of "adepti" (including those to whom he was indebted). *The Aspiring Adept: Robert Boyle and His Alchemical Quest* (Princeton: Princeton University Press, 1998), 34.

34. In *Alchemy Tried in the Fire: Starkey, Boyle, and the Fate of Helmontian Chymistry* (Chicago: University of Chicago Press, 2002), Newman and Principe affirm that it is "the artifactual nature of Mercury, Sulphur, and Salt that Robert Boyle adopted as a key argument in his *Sceptical Chymist*" (85).

35. Boyle, *Sceptical Chymist*, 303.

36. Ibid., 324.

37. Ibid., 303.

38. Ibid., 294 and 293.

39. Ibid., 290 and 316.

40. Ibid., 290.

41. Ibid., 306.

42. Ibid., 354.

43. Ibid., 356.

44. Steven Shapin and Simon Schaffer, *Leviathan and the Air-Pump: Hobbes, Boyle, and the Experimental Life* (Princeton: Princeton University Press, 1985), 339.

45. Boyle, *Sceptical Chymist*, 357.

46. Boyle, *Forms and Qualities*, 387.

47. Marie Boas Hall argues that *The Sceptical Chymist* marks "the great project that occupied Boyle's scientific life, the substitution of mechanical explanations, in terms of matter and motion, for the current peripatetic and spagyrical substantial forms and innate qualities." Boas [Hall], "The Establishment of the Mechanical Philosophy," *Osiris* 10 (1952): 489. Thomas S. Kuhn cites Boyle's reductionism as a key determinant of Newton's physics: "Newton, like Boyle, denied the specificity of the elementary corpuscles." Thomas S. Kuhn, "Robert Boyle and Structural Chemistry in the Seventeenth Century," *Isis* 43: 1 (April 1952): 33.

48. Boyle, *Forms and Qualities*, 326. Newman discusses the provenance of this term: "scholastics concluded that there are *minima naturalia*—smallest natural parts—out of which living, and even inanimate things, are composed." William R. Newman, *Gehennical Fire: The Lives of George Starkey, an American Alchemist in the Scientific Revolution* (Cambridge, Mass.: Harvard University Press, 1994), 24–25.

49. Newman, *Atoms and Alchemy*, xi.

50. Ibid., xii.

51. Ibid., 223–24.

52. Robert Boyle, "An Introduction to the History of Particular Qualities," in *Works of Robert Boyle*, ed. Hunter and Davis, 6: 274.

53. Robert Boyle, *Experiments, Notes, &c. about the Mechanical Origine or Production of divers particular Qualities*, in *Works of Robert Boyle*, ed. Hunter and Davis, 8: 445.

54. Boyle, *Forms and Qualities*, 408.

55. Boyle, *Sceptical Chymist*, 272.

56. In *Exquisite Mixture*, Schmidgen affirms Boyle's reductionism: "his atomism hinges on a massive act of theoretical reduction—all natural phenomena, Boyle believes, can be explained by matter and motion" (42). Schmidgen cites Boyle's combining corpuscles but overlooks the particulate agency of attraction.

57. Boyle, *Origin of Forms and Qualities*, 396. For recapitulation of the development of scholastic notions of the form (or essence) of mixture—which involves Thomistic and other efforts to explain how scholastic elements persist while substantial forms are replaced—see Newman, *Atoms and Alchemy*, Chapters 1 and 2. Aristotle's claim that true mixtures must be "homoeomerous" or "identical in all their parts" (*Atoms and Alchemy*, 30) is found in *On Generation and Corruption*, which stipulates "that, if combination has taken place, the compound *must* be uniform—any part of such a compound being the same as the whole, just as any part of water is water." *The Complete Works of Aristotle*, ed. Jonathan Barnes (Princeton: Princeton University Press, 1995), 1: 536. On Aristotelian form in seventeenth-century matter theory, see Norma E. Emerton, *The Scientific Reinterpretation of Form* (Ithaca: Cornell University Press, 1984).

58. Boyle, *Forms and Qualities*, 396. Boyle retains the word "Form" but does not use it in a scholastic sense. By "Form and Nature" he means the object's particulate texture in concert with the sensible qualities that elicit its denomination.

59. Boyle, *Sceptical Chymist*, 272.

60. Boyle, *Forms and Qualities*, 358.

61. Boyle, *Sceptical Chymist*, 268.

62. Newman and Principe credit the disproof of Aristotelian mixture by the mass balance to Helmont. If the chymist accepts scholastic claims that radical elemental change correlates with qualitatively perceived mixture, then "conservation of *weight* becomes an obvious impossibility, even though no *matter* is destroyed. Under such circumstances the balance would be totally useless for comparing the initial and final products in the laboratory." *Alchemy Tried*, 70.

63. On the humanly perceived miscibility of some reagents, Newman and Principe write: "the dissolved substance has merely undergone a division into smaller particles, hence keeping its other qualities while becoming a liquid at the level of sense perception" (ibid., 295 n. 64). See also Newman on "mere mixtures *ad sensum*—corpuscular juxtapositions rather than real mixtures, which had only fooled the eye of the beholder" (*Atoms and Alchemy*, 120).

64. Boyle, *Forms and Qualities*, 425.

65. Newman, *Atoms and Alchemy*, xiii.

66. Boyle, *Sceptical Chymist*, 281.

67. Newman, *Atoms and Alchemy*, 23 and 24.

68. Ibid., 122.

69. Ibid., 152.

70. Ibid., 5.

71. Boyle, *Forms and Qualities*, 365.

72. Ibid., 423 and 419.

73. Ibid., 420.

74. Ibid., 420–21.

75. Lawrence M. Principe, *The Secrets of Alchemy* (Chicago: University of Chicago Press, 2013), 17.

76. Jole Shackelford, "Seeds with a Mechanical Purpose: Servinus' Semina and Seventeenth-Century Matter Theory," in *Reading the Book of Nature: The Other Side of the*

Scientific Revolution, ed. Allen G. Debus and Michael T. Walton (Kirksville, Mo.: Sixteenth Century Journal Publishers, 1998), 16.

77. Shackelford, "Seeds with a Mechanical Purpose," 26.

78. Boyle, *Forms and Qualities*, 422.

79. Newman, *Atoms and Alchemy*, 136.

80. Boyle, *Forms and Qualities*, both citations, 449.

81. Ibid., 421.

82. Newman, *Atoms and Alchemy*, xii. Newman and Principe argue that "Boyle intentionally suppressed acknowledgment of the corpuscular theoretical principles of Daniel Sennert" to make the important historiographical point: "Boyle's lack of explicit acknowledgment of such origins for his corpuscular theory gives the appearance of a greater break between him and the foregoing chymical tradition than really was the case" (*Alchemy Tried*, 22).

83. Newman, *Atoms and Alchemy*, 198. Newman continues: "it was chymistry that allowed him to distinguish the essential differences of bodies in a relatively certain fashion, and without such stable essences Boyle could not argue that the qualitative mutability of the phenomenal world was mostly a matter of alterations in texture imposed on fundamentally unchanged corpuscles by mechanical means" (198).

84. Maurice Mandelbaum's *Philosophy, Science, and Sense Perception* (Baltimore: Johns Hopkins University Press, 1964) defines the relation between mechanical affections and perceived qualities as "the problem of transdiction": "the question of how observed data can serve as grounds for inferences to objects or events which not only have not yet been observed, but which cannot in principle be observed" (63). The term is currently "transduction."

85. Robert Boyle, "About the Excellency and Grounds of the Mechanical Hypothesis," in *Works of Robert Boyle*, ed. Hunter and Davis, both citations, 8: 107.

86. Boyle, "Excellency," 107.

87. Daniel Tiffany, *Toy Medium: Materialism and Modern Lyric* (Berkeley: University of California Press, 2000), 49.

88. Tiffany, *Toy Medium*, 86.

89. Catherine Wilson, *The Invisible World: Early Modern Philosophy and the Invention of the Microscope* (Princeton: Princeton University Press, 1997), 230.

90. Bernadette Bensaude-Vincent and Isabelle Stengers, *A History of Chemistry* (Cambridge, Mass.: Harvard University Press, 1996), 37. Tiffany, Wilson, Bensaude-Vincent, and Stengers present the view of mechanical philosophy prevailing in eighteenth-century literary studies. This version of Boyle's mechanism is more relentlessly analogical and substantively bereft than the one endorsed by Kuhn and Boas Hall, who read Boyle too closely to define him as a totalizing mechanist. Newman and Principe recapitulate the historical problem posed by reductionism: "This notion, when carried to its logical extreme, undercuts the possibility of quantitative analysis altogether, because if specific chymical species are not necessary components of a given 'concrete,' then they are not necessary products of analysis either" (*Alchemy Tried*, 288).

91. Bacon, *New Organon*, 13.

92. Robert Boyle, *Certain Physiological Essays*, in *Works of Robert Boyle*, ed. Hunter and Davis, 2: 23–24.

93. Boyle, *Certain Physiological Essays*, 24.

94. Shackelford, "Seeds with a Mechanical Purpose," 42 and 43.

95. Ibid., 43.

96. Sylvia Berryman, *The Mechanical Hypothesis in Ancient Greek Natural Philosophy* (Cambridge: Cambridge University Press, 2009), 241. See Berryman's invaluable "Appendix: Ancient Mechanics and the Mechanical in the Seventeenth Century."

97. Berryman, *Mechanical Hypothesis*, 248 and 248–49.

98. Newman, *Atoms and Alchemy*, 206.

99. Judith Butler, *Bodies That Matter: On the Discursive Limits of "Sex"* (London: Routledge, 1993), both citations, 28.

100. Butler, *Bodies That Matter*, 28.

101. Ibid., 29.

102. Vicki Kirby, *Telling Flesh: The Substance of the Corporeal* (London: Routledge, 1997), 107.

103. Kirby, *Telling Flesh*, 127.

104. Ibid., 113 and 112.

105. Ibid., 114.

106. Barad, *Meeting the Universe Halfway*, 151.

107. Ibid., 64.

108. Ibid., both citations, 353.

109. Luce Irigaray, *Speculum of the Other Woman*, trans. Gillian C. Gill (Ithaca: Cornell University Press, 1985), 265 and passim.

110. Irigaray, *Speculum*, 299.

111. Butler, *Bodies That Matter*, 47.

112. Ibid., 45.

113. Ibid., 50.

114. Ibid., 52 and 51.

115. Ibid., both citations, 54.

116. Ibid., 51.

117. See Jean-Paul Sartre, *Being and Nothingness*, trans. and intro. Hazel E. Barnes (New York: Washington Square Press, 1956), 770–80. Sartre theorizes "that particular quality which we call 'slimy'" (770) as the vertiginous and nauseating "possibility that the In-itself [unconscious facticity] might absorb the For-itself [conscious personhood]" (776), experienced as "a soft, yielding action, a moist and feminine sucking" (776).

118. Principe, *Secrets of Alchemy*, 125. For further testimony to alchemical figures of penetration, see Newman's *Gehennical Fire*: the medieval alchemist "Geber" sublimes and dissolves his would-be alchemical agent to "make its particles attain such minuteness that they will be able to penetrate through the pores of a base metal and transmute it into gold" (*Gehennical Fire*, 97); Helmont's putrefactive model of transmutation, Newman writes, is "predicated on a belief that the philosophical mercury could radically penetrate gold and break up its 'compact' substance" (*Gehennical Fire*, 140). Principe notes: "Sexual intercourse and reproduction are common elements of alchemical imagery, both textual and graphic" (*Secrets of Alchemy*, 75). While the first plate Principe reproduces from the fourteenth-century *Rosarium philosophorum* (Frankfurt, 1550) represents heterosexual intercourse, the resulting two images of male and female heads attached to the same (not distinctly sexed) body pose an outcome of penetration that does not affirm the penetrating agent's superior agency or immateriality.

119. Boyle, *Mechanical Origine*, 367.

120. Ibid., 368.

121. Isaac Newton, *De Natura Acidorum*, in *The Correspondence of Isaac Newton*, ed. H. W. Turnbull (Cambridge: Cambridge University Press, 1961), 3: 210.

122. Newton, *De Natura Acidorum*, 210.

123. Isaac Newton, *Opticks, or A Treatise of the Reflections, Refractions, Inflections & Colours of Light*, foreword Albert Einstein, intro. Edmund Whittaker, pref. I. Bernard Cohen (New York: Dover, 1979), 385–86.

124. In *Gender and Boyle's Law of Gases* (Bloomington: Indiana University Press, 2001), Elizabeth Potter argues for "[Boyle's] construction of the womanly woman at the margins of experimental science, serving as a foil for the new experimental scientist" (16).

125. Robert Hooke, *Micrographia: or, Some Physiological Descriptions of Minute Bodies* (1665), reprint (New York: Dover, 1961), 121, 125.

126. Hooke, *Micrographia*, 127, 122, 123.

127. Shackelford, "Seeds with a Mechanical Purpose," 44. Francesco Redi disproved the spontaneous generation of flies from rotting flesh in 1668. But as evidenced by later theories of the vegetable origin of insects, as well as seminal effluvia of minerals and gems, Redi did not preclude the ongoing confusion of mechanical and seminal agency. On Boyle and Redi, see Peter R. Anstey, "Boyle on Seminal Principles," *Studies in History and Philosophy of Biological and Biomedical Sciences* 33 (2002): 597–630.

128. Boyle, *Forms and Qualities*, 385.

129. In *On the Nature of Things*, Lucretius argues that because animals are manifestly not made up of foodstuffs, and foodstuffs do not spurt animal blood, food must be broken into atomic constituents before it augments the body. See book 2, ll. 711–12: "Foods are absorbed; from these the proper atoms spread to the body's parts." *The Nature of Things*, trans. Frank O. Copley (New York: Norton, 1977), 45.

130. Boyle, *Forms and Qualities*, 386.

131. See Kenelm Digby, *Of Bodies, and of Mans Soul* (1669), reprint (Kessinger Publishing's Rare Mystical Reprints, n.d.), which images the plant as a "Machine" (256) but posits a "*Seed*" that is "like a tincture extracted out of the whole plant and . . . at last dried up into a kind of magistery" (262). Digby's vegetable magistery, Principe explains of alchemical treatments of mercury, or "substance prepared from another without any separation of parts" (*Aspiring Adept*, 54), shows the entanglement of chymical and seminal causation in plant growth.

132. Boyle, *Sceptical Chymist*, 368–69.

133. Boyle, *Forms and Qualities*, 457.

134. Newman, *Atoms and Alchemy*, 151.

135. Boyle, *Sceptical Chymist*, 264. Newman and Principe remark of the agency of *semina* in Helmontian theory: "this is how water can be the basic substratum of all other matter: the action of *semina* radically converts water into all the various substances in the world" (*Alchemy Tried*, 66).

136. Boyle, *Forms and Qualities*, 457.

137. Robert Boyle, *Some Considerations Touching the Usefulness of Experimental Natural Philosophy*, in *Works of Robert Boyle*, ed. Hunter and Davis, both citations, 3: 401.

138. Hooke, *Micrographia*, 126, 127.

139. Ibid., 130.

140. Ibid.

141. See Cathy Cobb, Monty L. Fetterolf, and Harold Goldwhite, *The Chemistry of Alchemy from Dragon's Blood to Donkey Dung: How Chemistry Was Forged* (Amherst, N.Y.:

Prometheus Books, 2014): "It was reactions such as the Tree of Diana that led alchemists to believe that metals—such as gold—had a life of their own and might be made to grow" (134).

142. Hooke, *Micrographia*, 128.

143. Principe, *Secrets of Alchemy*, 166.

144. Hooke, *Micrographia*, 126.

145. Ibid., both citations, 127.

146. Ibid., 122.

147. Ibid., 123.

148. Ibid., 124.

149. Ibid., 124.

150. See Newman's discussion of Starkey on "insect generation and metamorphosis": "the same celestial influx that played at producing insects was also busily generating minerals within the earth" (*Gehennical Fire*, 17).

151. Hooke, *Micrographia*, 130.

152. Important in this regard, insofar as it applies to Epicurus as well as Aristotle, is Monte Johnson's suggestion that necessity and teleology must be interlaced to explicate motion like the ship's, whose sails, after all, are oriented in a direction. As Johnson notes: "neither the materialist or formalist theories on their own can explain arrangement or order" (*Aristotle on Teleology*, 120). See also Berryman, *Mechanical Hypothesis*: "Sometimes the term 'mechanistic' is treated as though it simply means the rejection of teleology. I have suggested . . . that the two categories are not exhaustive" (15).

CHAPTER 2

Epigraph: John Locke, *An Essay Concerning Human Understanding*, ed. Peter H. Nidditch (Oxford: Oxford University Press, 1975), 647. All citations from this edition will be noted parenthetically above by book, chapter, and section, as follows: 4.12.§12.

1. On the biographical details of this connection, see Richard Ithamar Aaron, *John Locke* (Oxford: Oxford University Press, 1937); Peter R. Anstey, *John Locke and Natural Philosophy* (Oxford: Oxford University Press, 2011); Stephen Gaukroger, "The Role of Natural Philosophy in the Development of Locke's Empiricism," *British Journal for the History of Philosophy* 17: 1 (2009): 55–83; Michael Hunter, *Boyle: Between God and Science* (New Haven: Yale University Press, 2009); J. R. Milton, "Locke at Oxford," in *Locke's Philosophy: Content and Context*, ed. G. A. J. Rogers (Oxford: Clarendon Press, 1994): 29–48; J. R. Milton, "Locke's Life and Times," in *The Cambridge Companion to Locke*, ed. Vere Chappell (Cambridge: Cambridge University Press, 1994), 5–25; G. A. J. Rogers, "The Intellectual Setting and Aims of the *Essay*," in *The Cambridge Companion to Locke's* Essay Concerning Human Understanding, ed. Lex Newman (Cambridge: Cambridge University Press, 2007), 7–32; Patrick Romanell, *John Locke and Medicine: A New Key to Locke* (Buffalo, N.Y.: Prometheus Books, 1984); Rose-Mary Sargent, *The Diffident Naturalist: Robert Boyle and the Philosophy of Experiment* (Chicago: University of Chicago Press, 1995).

2. Anstey writes of the chymical recipes contained in Locke's "chymical notebooks": "Boyle remains the single most important source of chymical advice and opinion in all of Locke's notebooks; many of Locke's chymical connections triangulate in various degrees with Boyle" (*Locke and Natural Philosophy*, 173). Boyle's attempts to convert mercury into Philosophical

Mercury, a precursor to the transmutation of metals into gold, were, Anstey demonstrates, known by Locke, as evidenced in the letter of a third party (dated May 20, 1660) to Locke: "this is the earliest known connection between Locke and Boyle and it suggests that they were already discussing chymistry together by May 1660" (171). See Kenneth Dewhurst's argument that "these early notebooks clearly show that Locke had already developed an interest in chemistry and medicine before the summer of 1660 when he met Robert Boyle who greatly accelerated these studies." *John Locke (1632–1704), Physician and Philosopher: A Medical Biography* (London: Wellcome Historical Medical Library, 1963), 7. In *The Aspiring Adept: Robert Boyle and His Alchemical Quest* (Princeton: Princeton University Press, 1998), Lawrence Principe relates of an item listed in Boyle's will as "the Animated Mercury" (or Philosophical Mercury): "Apparently, in the last year of his life, Boyle entrusted the preparation of this precious material to one or more of his friends. One of these friends was the philosopher John Locke, who in 1691 was compiling Boyle's notes for a *General History of the Air* into a publishable form. Locke had a long-standing interest in chymistry, including chrysopoeia . . . among the Boyle Papers there exists a receipt for 'Mercurius antimonii Lockii" (*Aspiring Adept*, 175).

3. Locke was exposed to scholastic pedagogy while a student at Westminster School from 1647 until he left for Christ Church in August 1652.

4. Anstey, *Locke and Natural Philosophy*, 51. J. R. Milton has established that between 1658 and March 1667, medical texts comprised 46.6 percent and natural philosophical texts 16.4 percent of Locke's reading ("Locke at Oxford," 36). Milton bases his claim that Locke read the approximately 350 books at issue on the excerpts in Locke's notebooks, although Milton concedes that such calculations are "inherently imprecise" (35) because of the challenges of disciplinary classification and Locke's possession of multiple copies of single texts.

5. John W. Yolton, *Locke and the Compass of Human Understanding: A Selective Commentary on the 'Essay'* (Cambridge: Cambridge University Press, 1970), 35 n. 2.

6. John Harrison and Peter Laslett, *The Library of John Locke* (Oxford: Oxford University Press, 1971), 16. See also Peter Laslett, "The Recovery of Locke's Library," in *The Philosophical Canon in the 17th and 18th Centuries: Essays in Honour of John W. Yolton*, ed. G. A. J. Rogers and Sylvana Tomaselli (Rochester, N.Y.: University of Rochester Press, 1996), for a gripping account of his personal endeavor to recover the King moiety of Locke's books, including the discovery of Locke's catalog, "the two huge vellum bound folios containing the flowers gathered by Locke in the 1660s and pressed by him between the exercises submitted by his Christ Church undergraduate pupils . . . These I had discovered at the bottom of a cupboard in the gun room smothered in the remains of a directory of Surrey of about 1910, a volume reduced to tatters by being shot through by a sporting gun" (75).

7. Among the authors discussed below, Locke's library contained works by John Amos Comenius (*Janua linguarum reserata aurea* [1668], *Porta linguarum trilinguis reserata et aperata* [1637], *Janua linguarum trilinguis* [1680]), Robert Hooke (*Micrographia* [1665], *An attempt for the explication of the phænomena observable in Mr Boyles 35 experiments touching the aire* [1661]), George Starkey / Eirenius Philalethes (*Secrets Reveal'd: or, An Open Entrance to the Shut-Palace of the King* [1669], *Enarratio methodica trium Gebri medicinar* [1678]), and the major works of René Descartes and Isaac Newton.

8. Ian Watt, *The Rise of the Novel: Studies in Defoe, Richardson, and Fielding* (Berkeley: University of California Press, 1957), 14.

9. Watt, *Rise of the Novel*, 32.

10. See Mark Blackwell, ed., *The Secret Life of Things: Animals, Objects, and It-Narratives in Eighteenth-Century England* (Lewisburg: Bucknell University Press, 2007); Jonathan Kramnick, *Actions and Objects from Hobbes to Richardson* (Stanford: Stanford University Press, 2010); Jonathan Lamb, *The Things Things Say* (Princeton: Princeton University Press, 2011); Deidre Lynch, *The Economy of Character: Novels, Market Culture, and the Business of Inner Meaning* (Chicago: University of Chicago Press, 1998); Sandra Macpherson, *Harm's Way: Tragic Responsibility and the Novel Form* (Baltimore: Johns Hopkins University Press, 2010); Julie Park, *The Self and It: Novel Objects in Eighteenth Century England* (Stanford: Stanford University Press, 2010); and Wolfram Schmidgen, *Eighteenth-Century Fiction and the Law of Property* (Cambridge: Cambridge University Press, 2002).

11. See Nancy Armstrong, *Desire and Domestic Fiction: A Political History of the Novel* (Oxford: Oxford University Press, 1987); and John Bender, *Imagining the Penitentiary: Fiction and the Architecture of Mind in Eighteenth-Century England* (Chicago: University of Chicago Press, 1989).

12. See Fredric Jameson, *The Political Unconscious: Narrative as a Socially Symbolic Act* (Ithaca: Cornell University Press, 1981). The most sustained alignment of Marxist and Jamesonian methodology with eighteenth-century literary studies is Michael McKeon's *The Origins of the English Novel, 1600–1740* (Baltimore: Johns Hopkins University Press, 1987).

13. John Amos Comenius, *The Great Didactic of John Amos Comenius*, trans. and intro. M. W. Keating (London: Adam and Charles Black, 1896), 337. On "universal knowledge," see Comenius, *A Patterne of Universall Knowledge, In a plaine and true Draught*, trans. Jeremy Collier (London, 1651).

14. Comenius, *Patterne of Universall Knowledge*, sig.B2r, 46, and 86.

15. John Amos Comenius, *Orbis Sensualium Pictus . . . Joh. Amos Comenius's Visible World*, trans. Charles Hoole (London, 1685), sig. A5. Emphasis in original.

16. Ibid.

17. Hezekiah Woodward, *A Light to Grammar* (London, 1641), sig.A2r and sig.A2v. Final word italicized in original.

18. See Charles Webster, *The Great Instauration: Science, Medicine, and Reform, 1626–1660* (London: Duckworth, 1975), 100–115. Webster writes: "Hartlib and his associates completely dominated the educational literature of the Puritan Revolution, being responsible for at least fifty educational works between 1640 and 1660" (112). He continues: "Comenian pansophia provided an explicit biblical sanction for Baconian empiricism, so cementing the integration of Bacon's philosophy into the puritan worldview and underlining the relevance of that philosophy to educational reform. Furthermore, Comenius's work demonstrated that educational reform was an integral element in the eschatological scheme" (113–14). Webster illuminates the affinity of Baconian natural history, "educational reform," and an eschatological program for the recovery of prelapsarian knowledge. "PANSOPHY," Comenius writes in *A Patterne of Universall Knowledge*, "teacheth us that onely solid things be solidly handled" (34).

19. Comenius, *Orbis Sensualium Pictus*, 163.

20. For the word "familiar," I am indebted to Michael McKeon's *The Secret History of Domesticity: Public, Private, and the Division of Knowledge* (Baltimore: Johns Hopkins University Press, 2005), which defines "domestication" as the accommodation of public knowledge to private modes of knowing: "the positive, protoscientific revaluation of sense impressions . . . bespeaks both a traditionalistic, rhetorically oriented deference to unsophisticated minds and an emergent, epistemologically oriented advocacy of empiricist cognition" (342).

21. Lamb, *The Things Things Say*, 136. Lamb draws on Shapin and Schaffer's historiography: "The real position was defined by Newton as mathematical demonstration, in which ideas are communicated directly as thought, without the obscurity caused by figurative language" (132). Lamb's account of empirical epistemology is accordingly polarized: the real is susceptible to mathematical abstraction, while "figurative language" claims no intelligible relation to matter.

22. Boyle does not consistently or frequently use the modifiers "primary" and "secondary." However, he does employ the former adjective when he leans towards maximally mechanistic stress on the adequacy of the extra-seminal agency of the "Divine Architect's Geometry" to produce the quality of some things: for example, "the *primary* and insensible Corpuscles of Salts and Metals [are] of such determinate, curious, and exact Shapes, that, as they happen to be associated together, they should naturally produce Concretions." Boyle, *The Origin of Forms and Qualities*, in *Works of Robert Boyle*, ed. Michael Hunter and Edward B. Davis (London: Pickering and Chatto, 1999), 5: 363.

23. John Locke, *An Examination of P. Malebranche's Opinion of Seeing all things in God*, in *Posthumous Works of Mr. Locke* (London, 1706), 148.

24. Romanell, *Locke and Medicine*, 91. Emphasis in original.

25. Anstey, *Locke and Natural Philosophy*, 166, 167.

26. Robert Hooke, *Micrographia: or Some Physiological Descriptions of Minute Bodies* (1665), reprint (New York: Dover, 1961), 54.

27. Hooke, *Micrographia*, 68. Emphasis in original.

28. To suggest that primary matter is heuristically mapped by Hooke is not, contra Karen Barad, equivalent to suggesting that matter is not real. It is to underscore sensory understanding that is produced by texture.

29. Hooke, *Micrographia*, 153.

30. See Cynthia Sundberg Wall, *The Prose of Things: Transformations of Description in the Eighteenth Century* (Chicago: University of Chicago Press, 2006) for discussion of the empirical character of *Micrographia*: "verbal visualization is matched with narration to try to create an *experiential* description of something not just seen but imaginatively inhabited" (74).

31. Jonathan Swift, *Gulliver's Travels*, ed. Robert DeMaria (London: Penguin, 2001), 105.

32. Lisa Downing, "The Status of Mechanism in Locke's *Essay*," *Philosophical Review* 107: 3 (July 1998): 400.

33. Downing, "Status of Mechanism," 403.

34. See Locke's *Essay*, 2.8.§13 on the analogical likeness of body to smaller body: "For it being manifest, there are Bodies, and good store of Bodies, each whereof is so small, that we cannot, by any of our Senses, discover either their bulk, figure, or motion, as is evident in the Particles of the Air and Water, and other extremely smaller than those, perhaps, as much smaller than the Particles of Air, or Water, are smaller than Pease or Hail-Stones." See also 2.4.§1 on solidity: "though our Senses take no notice of it, but in masses of matter, of a bulk sufficient to cause a Sensation in us; Yet the Mind, having once got this *Idea* from such grosser sensible Bodies, traces it farther; and considers it, as well as Figure, in the minutest Particle of Matter, that can exist."

35. See Richard S. Westfall, *The Construction of Modern Science: Mechanisms and Metaphors* (Cambridge: Cambridge University Press, 1977), for discussion of Lémery as an epitome of the reliance of reductionist mechanism on particulate shape.

36. Nicholas Lémery, *Cours de Chymie* (1675), reprint (Paris: Les Introuvables / Éditions d'aujourd'hui, 1981), 55. My translation.

37. The location of Locke's secondary qualities marks a sustained source of philosophical dispute. Significant contributions include: Peter Alexander, *Ideas, Qualities, and Corpuscles: Locke and Boyle on the External World* (Cambridge: Cambridge University Press, 1985); Michael Ayers, *Locke: Volume I: Epistemology* (London: Routledge, 1991); E. M. Curley, "Locke, Boyle, and the Distinction Between Primary and Secondary Qualities," *Philosophical Review* 81: 4 (October 1972): 438–64; Downing, "The Status of Mechanism in Locke's *Essay*"; James Hill, "Primary Qualities, Secondary Qualities and Locke's Impulse Principle," *British Journal for the History of Philosophy* 17: 1 (2009): 85–98; J. L. Mackie, *Problems from Locke* (Oxford: Clarendon Press, 1976); Samuel C. Rickless, "Locke on Primary and Secondary Qualities," *Pacific Philosophical Quarterly* 78 (1997): 297–319; John W. Yolton, *John Locke and the Way of Ideas* (Oxford: Clarendon Press, 1968). One strand of philosophical work binds the primary-secondary distinction to the "veil of perception" thesis, which presumes that Locke endorses a representational theory of perception. I hew to Yolton's, Downing's, and Martha Brandt Bolton's position that while perceptible qualities do not mimetically represent particulate reality, they are constitutively linked to the reality that produces them. (Locke argues for the vividness and involuntariness of ideas of things as testimony to their material liaison to the real world, as well as the divinely ordained consistency of the link between any given physiological trigger and the idea it stimulates in the mind.) See *Pacific Philosophical Quarterly* 85 (2004), especially Lex Newman, "Locke on Sensitive Knowledge and the Veil of Perception—Four Misconceptions," 273–300, and Martha Brandt Bolton, "Locke on the Semantic and Epistemic Role of Simple Ideas of Sensation," 301–21. Bolton argues, and I agree, that "Although Locke strips knowledge down to the bare existence of things with powers to produce our ideas, he nevertheless maintains that we know these things are, in reality, articulated in accord with idea-collections given in experience" (306).

38. See G. A. J. Rogers, who writes that although "Locke rejected completely the [inductive] Cartesian route to knowledge of the essence of self and matter," Locke is indebted to Descartes in a way that Boyle is not: "Locke adopted the Cartesian language of ideas to characterise our experience." Rogers continues: "[Locke] came to accept that clear and distinct ideas provide our best criterion of truth." "The Intellectual Setting and Aims of the *Essay*," both citations, 15. See also Stephen Gaukroger, who writes of Locke's response to the Cartesian Nicolas Malebranche's *The Search for Truth* (1674–75) (a text whose occasionalist claim that God effects human ideas of things Locke, following Antoine Arnauld, refutes): "It is in reaction to the epistemological issues raised in Malebranche that Locke was able to refine his understanding of how to respond to the problem of how one can advocate the autonomy of phenomenal explanations, while at the same time not denying that there is a micro-corpuscularian realm underlying physical processes" ("The Role of Natural Philosophy in the Development of Locke's Empiricism," 60). See my " 'In *Idea*, a thousand nameless Joys': Secondary Qualities in Arnauld, Locke, and Haywood's *Lasselia*," *Eighteenth Century: Theory and Interpretation* 48: 3 (Fall 2007): 225–44, on Locke's agreement with Arnauld's claim for the identity of perception and idea.

39. Robert Boyle, *Experiments and Considerations Touching Colours*, in *Works of Robert Boyle*, ed. Hunter and Davis, 4: 36.

40. Boyle, *Colours*, 73–74.

41. Ibid., 5. Boyle's text was, he states, "written to a private Friend" (5), who is, the editors explain, "probably . . . his nephew Richard Jones (1641–1712), 3rd Viscount and 1st Earl of Ranelagh, the son of his sister Katherine, at whose home Boyle lived from 1668 until his death in 1691" (5 n.a).

42. Ibid., 25.

43. See *Oxford English Dictionary*, s.v. "freeze" (v.), definitions 5c, 5d, and 6. The citation from Milton's *Comus* is given under definition 6.

44. John Milton, *Comus, A Mask*, ll. 447–49, in *Complete Poems and Major Prose*, ed. Merritt Y. Hughes (New York: Macmillan, 1957), 100.

45. Robert Boyle, "Experimental Notes of The Mechanical Origin or Production of Fixedness" (1675–76), in *Works of Robert Boyle*, ed. Hunter and Davis, 8: 453, 450.

46. Robert Boyle, *Certain Physiological Essays*, in *Works of Robert Boyle*, ed. Hunter and Davis, 2: 163–64.

47. After reading Isaac Newton's *Principia* (1687), Locke revised his attribution of material causality to "nothing else but Modifications of Motion." For discussion of Locke's acknowledgement of gravity as a legitimate mode of inanimate action, see Anstey, *Locke and Natural Philosophy*, 111. However, I disagree with Anstey's claim that Locke's acceptance of gravity "reinforced" what Anstey takes as the tendency of the *Essay* from Draft A forward: "[Locke] envisaged a kind of corpuscular metric that would enable us to develop a demonstrative science of material bodies, a science of nature founded on a quantitative analysis of their constituent parts" (223 and 222). "Quantitative analysis" never surfaces in the *Essay* as a potential rationalizer of the link between primary and secondary qualities, as evidenced by the fact that primary ideas of geometrical attributes of matter fall by the wayside.

48. Margaret D. Wilson, "Superadded Properties: The Limits of Mechanism in Locke," *American Philosophical Quarterly* 16: 2 (April 1979): 147.

49. In several places Locke disavows any knowable connection between primary and secondary qualities, disavowals that threaten his claims for the latter's dependency on the former: "nor, did we know them [particles], could we discover any necessary *connexion* between them, and any of the *secondary Qualities*: which is necessary to be done, before we can certainly know their *necessary co-existence*" (4.3.§14).

50. I echo the title of Lisa Downing's article, "Are Corpuscles Unobservable in Principle for Locke?," *Journal of the History of Philosophy* 30: 1 (January 1992): 33–52. Downing cites Yolton's claim: "Locke did not think the particles were unobservable in principle" (*Locke and the Compass of Human Understanding*, 45). Downing affirms the observability of corpuscles by insisting upon Locke's reliance on analogy: "The great virtue of corpuscularianism, which was thought to render it uniquely intelligible and explanatory, was that the theoretical entities of its explanations are tiny *bodies* . . . The unperceivable causes [i.e. primary texture] argument destroys the analogy, and with it the corpuscular theory's claim to intelligibility" (47). Marking a blind spot in her own argument, Downing never acknowledges the experimental proof of tiny parts by chemical processes like sublimation. By contrast, Alexander argues that "corpuscles are in principle unobservable" (*Ideas, Qualities and Corpuscles*, 184) to claim of Locke's microscopical eyes: "That a man so equipped could 'come nearer' the discovery of the inner constitution of bodies and even 'get *Ideas*' of them does not imply that he would be able to see them" (186). This claim is unpersuasive because Locke ends the passage (cut from Alexander's excerpt) by stating that his microscopical man is "sharp-sighted enough to *see* the Configuration of the minute Particles" (emphasis mine).

51. See Neal Wood, *The Politics of Locke's Philosophy: A Social Study of "An Essay Concerning Human Understanding"* (Berkeley: University of California Press, 1983), for the claim that the militarization of religious politics during the Civil War era defined the corrective project

of the *Essay*: "Locke's own project for a new 'instauration' . . . would free men from error and prejudice, allowing them to fulfill their rational capacities" (95).

52. Martha Brandt Bolton, "The Real Molyneux Question and the Basis of Locke's Answer," in *Locke's Philosophy*, ed. Rogers, both citations, 77. Emphasis mine. See J. L. Mackie's discussion of two-dimensional versus three-dimensional shape: in the case of two-dimensional shape without prior understanding of shadowing and other visual signifiers, "a circle as seen and a circle as felt share the genuinely common feature of being the same all round, whereas a square, whether seen or felt, has the contrasting feature that its four corners are different from its four sides" (*Problems from Locke*, 31). With Bolton, Mackie confirms that the Molyneux outcome asserted by Locke "force[s] us to qualify Locke's claim that our ideas of shapes resemble the intrinsic qualities of objects which causally produce those ideas" (32).

53. Bolton, "Real Molyneux Question," both citations, 79.

54. Ibid., 98. Bolton continues: "However, this is not clearly consistent with Locke's doctrine that the 'reality' and 'adequacy' of simple ideas are guaranteed" (98). Figure would thus have the paradoxical status of both a simple and a complex (that is, recombined or aggregated) idea, violating a core distinction of Locke's *Essay*.

55. Texts like *L'Ecole des Filles* (1655) and *L'Académie des dames* (1680) assert the verbal transmission of erotic sensation that precedes stimulation at the so-called proper inlet. In this, the virgin resembles Locke's blind man. These texts also include as part of the virgin's sexual tutelage the rigorously mechanical explanation of activities like "frotter," "remuer," and "s'échauffer." *L'Ecole des filles ou La Philosophie des dames* (Paris: Éditions Allia, 1997), 31. Harrison and Laslett state that Locke read "French romance, some of it salacious" (*Library of John Locke*, 29).

56. Comenius, *Orbis Sensualium Pictus*, 74–89.

57. Ibid., 87.

58. In his manuscript "Anatomia" (1668), written three years before Drafts A and B of the *Essay*, Locke asserts the inscrutability of particulate matter: "it is certaine & beyond controversy that nature performs all her operations in the body by parts soe minute & insensible that I thinke noe body will ever have or pretend even by the assistance of glasses or any other invention to come to a sight of them." Cited by Anstey, *John Locke and Natural Philosophy*, 40. See Romanell, *John Locke and Medicine*, 111, for a slightly different transcription of this passage.

59. Comenius, *Orbis Sensualium Pictus*, 90–91.

60. Ibid., 91.

61. Robert Boyle, *The Sceptical Chymist*, in *Works of Robert Boyle*, ed. Hunter and Davis, 2: 300.

62. Boyle, *Sceptical Chymist*, 272.

63. Ibid., 306.

64. Ibid., 367.

65. Ibid., 299.

66. Ibid., 255.

67. Michael Ayers notes that Locke relies upon an "Aristotelian theory of definition." Whereas Aristotle distinguishes nominal from "simple" definition, which refers to a thing's "unitary essence," "Locke wanted to say that all candidate definitions of essences are in fact nominal." "The Foundations of Knowledge and the Logic of Substance: The Structure of Locke's General Philosophy," in *Locke's Philosophy*, ed. Rogers, 63.

68. Boyle, *Sceptical Chymist*, 357.

69. Ibid., 354.

70. Ibid., 357, 255, 307.

71. Ibid., 290.

72. Ibid., 316.

73. Ibid., 303.

74. Ibid., 302.

75. Daniel Defoe, *Roxana: The Fortunate Mistress*, ed. John Mullan (Oxford: Oxford University Press, 1996), 171.

76. Roxana's lapses of maternal obligation illuminate the difference between a chymical and a sentimental genealogy of human rights. For the latter, see Lynn Hunt, *Inventing Human Rights: A History* (New York: Norton, 2007): "What might be termed 'imagined empathy' serves as the foundation of human rights . . . Novels generated it by inducing new sensations about the inner self" (32). For a significant challenge to Hunt, see Markman Ellis's discussion of sentimentalism as an accessory to plantation slavery in *The Politics of Sensibility: Race, Gender and Commerce in the Sentimental Novel* (Cambridge: Cambridge University Press, 1996).

77. Monique Wittig, "On the Social Contract," in *The Straight Mind and Other Essays* (Boston: Beacon Press, 1992), 44. Wittig cites Locke as a theorist of the social contract (36).

78. Robert Boyle, "An Introduction to the History of Particular Qualities," in *Works of Robert Boyle*, ed. Hunter and Davis, 6: 282–83.

79. Eliza Haywood, *Fantomina: or, Love in a Maze*, in *Popular Fiction by Women, 1660–1730: An Anthology*, ed. Paula Backscheider and John Richetti (Oxford: Oxford University Press, 1996), 227.

80. Eliza Haywood, *Love in Excess*, ed. David Oakleaf (Peterborough, Ontario: Broadview Press, 2000), 142.

81. Haywood, *Love in Excess*, 144.

82. Both citations, Haywood, *Fantomina*, 247.

CHAPTER 3

Epigraph: George Thomson, *Loimotomia: or the Pest Anatomized* (London, 1666), 53.

1. George Thomson, *Galeno-pale: or, a Chymical Trial of the Galenists* (London, 1665), 62 (erroneously typeset as 26).

2. Robert Boyle, *Some Considerations Touching the Usefulness of Experimental Natural Philosophy* (1663), in *The Works of Robert Boyle*, ed. Michael Hunter and Edward B. Davis (London: Pickering and Chatto, 1999), 3: 223–24.

3. Robert Boyle, *Experiments and Considerations about the Porosity of Bodies*, in *Works of Robert Boyle*, ed. Hunter and Davis, 10: 107.

4. Francis Bacon, *The New Organon*, ed. Lisa Jardine and Michael Silverthorne (Cambridge: Cambridge University Press, 2000), 222.

5. Boyle, *Porosity*, 109. For an invaluable account of the milieu in which Harvey's ideas were elaborated at Oxford, see Robert G. Frank, *Harvey and the Oxford Physiologists: Scientific Ideas and Social Interaction* (Berkeley: University of California Press, 1980).

6. Charles Webster writes: "Boyle met Hartlib shortly after his return from the continent in 1644. They were probably introduced by Boyle's sister Katherine, Lady Ranelagh, one of Hartlib's major patrons. Boyle's interest in chemistry probably resulted from his contacts with the

Hartlib circle." "English Medical Reformers of the Puritan Revolution: A Background to the 'Society of Chymical Physitians,'" *Ambix* 14: 1 (February 1967): 30 n. 60. On Boyle's contact with Samuel Harlib and his circle, see also Betty Jo Teeter Dobbs, *The Foundations of Newton's Alchemy or "The Hunting of the Greene Lyon"* (Cambridge: Cambridge University Press, 1975); Michael Hunter, *Boyle: Between God and Science* (New Haven: Yale University Press, 2009); William R. Newman and Lawrence M. Principe, *Alchemy Tried in the Fire: Starkey, Boyle, and the Fate of Helmontian Chymistry* (Chicago: University of Chicago Press, 2002); and Charles Webster, *The Great Instauration: Science, Medicine and Reform, 1626–1660* (London: Duckworth, 1975).

7. Boyle, *Porosity*, 113.

8. Ibid., 116. This term and its cognates are ubiquitous in the Great Plague literature.

9. Ibid., 119.

10. For discussions of the chemical physicians' attempt to found their own college, see P. M. Rattansi, "The Helmontian-Galenist Controversy in Restoration England," *Ambix* 12: 1 (1964): 1–23; Webster, "English Medical Reformers"; and Harold J. Cook, "The Society of Chemical Physicians, the New Philosophy, and the Restoration Court," *Bulletin of the History of Medicine* 61: 1 (Spring 1987): 61–77; George Thomson also includes an account in *Galenopale*. On Helmontian medicine in England, see Harold J. Cook, *The Decline of the Old Medical Regime in Stuart London* (Ithaca: Cornell University Press, 1986), and Andrew Wear, *Knowledge and Practice in English Medicine, 1550–1680* (Cambridge: Cambridge University Press, 2000). William R. Newman shows that chemical physicians were also challenged by "empirics," practitioners who touted miracle cures and usurped the chemists' claims of therapeutic superiority: "The Society of Chymical Physicians was not merely an organ for the protection of iatrochemists against reprisals from the College: in the true Helmontian spirit it was meant to weed out all those who were not true 'sons of art.'" *Gehennical Fire: The Lives of George Starkey, an American Alchemist in the Scientific Revolution* (Cambridge, Mass.: Harvard University Press, 1994), 202.

11. *London's Dreadful Visitation: Or, A Collection of All the Bills of Mortality For this Present Year: Beginning the 20th of December 1664. and ending the 19th of December following* (London, 1665) provides a weekly breakdown of deaths from all causes as they occurred in parishes within the London walls (ninety-seven total), parishes outside the London walls (sixteen total), parishes in Middlesex and Surrey (twelve total), parishes in the City and Westminster (five total), and the pesthouse. For historical and biological synopsis, see A. Lloyd Moote and Dorothy C. Moote, *The Great Plague: The Story of London's Most Deadly Year* (Baltimore: Johns Hopkins University Press, 2004); Stephen Porter, *The Great Plague* (Gloucestershire: Amberley, 2009); Paul Slack, *The Impact of Plague in Tudor and Stuart England* (London: Routledge, 1985); Graham Twigg, *The Black Death: A Biological Reappraisal* (London: Batsford Academic and Educational, 1984).

12. Moote and Moote, *Great Plague*, 11. During the plague, about two hundred thousand of a total metropolitan population of about five hundred thousand fled the city.

13. Webster, "English Medical Reformers," 18.

14. Ibid.

15. Rattansi, "Helmontian-Galenist Controversy," 8. Rattansi notes: "At least one reason [for this surge of publication] was the temporary lapse of the Crown authority over printing" during the Puritan Revolution (8).

16. "[I]t was predominantly through Starkey that Boyle acquired a firm grounding in Helmontian theory and practice." Newman and Principe, *Alchemy Tried*, 222.

17. Thomson, *Loimotomia*, "To the Reader," n.p. Italicized in original.

18. George Starkey, *Natures Explication and Helmont's Vindication* (London, 1657), 304.

19. Starkey, *Helmont's Vindication*, 217, 226, 252.

20. George Thomson, *A Gag for Johnson, that Published Animadversions upon Galeno-pale* (London, 1665), 15.

21. Thomson, *Gag*, 21.

22. Thomson, *Loimotomia*, 175.

23. Thomson, *Galeno-pale*, "To the Reader," n.p.

24. George Starkey, *An Epistolar Discourse to the Learned and Deserving Author of Galeno-pale* (London: R. Wood, 1665), 41.

25. Thomson, *Gag*, 6.

26. Ibid., 27.

27. Thomson, *Galeno-pale*, 83.

28. Ibid., 89.

29. Thomson, *Gag*, 5.

30. Ibid., 24.

31. Thomson, *Galeno-pale*, 105.

32. Starkey, *Epistolar Discourse*, 53.

33. Thomson, *Gag*, 27–28.

34. George Thomson, *Loimologia: A Consolatory Advice, And some brief Observations Concerning the Present Pest* (London: L. Chapman, 1665), 2.

35. Thomson, *Loimologia*, 4.

36. Thomson, *Loimotomia*, 70.

37. Ibid., 77.

38. Ibid., 78 and 10.

39. Ibid., 78–79.

40. Ibid., 95.

41. Ibid., 133.

42. On Paracelsus and Helmont, see Allen G. Debus, *The English Paracelsians* (London: Oldbourne Book Co., 1965); Walter Pagel, *Joan Baptista Van Helmont: Reformer of Science and Medicine* (Cambridge: Cambridge University Press, 1982), and *Paracelsus: An Introduction to Philosophical Medicine in the Era of the Renaissance*, 2nd rev. ed. (Basel: Karger, 1982); and Charles Webster, *Paracelsus: Medicine, Magic and Mission at the End of Time* (New Haven: Yale University Press, 2008).

43. Audrey B. Davis, *Circulation Physiology and Medical Chemistry in England, 1650–1680* (Lawrence, KS: Coronado Press, 1973), 14.

44. Thomson, *Loimotomia*, 65.

45. Ibid., 64.

46. Ibid., 141–42.

47. The fact that Charleton and his Helmontian compatriot Sir Kenelm Digby (1603–1665) were royalists shows that the political affiliations of Helmontian scientists were not unilaterally puritan. Thomson reports of the chemists' failed attempt to found a Society of Chemical Physicians: "seldom hath there been any Design more generally countenanced and animated by the learned Nobility, and ingenious Gentry of *England*, then this" (*Galeno-pale*, 104). For a corrective to Webster's claim for the consistently republican agenda advanced by chemical physicians, see Cook, "Society of Chemical Physicians," which argues that chemistry was

simultaneously a craft, bourgeois, and aristocratic avocation. On Charleton's Helmontianism, see Antonio Clericuzio, "From van Helmont to Boyle: A Study of the Transmission of Helmontian Chemical and Medical Theories in Seventeenth-Century England," *British Journal for the History of Science* 26: 3 (1993): 303–34. Clericuzio notes: "Charleton's adherence to van Helmont's doctrines . . . was a peculiar one" (306).

48. Walter Charleton, *A Ternary of Paradoxes: Of The Magnetic Cure of Wounds. Nativity of Tartar in Wine. Image of God in Man* (London, 1650), 16–17, 17.

49. Walter Pagel, *Helmont*, 10. Of material force that is, for Helmont, also spiritual, Pagel writes: "This . . . does not imply a dualistic view in which spirit is imposed on matter; the spiritual and the material are rather seen as the two convertible faces of the same coin, the individual unit, in which they are inseparably interwoven" (10).

50. Ibid., 197 and 196.

51. Charleton, *Ternary*, 18.

52. Ibid., 19.

53. See Pagel, *Helmont*, Chapter 5, "The Ontological Conception of Disease." I am indebted to Clericuzio, "From van Helmont to Boyle."

54. Charleton, *Ternary*, both citations, 20.

55. Ibid., 22.

56. Thomson, *Loimotomia*, 154–55.

57. Ibid., 159–60.

58. Thomas Willis, *A Plain and Easie Method for Preserving [by God's Blessing] those that are Well from the Infection of the Plague . . . Written in the Year 1666* (London: W. Crook, 1691), 27.

59. Thomson, *Loimotomia*, 13.

60. Ibid, 14 and 15–16.

61. Ibid., 16.

62. Ibid., 13.

63. Ibid., 11. Thomson's employment of the Helmontian word "consentaneous" marks a resonant incidence of its use before it is adopted by Samuel Richardson's *Clarissa* (1748–49), since Clarissa also deploys it to refer to the congruence of a person's spiritual and bodily attributes. On metaphysics in *Clarissa*, see Chapter 6.

64. Starkey, *Helmont's Vindication*, 264.

65. Thomson, *Loimotomia*, 24–25.

66. Ibid., 31–32.

67. Ibid., 29.

68. Ibid., 46.

69. Newman and Principe, *Alchemy Tried*, 63–64.

70. Jole Shackelford, "Seeds with a Mechanical Purpose: Servinus' Semina and Seventeenth-Century Matter Theory," in *Reading the Book of Nature: The Other Side of the Scientific Revolution*, ed. Allen G. Debus and Michael T. Walton (Kirksville, Mo.: Sixteenth Century Journal Publishers, 1998), 43.

71. Thomson, *Galeno-pale*, 90.

72. Kenneth D. Keele, "The Sydenham-Boyle Theory of Morbific Particles," *Medical History* 18 (1974): 247.

73. Vivian Nutton, "The Reception of Fracastoro's Theory of Contagion: The Seed that Fell Among Thorns?" *Osiris* (2nd series) 6 (1990): 233. While Nutton discusses the Veronese

doctor and poet Girolamo Fracastoro (c. 1478–1553), whose disease theory influenced Helmont, his historical caution holds "until the bacteriological and microscopic revolution of the late nineteenth century, [before which] there was no convincing reason why one should interpret Fracastoro's seeds of disease literally, and much to be said against it" (233). See also Nutton's "The Seeds of Disease: An Explanation of Contagion and Infection from the Greeks to the Renaissance," *Medical History* 27 (1983): 1–34, which situates medical thinking about contagion in "a long and vigorous debate among philosophers and others interested in the natural world on the problem of causation and, in particular, of action at a distance" (30).

74. Nutton, "Fracastoro's Theory of Contagion," 232.

75. Starkey, *Helmont's Vindication*, 268–69.

76. Ibid., 291.

77. Willis, *Plain and Easie Method*, 7.

78. Starkey, *Helmont's Vindication*, 277.

79. Thomson, *Loimotomia*, 110–11.

80. Starkey, *Helmont's Vindication*, 304–5. Newman cautions of Starkey's "Sope": "this comparison . . . was not just a homey analogy, for the very technology by which Starkey thought to obtain his volatile alkali was that of soap-making" (*Gehennical Fire*, 176).

81. Thomson, *Galeno-pale*, 31.

82. Robert Boyle, *Of the Reconcileableness of Specifick Medicines to the Corpuscular Philosophy*, in *Works of Robert Boyle*, ed. Hunter and Davis, 10: 368.

83. Boyle, *Specifick Medicines*, 369.

84. Ibid., 365.

85. Ibid.

86. Thomson, *Galeno-pale*, 44–45.

87. Ibid., 75.

88. Ibid., 16.

89. Ibid., 68.

90. Ibid., both citations, 61.

91. Boyle, *Specifick Medicines*, 369.

92. Boyle, *Usefulness*, 438.

93. Boyle, *Specifick Medicines*, 390.

94. Letter to John Mallet, cited in Newman, *Gehennical Fire*, 75.

95. Boyle, *Usefulness*, 439, 440.

96. Ibid., 438.

97. Ibid.

98. See Wear, *Knowledge and Practice*, on patients' evacuation of, for example, "twenty Chamber-pots full of vomit . . . and eight stools so great, as to fill a Close-stool-pan twice full" (Thomas O'Dowde, *The Poor Man's Physician, Or the True Art of Medicine* [1665], cited by Wear, 422).

99. Wear, *Knowledge and Practice*, 367, 392, 419.

100. Robert Boyle, *An Essay about the Origin & Virtues of Gems*, in *Works of Robert Boyle*, ed. Hunter and Davis, 7: 67, 47.

101. Ibid., 52, 67, 70.

102. Ibid., 70.

103. Ibid., 52, 47.

104. Ibid., 52.

105. Thomas Sydenham, *Collections of Acute Diseases. The Second and Third Part* (London, 1688), 7.

106. Sydenham, *Acute Diseases*, 29.

107. Ibid., 4.

108. See Kenneth Dewhurst, *John Locke (1632–1704), Physician and Philosopher: A Medical Biography* (London: Wellcome Historical Medical Library, 1963), 35.

109. Cited in Dewhurst, *Physician and Philosopher*, 39. While Dewhurst notes that the essay "is in Locke's handwriting, but contemporary evidence suggests that Sydenham was the author" (38), Patrick Romanell cites it as "inspired by Sydenham" in *John Locke and Medicine: A New Key to Locke* (Buffalo, N.Y.: Prometheus Books, 1984), 115.

110. Romanell, *John Locke and Medicine*, 119.

111. Locke to William Molyneux, January 20, 1682/3. Cited in Dewhurst, *Physician and Philosopher*, 310.

112. Cook, *Decline of the Old Medical Regime*, 185.

113. Sydenham, *Acute Diseases*, 97.

114. The tenacity with which Sydenham defends bleeding in the face of "great obstruction," "friends violently oppos'd [to] the taking away any more blood," and "frequently met . . . obstacles" (*Acute Diseases*, 23, 24, 24) belies Wear's claims for a unilateral cultural attachment to phlebotomy. Sydenham proposes "bleed[ing] moderately with respect to the strength and temperament of the sick" as a compromise aimed to placate demands for treatment that is "not so displeasing" (*Acute Diseases*, 25, 24).

115. Sydenham, *Acute Diseases*, 8.

116. Ibid., 26.

117. Ibid., 10.

118. Ibid., 8.

119. Ibid., 91, 92.

120. Daniel Defoe, *A Journal of the Plague Year*, ed. Cynthia Wall (London: Penguin, 2003), 9. All citations from this edition will be noted parenthetically.

121. John Locke, *An Essay Concerning Human Understanding*, ed. Peter H. Nidditch (Oxford: Oxford University Press, 1975), 1.2.§15; 55.

122. Locke, *Essay*, 2.1.§23; 117.

123. Robert Boyle, "An Experimental Discourse of Some Unheeded Causes of the Insalubrity and Salubrity of the Air," in *Works of Robert Boyle*, ed. Hunter and Davis, both citations, 10: 335.

124. Boyle, "Salubrity of the Air," 313, 313–14.

125. Thomson, *Loimotomia*, 61.

126. For a synopsis of recent critical deployments of form, see Marjorie Levinson, "What Is New Formalism?" *PMLA* 122: 2 (2007): 558–68.

127. See Frances Ferguson, *Pornography, the Theory: What Utilitarianism Did to Action* (Chicago: University of Chicago Press, 2004), especially her discussion of tort law, utilitarianism, and pornography; and Sandra Macpherson, *Harm's Way: Tragic Responsibility and the Novel Form* (Baltimore: Johns Hopkins University Press, 2010), which argues that eighteenth-century law and the novel construct "a formal person—a person whose content as represented by the state of her interior (or mind) is irrelevant to the question of her responsibility" (23).

128. For treatments of the *Journal* that argue for Defoe's representation of persons who cannot be assimilated to the model of the liberal individual, see Macpherson, *Harm's Way*,

40–49, and Nancy Armstrong, "The Other Side of Modern Individualism: Locke and Defoe," in *Individualism: The Cultural Logic of Modernity* (Lanham, Md.: Lexington Books, 2011). In a suggestive account of Defoe's *Robinson Crusoe* (1719), Jayne Lewis argues that Crusoe's weather-writing "blur[s] . . . the line of demarcation between a character's inside and his outside." *Air's Appearance: Literary Atmosphere in British Fiction, 1660–1794* (Chicago: Chicago University Press, 2012), 107.

129. Ernest B. Gilman, "The Subject of the Plague," *Journal for Early Modern Cultural Studies* 10.2 (Fall/Winter 2010), 29.

130. Boyle, "Salubrity of the Air," 340, 339, 340.

131. Ibid., all citations, 341.

132. Paula R. Backscheider, *Daniel Defoe: His Life* (Baltimore: Johns Hopkins University Press, 1989), 15–16.

133. Charles Morton, *Compendium Physicae*, intro. Samuel Eliot Morison, in *Publications of the Colonial Society of Massachusetts* (Boston: Published by the Society, 1940), 33: 11. See Morison's introduction on Morton's debt to Boyle. See also Ilse Vickers, *Defoe and the New Sciences* (Cambridge: Cambridge University Press, 1996): "Whether writing on the circumference of the earth, on cold, on fluids, on form, Morton always cites Boyle as the decisive authority" (26).

134. Morton, *Compendium*, 134, 121.

135. Daniel Defoe, *The History Of the Principal Discoveries and Improvements, In the Several Arts and Sciences* (London: Printed for W. Mears et al., 1727), 251. See Vickers, *Defoe and the New Sciences*, 76–79, for an exhaustive demonstration that Defoe "copies from" (76) Boyle's *Certain Physiological Essays* in his discussion of magnetism. For reference to *Discoveries and Improvements*, I am grateful to Vickers and Backscheider. Vickers's study does not mention Defoe's *Journal*.

136. Defoe, *Discoveries and Improvements*, 255.

137. Robert Boyle, *Experiments and Considerations Touching Colours*, in *Works of Robert Boyle*, ed. Hunter and Davis, 4: 30.

138. Robert Boyle, *New Experiments Physico-Mechanicall, Touching the Spring of the Air*, in *Works of Robert Boyle*, ed. Hunter and Davis, both citations, 1: 203.

139. Locke, *Essay*, 2.8.§17, 138.

140. Boyle, *Usefulness*, 446.

141. Daniel Defoe, *Due Preparations for the Plague, as well for Soul as Body*, in *Romances and Narratives by Daniel Defoe*, ed. George A. Aitken (London: J. M. Dent and Co., 1895), 15: 88.

142. Defoe, *Due Preparations*, 10.

143. Ibid., 88.

144. Ibid., 81.

145. See Moote and Moote, 67–71, for discussion of the difference between "miasmist" and "contagionist" etiologies of the plague, which were "[m]ost often . . . combined" (70). Now generally accepted as the cause of the great plague, the bacillus *Yersinia pestis* is transmitted by a bacillus-containing or "blocked" flea (*Xenopsylla cheopis*) that has fed on the blood of the infected Black rat (*Rattus rattus*) or Brown rat (*Rattus norvegicus*). However, variant human cases of "pneumonic plague" may be spread by human sputum "without the intermediation of fleas" (Twigg, *The Black Death*, 20). Another variant, "intense septicaemia" (ibid., 19) that bypasses the lymph nodes, explains the difference between victims who seem to perish instantaneously and those who harbor the infection, develop symptoms like buboes, and then die.

146. Defoe, *Due Preparations*, 16–17.

147. Macpherson's *Harm's Way* argues that Defoe's characters are exculpated from guilt for the course of the plague. But Defoe refines modalities of moral agency to correct for the obfuscated self-knowledge of all not openly sick persons: most notably, he applauds people who voluntarily curtail their conversation with others.

148. Thomson, *Loimotomia*, all citations, 110. Italicized in original.

149. Wolfram Schmidgen's *Exquisite Mixture: The Virtues of Impurity in Early Modern England* (Philadelphia: University of Pennsylvania Press, 2012) suggests that natural or secondary as opposed to immediate causes constitute a flexible mode of providential explanation in late seventeenth-century Britain, in part to counter an upsurge of predestinarian thought.

150. Boyle, *Salubrity of the Air*, 329–30.

151. Ibid., 312.

152. Defoe, *Due Preparations*, 56.

153. Boyle, *Salubrity of the Air*, 330.

154. Defoe, *Due Preparations*, 33.

CHAPTER 4

Epigraphs: Penelope Aubin, *The Life of Charlotta Du Pont, An English Lady* (London: Printed for A. Bettesworth, 1723), 164–65. *A New Canting Dictionary* (London: Booksellers of London and Westminster, 1725), n.p.

1. William J. Burling, "Chetwood, William Rufus (d. 1766)," *Oxford Dictionary of National Biography*, online ed., May 2005. Maximillian E. Novak includes Chetwood among the "members of the conger who would publish most of his [Defoe's] future work [after *Robinson Crusoe*]: A. Bettersworth, C. King, J. Brotherton, W. Meadows, W. Chetwood, and William Boreham." *Daniel Defoe: Master of Fictions* (Oxford: Oxford University Press, 2001), 580.

2. Eliza Haywood, *Love in Excess*, ed. David Oakleaf (Peterborough, Ontario: Broadview Press, 2000), 142.

3. Haywood, *Love in Excess*, both citations, 144.

4. William Chetwood, *The Voyages, Dangerous Adventures And imminent Escapes of Captain Richard Falconer* (London: Printed for W. Chetwood, 1720), 3: 139, 141.

5. Chetwood, *Falconer*, 3: 141, 142.

6. While aware of the historical anachronism of the modern term "race," which did not signify a coherent category of either genetic, phenotypic, or ethnographic identity in the eighteenth century, I have chosen to use the word due to pressures of expedience. "Race" must thereby be understood as a signifier whose meaning is not consolidated. When the term applies in its eighteenth-century sense as family or starting generation, I underscore this meaning in the text.

7. H. R. Luhning, "A Crafted Debut: Haywood's *Love in Excess* and the Literary Marketplace," *Lumen* 28 (2009): 97–110.

8. The daughter of a mixed-race mother and black father, "Melanctha" is a titular personage in Gertrude Stein's *Three Lives* (1909). For this connection, I am grateful to the neon sculpture of Glenn Ligon. See *La Biennale di Venezia, All the World's Futures*, Short Guide (Venice: Marsilio Editori, 2015), 30–31.

9. Robert Boyle, *Experiments and Considerations Touching Colours*, in *The Works of Robert Boyle*, ed. Michael Hunter and Edward B. Davis (London: Pickering and Chatto, 1999), 4: 40.

10. Boyle, *Colours*, 40–41.

11. Ibid., 42.

12. Ibid., 28.

13. Ibid.

14. Ibid., 52.

15. Ibid., 53.

16. Boyle's editors note that a "preliminary version of this work [*Experiments and Considerations Touching Colours*], entitled 'The History of Colours Begun' . . . was printed at London in 1663" (Boyle, *Colours*, xii). They continue: "The reader of the work whose reaction is perhaps best known is Isaac Newton, who made extensive notes on it and who was apparently stimulated by it to record his first known optical experiments" (xv–xvi). Betty Jo Teeter Dobbs includes the 1663 publication among the texts by Boyle included in "Newton's student notebook of 1661–65." *The Foundations of Newton's Alchemy or "The Hunting of the Greene Lyon"* (Cambridge: Cambridge University Press, 1975), 123. Alan E. Shapiro writes: "Newton's interest in the colors of bodies was stimulated and perhaps first aroused by his encounter with Boyle's *Touching Colours* in late 1664 or early 1665." *Fits, Passions, and Paroxysms: Physics, Method, and Chemistry and Newton's Theories of Colored Bodies and Fits of Easy Reflection* (Cambridge: Cambridge University Press, 1993), 99.

17. Dennis L. Sepper, *Newton's Optical Writings: A Guided Study* (New Brunswick, N.J.: Rutgers University Press, 1994), 39. I am indebted to Sepper throughout this section.

18. Isaac Newton, *Opticks; or A Treatise of the Reflections, Refractions, Inflections & Colours of Light* (1730), foreword Albert Einstein, intro. Edmund Whittaker, pref. I. Bernard Cohen (New York: Dover, 1952), 113. Italicized in original.

19. Sepper lists Hooke and Christiaan Huygens (1629–1695) among "Newton's chief critics" (*Newton's Optical Writings*, 51). Edmund Whittaker states that the 1672 letter "gave rise to an acute controversy. Hooke in particular assailed it vehemently; and the unpleasant consequences which followed its announcement had much to do with the reluctance with which Newton ever afterwards showed to make known his results to the world." In a subsequent letter to Henry Oldenberg, secretary of the Royal Society, Newton declares his intention "to be no farther solicitous about matters of Philosophy" (*Opticks*, intro., both citations, lxvii).

20. Newton, 1672 letter, in Sepper, *Newton's Optical Writings*, 37. I rely upon Sepper's version of the letter, taken from Newton's *Correspondence*, throughout.

21. See Sepper, *Newton's Optical Writings*, 39.

22. Newton, 1672 letter, in Sepper, *Newton's Optical Writings*, 40.

23. Ibid., 43–44.

24. I follow Newton's published terminology in referring to these as "Rays." See Shapiro, "Newton's Optics and Atomism," in *The Cambridge Companion to Newton*, ed. I. Bernard Cohen and George E. Smith (Cambridge: Cambridge University Press, 2002), which shows the extent to which Newton avoided references to the corpuscularity of light in the *Opticks* while affirming it in other venues.

25. Newton, 1672 letter, in Sepper, *Newton's Optical Writings*, 44.

26. Ibid., 47–48.

27. Ibid., 48.

28. For Curran, Arbuthnot "updates Hippocrates's understanding of air-based national dispositions with new information on the 'mechanical' results of different climates on human

fibers and character." Curran does not discuss particulate science. Andrew S. Curran, *The Anatomy of Blackness: Science & Slavery in an Age of Enlightenment* (Baltimore: Johns Hopkins University Press, 2011), 134.

29. Roxann Wheeler, *The Complexion of Race: Categories of Difference in Eighteenth-Century British Culture* (Philadelphia: University of Pennsylvania Press, 2000), 21–22, 26.

30. Wheeler, *Complexion of Race*, 26.

31. Ibid., 32. Wheeler writes: "In the nineteenth century, Foucault argues, a science of surfaces is transformed into a science that promotes a relatively stable relationship between surface and depth. In new scientific practices, natural historians made the visible correspond to the invisible structure beneath . . . To be sure, this methodology can be traced in its incipient form to [Francis] Bacon, [William] Harvey, and a few other seventeenth-century representatives of the new science" (32). This synopsis of eighteenth-century British "science of surfaces" denies particulate explanation. Wheeler's "few seventeenth-century representatives" aggravate her omission of Boyle and Locke, as well as other physicians, chemists, and matter theorists. Her reliance on Foucault's "*The Order of Things*" (217) is anticipated in Mary Louise Pratt, *Imperial Eyes: Travel Writing and Transculturation* (London: Routledge, 1992), which offers a confusing account of how "depth" might be added to "the surface blanket" of eighteenth-century classificatory discourse theorized by Foucault (60).

32. John Arbuthnot, *An Essay Concerning the Effects of Air on Human Bodies* (London: J. Tonson, 1733), vi–vii. All subsequent citations will be noted parenthetically.

33. Robert Boyle, *New Experiments Physico-Mechanicall, Touching the Spring of the Air and its Effects*, in *Works of Robert Boyle*, ed. Hunter and Davis, 1: 165.

34. Boyle, *Spring of the Air*, 165.

35. Ibid.

36. Stephen Hales, *Statical Essays: Containing Hæmastaticks; or, An Account of some Hydraulick and Hydrostatical Experiments made on the Blood and Blood-Vessels of Animals* (1733), reprint (New York: Hafner Publishing Co., 1964), 108.

37. Arbuthnot attributes the state of slavery to oscillatory difference that aligns "activity" with the difficulty of farming in northern lands: "in Countries which do not produce without much Labour . . . the Workman ought to have a certain Title to the Fruits of it. There are Degrees of Slavery, and, generally speaking, it is most extreme in some hot and fruitful Countries" (153).

38. For correlations of race and culture, see Walter Benn Michaels, "Race into Culture: A Critical Genealogy of Cultural Identity," *Critical Inquiry* 18: 4 (Summer 1992): 655–85, and *Our America: Nativism, Modernism, and Pluralism* (Durham: Duke University Press, 1995). See also Étienne Balibar, "Is There a 'Neo-racism'?" in *Race, Nation, Class: Ambiguous Identities*, ed. Balibar and Immanuel Wallerstein (London: Verso, 1991).

39. Anthony Appiah, "The Uncompleted Argument: Du Bois and the Illusion of Race," in *'Race,' Writing, and Difference*, ed. Henry Louis Gates (Chicago: Chicago University Press, 1986): 21–37. Drawing on genetic research, Appiah shows the vanishingly small numerical "extent to which members of these human populations we call races differ more from each other than they do from members of the same race." He concludes: "race is relatively unimportant in explaining biological differences between people, where biological difference is measured in the proportion of differences in loci on the chromosome" (both citations, 31).

40. See Locke's *Essay*, Book 2, Chapter 21, "On Power."

41. Curran, *Anatomy of Blackness*, 122.

42. John Mitchell and Peter Collinson, *An Essay upon the Causes of the different Colours of People in different Climates; by John Mitchell, M. D. Communicated to the Royal Society by Mr. Peter Collinson, F. R. S. Philosophical Transactions of the Royal Society* (London, 1744), 43: 102–50. All further citations will be noted parenthetically.

43. James Delbourgo, "The Newtonian Slave Body: Racial Enlightenment in the Atlantic World," *Atlantic Studies: Literary, Cultural and Historical Perspectives* 9: 2 (2012): 200.

44. Delbourgo, "Newtonian Slave Body," 196.

45. Ibid., 202, 197.

46. As Delbourgo underscores, Mitchell's *Colours of People* is cited by the abolitionists Olaudah Equiano and Thomas Clarkson, among others. See Olaudah Equiano, *The Interesting Narrative of the Life of Olaudah Equiano, or Gustavus Vassa, the African*, ed. Werner Sollors (New York: Norton, 2001), where Equiano credits "the difference of colour between the Eboan Africans and the modern Jews" to "a fact as related by Dr. Mitchell. 'The Spaniards, who have inhabited America, under the torrid zone, for any time, are become as dark coloured as our native Indians of Virgina; of which *I myself have been a witness*' " (30, emphasis Equiano's).

47. I am indebted to Alan E. Shapiro's discussion of the pervasive—and misguided—mechanism of Newton's attempt at a "new, mathematical science of color" that extends his findings for thin films to three-dimensional bodies (*Fits, Passions*, 119).

48. Boyle, *Touching Colours*, 51, 51–52.

49. Ibid., 52.

50. Ibid., 50.

51. Shapiro, *Fits, Passions*, 102. Shapiro notes that "there is strong evidence that he [Newton] drew the less complex compositional theory that he was using in his early optical theory from Boyle's *Origine of Formes and Qualities, (according to the Corpuscular Philosophy)* (1666) . . . In the 1672 version of the 'Hypothesis,' Newton appears to have adopted Boyle's terms 'corpuscle' and 'clusters' for the compound particles" (*Fits, Passions*, 88).

52. Newton, *Opticks*, 193 and passim. Italicized in original.

53. Robert Hooke, *Micrographia; or, Some Philosophical Descriptions of Minute Bodies* (1665), reprint (New York: Dover, 1961), 49, 49, 47.

54. Hooke, *Micrographia*, all citations, 50.

55. Ibid., 51.

56. Newton, *Opticks*, 194.

57. Ibid., 255.

58. Ibid., all citations, 256.

59. Ibid.

60. Ibid., 255.

61. Ibid., 245.

62. Ibid., 255. Italicized in original. Mitchell's *Colours of Persons* notes: "*Opt.* lib. II. *Part* 3. PROP. 7" (119).

63. Ibid., 250. Italicized in original. Mitchell's *Colours of Persons* footnotes: "Newton *Opt.* lib. II. Part 3. PROP. IV" (117).

64. Ibid., 250.

65. Ibid., 251.

66. Ibid.

67. Ibid., all citations, 206.

68. On the problem of opaque black bodies, I am indebted to Shapiro's *Fits, Passions*, Chapter 3. However, I am perplexed by his discussion of transduction and transparency, defined by Newton in terms of the contiguity of two "*equally refracting Mediums*" (*Opticks*, 246). Shapiro claims that Newton's contiguous, identically refracting media are not porous: "contiguous parts have no pores" (Shapiro, *Fits, Passions*, 115 n. 46). Shapiro proceeds to argue that nonporous media would violate Newton's Proposition 4, which predicates transparency on parts that fall below a certain definite bigness, to suggest: "Newton seems to have carelessly and unnecessarily carried his transduction to too fine a level" (ibid.). But Newton never specifies that his contiguous transparent media (e.g., glass, water, crystal) "have no pores" and never predicates the transmission of light on nonporosity.

69. Newton *Opticks*, 265.

70. Ibid., 262. Italicized in original.

71. Ibid., 266–67.

72. Ibid., 267. Newton cites the surprising porosity of "Gold," not a black body, to make his point.

73. Ibid., both citations, 248. Proposition 2 italicized in original.

74. Shapiro, *Fits, Passions*, 125.

75. Thomas Melvill, cited in ibid., 125 n. 76. Delivered as a talk in 1752, *Observations on Light and Colours* was posthumously published in *Essays and Observations, Physical and Literary*, vol. 2 (Edinburgh, 1756). Citation, 82.

76. Shapiro notes that "the primordials would be black, though Newton never acknowledged this consequence." *Fits, Passions*, 133.

77. Mitchell refers to what is below "the Cuticle," a synonym for the epidermal or outermost level of skin.

78. William Cowper, *Anatomy of Humane Bodies*. Digital source: http://www.biusante.parisdescartes.fr/histmed/medica/cote?02042. Page: http://www.biusante.parisdescartes.fr/histmed/medica/page?02042&p=27.

79. Cowper, *Anatomy*, http://www.biusante.parisdescartes.fr/histmed/medica/page?02042&p=27.

80. Ibid.

81. Ibid.

82. Mitchell repeats: "there is not so great, unnatural, and unaccountable a Difference between Negroes and white People, on account of their Colours, as to make it impossible for both ever to have been descended from the same Stock" (131).

83. An example is Katy L. Chiles, *Transformable Race: Surprising Metamorphoses in the Literature of Early America* (Oxford: Oxford University Press, 2014), which asserts the mutability of eighteenth-century race while considering countervailing ontological tendencies only as the harbinger of an essentialist nineteenth century.

84. Wheeler, *Complexion of Race*, 21, 43.

85. Delbourgo, "Newtonian Slave Body," 201.

86. Dror Wahrman, *The Making of the Modern Self: Identity and Culture in Eighteenth-Century England* (New Haven: Yale University Press, 2004), Chapter 3. Wahrman equates climatic influence and malleability: "climatic understanding of human diversity was intertwined with an acceptance of the mutability of those physical traits" (88).

87. Catherine Molineux, *Faces of Perfect Ebony: Encountering Atlantic Slavery in Imperial Britain* (Cambridge, Mass.: Harvard University Press, 2012), 95. Molineux's gloss of seventeenth-

century color theory stresses hue whose "essential nature might ultimately reside in the structure of the skin and the particular way in which that object formed an image in the viewer's mind" (96).

88. Chetwood, *Falconer*, both citations, 1: 11.

89. Ibid., 1: vii.

90. Michael McKeon, *The Origins of the English Novel, 1600–1740* (Baltimore: Johns Hopkins University Press, 1987), 21.

91. Chetwood, *Falconer*, 32. On Locke and pineapples, see Deidre Shauna Lynch, *The Economy of Character: Novels, Market Culture, and the Business of Inner Meaning* (Chicago: University of Chicago Press, 1998): "He finds out new ideas for the inquiring mind to collect" (50).

92. Chetwood, *Falconer*, 1: 32–36.

93. Ibid., 1: 38.

94. John Locke, *An Essay Concerning Human Understanding*, ed. Peter H. Nidditch (Oxford: Oxford University Press, 1975), 2. 33 (italicized as chapter title "*Of the Association of* Ideas"); 2.33.§6.

95. Locke, *Essay*, 2.10.§2.

96. Ibid., 2.33.§6, 2.33.§7.

97. William Chetwood, *The Voyages and Adventures of Captain Robert Boyle, In several Parts of the World. Intermix'd with the Story of Mrs. Villars, an English Lady with whom he made his surprizing Escape from Barbary* (London: Printed for John Watts, 1726), 2.

98. Chetwood, *Boyle*, 25.

99. Ibid., 46, 50.

100. Ibid., 67.

101. On European enslavement in North Africa, which affirms the non-binary treatment of color evidenced in Chetwood and Aubin's novels, see Adam R. Beach, "African Slaves, English Slave Narratives, and Early Modern Morocco," in *Eighteenth-Century Studies* 46: 3 (Spring 2013): 333–48. While, as Beach recounts, "the European *aseer* slaves were fundamentally different from black African slaves, who hailed from what were considered to be more primitive societies" (339), he affirms the necessity of critical attention to "intra-slave dynamics" (345) like those mobilized by Chetwood and Aubin. On the history of white enslavement in north Africa and ensuing complexities for a black-white schematics of race, see Srinivas Aravamudan, *Enlightenment Orientalism: Resisting the Rise of the Novel* (Chicago: University of Chicago Press, 2012); Linda Colley, *Captives* (New York: Pantheon Books, 2002); Robert C. Davis, *Christian Slaves, Muslim Masters* (London: Palgrave, 2003); Nabil Matar, *Britain and Barbary, 1589–1689* (Gainesville: University Press of Florida, 2005), and *Turks, Moors, and Englishmen in the Age of Discovery* (New York: Columbia University Press, 1999); Felicity Nussbaum, "Between 'Oriental' and 'Blacks So Called', 1688–1788," in *The Postcolonial Enlightenment: Eighteenth-Century Colonialism and Postcolonial Theory*, ed. Daniel Carey and Lynn Festa (Oxford: Oxford University Press, 2009); and Roxann Wheeler's invaluable chapter on Defoe's Friday in *Complexion of Race*.

102. Chetwood, *Falconer*, 1: 65.

103. Chetwood, *Boyle*, 115.

104. Ibid., 307.

105. Ibid., 318.

106. William Chetwood, *The Voyages, Travels and Adventures, of William Owen Gwin Vaughan, Esq.* (London: Printed for J. Watts, 1736), 2: 250, 251.

107. Chetwood, *Vaughan*, 1: 300. Italicized in original.

108. Chetwood, *Boyle*, 229.

109. Ibid., 234.

110. Ibid., both citations, 240.

111. Ibid., 241.

112. Ibid., 242.

113. Ibid., 59.

114. Chetwood, *Vaughan*, 2: 126.

115. See Alan Grosrichard, *The Sultan's Court: European Fantasies of the East*, trans. Liz Heron (London: Verso, 1998): "Now it is precisely because royal power is rooted in domestic power that its abuse can be denounced by describing it as *despotic*" (5). See my *Ingenuous Subjection: Compliance and Power in the Eighteenth-Century Domestic Novel* (Philadelphia: University of Pennsylvania Press, 2005) on wives' simulation of liberal consent. On the modalities of subservience exacted from British captives in early modern North Africa, see *Invoking Slavery in the Eighteenth-Century British Imagination*, ed. Srividhya Swaminathan and Adam R. Beach (Burlington, Vt.: Ashgate, 2013).

116. Chetwood, *Vaughan*, 2: 129.

117. Boyle, *Colours*, 84.

118. Ibid., 85.

119. Ibid., 89.

120. Ibid.

121. Ibid., 92.

122. Ibid., 33, 47.

123. Ibid., 175.

124. Ibid., 90.

125. For a trenchant reevaluation of Aubin's biography and political inclinations, see Debbie Welham, "The Particular Case of Penelope Aubin," *Journal for Eighteenth-Century Studies* 31: 1 (2008): 63–76, and "The Political Afterlife of Resentment in Penelope Aubin's *The Life and Amorous Adventures of Lucinda* (1721)," *Women's Writing* 20: 1 (2013): 49–63. Welham musters key textual and biographical evidence that "Aubin was just that—a Tory and Anglican (probably High Anglican) woman" ("Political Afterlife of Resentment," 3). Aubin was also the royalist Helmontian doctor Walter Charleton's illegitimate granddaughter. In arguing for Aubin's retrenchment of white patrilineal femininity, I depart from Edward Kozaczka, whose "Penelope Aubin and Narratives of Empire," *Eighteenth-Century Fiction* 25: 1 (Fall 2012), claims both empire and heteronormative gender as "ideological frameworks that Aubin sets out to destabilize" (225).

126. McKeon, *Origins of the English Novel*, 171–74.

127. Penelope Aubin, *The Strange Adventures of the Count de Vinevil and his Family* (London: Printed for E. Bell et al., 1721), 7. Italicized in original.

128. Penelope Aubin, *The Life of Madam de Beaumount* (London: Printed for E. Bell et al., 1721), both citations, 102.

129. Aubin, *Vinevil*, 6. Italicized in original.

130. Ibid. Italicized in original.

131. Ibid., 12.

132. Ibid., 64–65.

133. Penelope Aubin, *The Life and Amorous Adventures of Lucinda, an English Lady* (London: Printed for E. Bell et al., 1722), 85.

134. Aubin, *Vinevil*, 27–28, 36.

135. Ibid., 72.

136. Aubin, *Charlotta Du Pont*, 282.

137. Penelope Aubin, *The Noble Slaves: or, The Lives and Adventures of Two Lords and Two Ladies* (London: Printed for E. Bell et. al, 1722), 26 and 29.

138. Aubin, *Noble Slaves*, 30.

139. Ibid., 31, 33.

140. Ibid., 198.

141. Ibid., 200.

142. Ibid., 202.

143. For political and literary context that illuminates the exceptionality of Aubin's treatment of seduction, see Toni Bowers, *Force or Fraud: British Seduction Fiction and the Problem of Resistance, 1660–1760* (Oxford: Oxford University Press, 2011). While Bowers stresses the paradoxical virtue of seduction as a model for Tories' ambivalent concession to Whig hegemony, Aubin refuses to endorse the irresistibility even of violent assault. A Tory sympathizer, Aubin would seem analogically to condemn even the political expedient of passive obedience.

144. Richardson may have written the preface to Aubin's posthumous collected works. See Wolfgang Zach, "Mrs. Aubin and Richardson's Earliest Literary Manifesto (1739)," *English Studies* 62 (1981): 271–85. Peter Sabor and Thomas Keymer affirm that Zach "convincingly" establishes Richardson's authorship in *'Pamela' in the Marketplace: Literary Controversy and Print Culture in Eighteenth-Century Britain and Ireland* (Cambridge: Cambridge University Press, 2005), 27. Aubin's preface affirms that novels must "imprint noble Principles in the ductile Souls of our Youth," but constitutes a rote endorsement, which gives me some pause. *A Collection Of Entertaining Histories and Novels . . . By Mrs. Penelope Aubin*, vol. 1 (London, 1739), n.p. An Englishman, Lovelace domesticates the seraglio-keeper in two ways: as a rake, he arrogates the scarce resource of feminine virginity to himself; as a future husband, his excessive despotism exposes his failure to legitimize his contractual right. See Letter 232: "I would have gone to war with the Great Turk, and the Persian, and the Mogul, for their seraglios; for not one of those Eastern monarchs should have had a pretty woman . . . till I had done with her." *Clarissa*, ed. Angus Ross (London: Penguin, 1985), 762. On orientalist tropes that install the East-West axis inside the metropole, see Saree Makdisi, *Making England Western: Occidentalism, Race, and Imperial Culture* (Chicago: University of Chicago Press, 2014). On Lovelace's enhanced despotism in Richardson's revised *Clarissa* (1751), see my "Secondary Qualities and Masculine Form in *Clarissa* and *Sir Charles Grandison*," *Eighteenth-Century Fiction* 24: 2 (Winter 2011–12): 195–226.

145. Edward Long, *The History of Jamaica. Or, General Survey of the Antient and Modern State of that Island* (London: T. Lowndes, 1774), 2: 336.

146. Long, *History of Jamaica*, all citations, 2: 260.

147. Angela Y. Davis, *Women, Race, and Class* (New York: Vintage, 1983), and bell hooks, *Ain't I A Woman?: Black Women and Feminism* (Boston: South End Press, 1981). See also Jennifer L. Morgan, *Laboring Women: Reproduction and Gender in New World Slavery* (Philadelphia: University of Pennsylvania Press, 2004).

148. Aubin echoes Aphra Behn's *Oroonoko: or, the Royal Slave* (1688), whose eponymous protagonist is African.

149. Aubin, *Noble Slaves*, 6–7.

150. Aubin, *Charlotta Du Pont*, both citations, 78.

151. Ibid., 79, 83.
152. Ibid., 83, 87.
153. Ibid., 88.
154. Ibid., 84.
155. Ibid., 83.
156. Ibid., 99.
157. Long, *History of Jamaica*, both citations, 2: 328.
158. Ibid., 2: 329.

CHAPTER 5

Epigraphs: Henry Fielding, *Contributions to The Champion and Related Writings*, ed. W. B. Coley (Oxford: Clarendon Press, 2003), 428. Henry Fielding, *The History of the Adventures of Joseph Andrews* and *An Apology for the Life of Mrs. Shamela Andrews*, ed. Douglas Brooks-Davies (Oxford: Oxford University Press, 2008), 158. All further references to *Joseph Andrews* and *Shamela* will be noted parenthetically.

1. Fielding, *Contributions*, 239.
2. Ibid., 241.
3. Ibid.
4. Nancy A. Mace claims that Fielding opposes his classical knowledge to "the absurd learning" of "pedants" and virtuosi of the Royal Society. *Henry Fielding's Novels and the Classical Tradition* (Newark: University of Delaware Press, 1996), 46.
5. Robert Boyle, *The Sceptical Chymist*, in *The Works of Robert Boyle*, ed. Michael Hunter and Edward B. Davis (London: Pickering and Chatto, 1999), 2: 306.
6. Fielding, *Contributions*, 431.
7. John Locke, *An Essay Concerning Human Understanding*, ed. Peter H. Nidditch (Oxford: Oxford University Press, 1975), 1.2.§15. All further citations will be noted parenthetically.
8. Fielding, *Contributions*, 324 and 327.
9. Ibid., all citations, 329.
10. Word changed from "sublimed" in original. George Cheyne, *An Essay on Regimen. Together with Five Discourses, Medical, Moral, and Philosophical* (London, 1740), 6.
11. Fielding, *Contributions*, 329.
12. Ibid., 330.
13. Walter Pagel, *Paracelsus: An Introduction to Philosophical Medicine in the Era of the Renaissance*, 2nd rev. ed. (Basel: Karger, 1982), 113. The words "fall" and "separation" are quoted from Paracelsus by Pagel.
14. Pagel, *Paracelsus*, 89.
15. See Betty Jo Teeter Dobbs, *The Foundations of Newton's Alchemy or "The Hunting of the Greene Lyon"* (Cambridge: Cambridge University Press, 1975), 35–39, for Dobbs's account of the influence of fifteenth- through seventeenth-century "Renaissance Neoplatonism" (36)—and especially "intermediate stages between matter and spirit" (36)—on alchemy.
16. For discussion of all three satires and their medico-historical context, see R. C. Jarvis, "The Death of Walpole: Henry Fielding and a Forgotten Cause Célèbre," *Modern Language Review* 41: 2 (April 1946): 113–30. Of the "Catalogue," Jarvis writes: "It is without doubt the work of Fielding, and is in his best ironical vein" (127). Of *A Project for the Advancement of*

Physic in this Island, Jarvis is unsure because it does not employ Fielding's preferred verb forms "hath" and "doth," but notes that "it is certainly after his characteristic ironical style" (128). In *Henry Fielding: A Life* (London: Routledge, 1989), Martin C. Battestin and Ruth Battestin attribute all three texts to Fielding, remarking that his intervention in defense of his friend Dr. John Ranby (1703–1773), who disputed the efficacy of Walpole's treatment, "was of a piece with the irreverent treatment [Fielding] generally accorded the medical establishment" (395). They remark: "Fielding filled out the pamphlet by adding a pair of other satires on the College of Physicians: one takes the form of an 'Advertisement' for the sale of a physician's library" (396). Battestin adds in a note that although *A Project* does not use Fielding's "hath" and "doth," "this essay bears every other mark of his style and appears to be Fielding's work" (665 n. 230).

17. Henry Fielding, *The Charge to the Jury: or, the Sum of the Evidence, on the Trial of A.B.C.D. and E.F.* (London: M. Cooper, 1745), 24.

18. Fielding, *Charge*, both citations, 23.

19. Ibid., 25.

20. Ibid., 23.

21. Ibid., 38.

22. Ibid., 25.

23. Ibid., 42.

24. See William R. Newman, *Gehennical Fire: The Lives of George Starkey, an American Alchemist in the Scientific Revolution* (Cambridge, Mass.: Harvard University Press, 1994), 196–203, for discussion of "[t]he competitive pressure exerted [on the Helmontians] by chemically given empirics who were masters of salesmanship" (202). Newman gives a brief but fascinating account of the empiric Lionel Lockyer (d. 1672), who relied upon "mass promotion" (197) to tout the benefit of Lockyer's Pill, which "was still being manufactured into the mid-eighteenth century" (197).

25. George Starkey, *Brief Examination and Censure of Several Medicines* (London, 1664), 8. Cited by Newman, *Gehennical Fire*, 198.

26. Fielding, *Charge*, 24.

27. For a discussion of *spagyria*, its etymology and its corpuscular interpretation by the anti-Paracelsian sixteenth- and early seventeenth-century chemist Libavius, see William R. Newman, *Atoms and Alchemy: Chymistry and the Experimental Origins of the Scientific Revolution* (Chicago: University of Chicago Press, 2006), Chapter 3.

28. Samuel Richardson, *Pamela: or, Virtue Rewarded*, ed. T. C. Duncan Eaves and Ben D. Kimpel (Boston: Houghton Mifflin, 1971), 12.

29. Samuel Richardson, *Pamela; or, Virtue Rewarded*, ed. Thomas Keymer and Alice Wakely (Oxford: Oxford University Press, 2001), 56. All further citations refer to this edition, which does not include Hill's letter, and are noted parenthetically.

30. For a comprehensive overview of the *Pamela* controversy, see William B. Warner, *Licensing Entertainment: The Elevation of Novel Reading in Britain, 1684–1750* (Berkeley: University of California Press, 1998).

31. See Michael Ayers, "Locke's Logical Atomism," in *John Locke: Critical Assessments*, ed. Richard Ashcraft (London: Routledge, 1991), 4: 21–36. Ayers remarks: "the impediment is not a veil of ideas as much as a veil of words which can entangle and impose on us especially with respect to the formation of complex ideas" (33–34).

32. *A New Canting Dictionary* (London, 1725), n.p. At the level of the phoneme, *Pamela* anticipates Fielding's satire with Pamela's "Fears of this Sham-marriage" (*Pamela*, 260).

33. Henry Fielding, *Tom Thumb. A Tragedy. As it is Acted at the Theatre in the Hay-Market. Written by Scriblerus Secondus* (London: J. Roberts, 1730), A2.

34. Glenn W. Hatfield, *Henry Fielding and the Language of Irony* (Chicago: University of Chicago Press, 1968), 67.

35. Jacques Derrida, "Différance," in *Margins of Philosophy*, trans. Alan Bass (Sussex: Harvester Press, 1982), 9.

36. Derrida, "Différance," 11.

37. Fielding, *Contributions*, 252–53.

38. Hatfield, *Language of Irony*, 43.

39. Henry Fielding, *The Author's Farce; and the Pleasures of the Town.* Written by *Scriblerus Secundus* (London: J. Roberts, 1730), 25.

40. See Thomas Hobbes, *Leviathan* (1651): "before the names of *just* and *unjust* can have place, there must be some coercive power to compel men equally to the performance of their covenants by the terror of some punishment greater than the benefit they expect by the breach of their covenant." *Leviathan*, ed. A. P. Martinich (Peterborough, Ontario: Broadview Press, 2002), 108.

41. Jonathan Swift, *Gulliver's Travels*, ed. Robert DeMaria (London: Penguin, 2001), 229.

42. Swift, *Gulliver's Travels*, 221.

43. See Toni Bowers, *Force or Fraud: British Seduction Stories and the Problem of Resistance, 1660–1760* (Oxford: Oxford University Press, 2011): "Henry Fielding's *Shamela* (1741), finds hidden, salacious motives in Pamela's resistance, self-serving strategies in her apparently most submissive acts" (255).

44. Fielding, *Contributions*, 132.

45. Ibid., 413–5.

46. Ibid., 132.

47. Hatfield, *Language of Irony*, 14, 151. Emphasis Hatfield's.

48. Ibid., 160, 165.

49. Sandra Macpherson, *Harm's Way: Tragic Responsibility and the Novel Form* (Baltimore: Johns Hopkins University Press, 2010), 125.

50. Citing Martin C. Battestin, Hatfield claims the "exalted concept" of charity as Fielding's seminal Christian virtue. *Language of Irony*, 176.

51. Hatfield, *Language of Irony*, all citations, 200.

52. Ibid., 211.

53. Michael McKeon, *The Origins of the English Novel, 1600–1740* (Baltimore: Johns Hopkins University Press, 1987), 405.

54. McKeon, *Origins of the English Novel*, 408.

55. Hatfield, *Language of Irony*, 23.

56. Fielding, *Contributions*, 304.

57. See Jürgen Habermas, *The Structural Transformation of the Public Sphere*, trans. Thomas Burger (Cambridge, Mass.: MIT Press, 1989).

58. Henry Fielding, *The Life of Mr. Jonathan Wild the Great*, ed. Hugh Amory and Claude Rawson (Oxford: Oxford University Press, 2003), 76. This text was originally published as vol. 3 of Fielding's *Miscellanies* (London, 1743). All references to the modern edition are noted parenthetically above.

59. Isaac Newton, *De natura Acidorum*, in *The Correspondence of Isaac Newton*, ed. H. W. Turnbull (Cambridge: Cambridge University Press, 1960), 3: 211.

60. Ian Watt marshals *Robinson Crusoe* and *Pamela* as exponents of a "modern individualism" that emerged in "Protestant societies." *The Rise of the Novel: Studies in Defoe, Richardson, and Fielding* (Berkeley: University of California Press, 1957), 136, 140. Pursuing this affinity, McKeon wittily evokes Pamela as " 'Beached' at the top of the hierarchy of domestic service" (*Origins of the English Novel*, 375).

61. See Dobbs, *Foundations of Newton's Alchemy*; Newman, *Atoms and Alchemy*; Lawrence M. Principe, *The Aspiring Adept: Robert Boyle and His Alchemical Quest* (Princeton: Princeton University Press, 1998); Richard S. Westfall, *The Life of Isaac Newton* (Cambridge: Cambridge University Press, 1994).

62. On Paul of Taranto's authorship and borrowing from the *70 Books*, as well as the text's descent from work proceeding from the eighth- to ninth-century alchemist and physician Rāzi, see Geber, Chapter 2 of the introduction to *The "Summa Perfectionis" of Pseudo-Geber*, trans. and ed. William R. Newman (Leiden: E. J. Brill, 1991). Citation, 86.

63. Geber, *Summa*, 768.

64. On the transmutation history as a genre and on "dispersion de la science" as a technique of secrecy, see Newman, *Gehennical Fire*, 115–35. Alchemical writers' recourse to "figurative language" (117), "*syncope*" (134) or elliptical description, a "specialized language of images and tropes" (116), and at least in the case of one of Philalethes's texts, the disguised flight of the adept-narrator, is suggestive for the history of the novel. See also Principe, *Aspiring Adept*, 93–98, where Principe remarks that "we can point to an entire genre of alchemical writings that may be termed 'transmutation histories' " (93). McKeon's argument for the "domestication" of arcane political knowledge in *The Secret History of Domesticity: Public, Private, and the Division of Knowledge* (Baltimore: Johns Hopkins University Press, 2005) briefly looks at pictorial images of alchemy to instance the familiarization of once-secret practice, but McKeon claims "growing skepticism about the figure of the alchemist" (215). Attention to the transmutation history would help address Dobbs's call for an "adequate history of the impact of alchemy on other aspects of European intellectual life" (*Foundations*, 25).

65. Dobbs, *Foundations*, 46.

66. Newman, *Gehennical Fire*, 239.

67. Dobbs, *Foundations*, 10.

68. John Harrison and Peter Laslett, *The Library of John Locke* (Oxford: Oxford University Press, 1971), 238.

69. The Act Against Multipliers or the Act Against Multiplication of Metals, in part implemented under Henry IV in 1404, was repealed in 1689, an appeal initiated by Boyle, according to Newton as well as the Royal Society Journal Book (Principe, *Aspiring Adept*, 105–6). In *The Secrets of Alchemy* (Chicago: University of Chicago Press, 2013), Principe states that in the eighteenth century "chrysopoeia was by no means dead. In fact, the last decades of the eighteenth century witnessed the first of several 'alchemical revivals' " (90).

70. Principe, *Aspiring Adept*, all citations, 48.

71. Dobbs, *Foundations*, 135.

72. Geber, *Summa*, 669.

73. Ibid., 715.

74. Newman, preface to *Summa*, iii.

75. Geber, *Summa*, 743.

76. Ibid., 670.

77. Newman, introduction to *Summa*, 27. See Newman on Starkey: "Philalethes's term 'primary quality' means the Aristotelian elemental qualities—rather than size, shape, position, and arrangement" (*Gehennical Fire*, 156).

78. Newman, introduction to *Summa*, 176.

79. The terms "body" and "spirit" are used by Geber and other alchemists to signify changes in state, not metaphysical difference.

80. Geber, *Summa* 681, 681–82.

81. Ibid., 704.

82. Ibid., 707 and 704. "Calcination" might be translated as oxidation, the heating of a substance to below its melting or fusing point.

83. Geber differentiates between extrinsically adduced principles and those deeply entrenched within a metal. The former can be easily removed, while the latter, as profound constituents of the matter of these metals, cannot: "it is therefore granted that there will be a double sulfureity in the bodies, one enclosed in the depth of quicksilver in the beginning of its mixture, the other added on. One of these is removed with labor, but it is not possible that our work can fittingly and usefully arrive at the other by any of the techniques brought about by fire, since it has already been united with that quicksilver during the creation of the same" (*Summa*, 732).

84. Newman, *Gehennical Fire*, 94.

85. Geber, *Summa*, 680.

86. Isaac Newton, *Opticks; or A Treatise of the Reflections, Refractions, Inflections & Colours of Light* (1730), foreward Albert Einstein, intro. Edmund Whittaker, pref. I. Bernard Cohen (New York: Dover, 1952), 385.

87. Newton, *Opticks*, 386.

88. Newman, *Gehennical Fire*, 234.

89. Dobbs, *Foundations*, 221.

90. On this passage, see Willian Newman, "The Background to Newton's Chymistry," in *The Cambridge Companion to Newton*, ed. I. Bernard Cohen and George E. Smith (Cambridge: Cambridge University Press, 2002): 358–69. Citation, 366.

91. Newman, *Gehennical Fire*, 95.

92. Ibid., Chapters 3 and 4. For histories that include other major adepts like Sendivogius, see Dobbs and Principe.

93. Newton to Boyle, February 28, 1678/9, in *Correspondence*, ed. Turnbull, 2: 292.

94. See Frederick G. Ribble and Anne G. Ribble, *Fielding's Library: An Annotated Catalogue* (Charlottesville: Bibliographical Society of the University of Virginia, 1996). The authors caution that many works Fielding obviously knew (like Cibber's *Life*) are not listed in the catalogue (xiii–xv), which lists only one text by Newton: *Newton's Observations on Daniel, and the Apocalypse* (1733). Fielding owned Locke's *Works* (1751) and *Two Treatises of Government* (1698), though his citations from the *Essay* indicate that he had access to a copy of this text before 1751. He also subscribed to William Chetwood's *Falconer* and *William Owen Gwin Vaughan*.

95. Henry Knight Miller suggests that Fielding wrote this essay "at about the same time" he contributed to the *Champion*, "during the period 1739–40." Henry Fielding, *Miscellanies by Henry Fielding, Esq.*, ed. Henry Knight Miller (Oxford: Wesleyan University Press, 1972), 1: 153 n. 1.

96. Fielding, *Miscellanies*, 1: 168.

97. Ibid., both citations, 1: 169.

98. On the topic of dietary law, the Gospel of Mark stipulates: "whatever goes into a person from outside cannot defile . . . For it is from within . . . that evil intentions come." *The New Oxford Annotated Bible*, ed. Bruce M. Metzger and Roland E. Murphy (New York: Oxford University Press, 1991), Mark 7: 18, 21.

99. Fielding, *Miscellanies*, 1: 173.

100. Geber, *Summa*, 663.

101. See McKeon, *Origins of the English Novel*, who elucidates the "crisis" entailed by this term: "Progressive ideology would . . . [destroy] genealogical inequities, thereby liberating merit to reestablish status consistency by winning its just deserts. Conservative ideology, also mindful of aristocratic injustice, is yet more mindful of its modern replacement by progressive ideology and the rise of the 'new aristocracy' of the undeserving" (174).

102. On Fielding's defense of the activity of judgment in time of crisis, see Vivasvan Soni, "Judging, Inevitably: Aesthetic Judgment and Novelistic Form in Fielding's *Joseph Andrews*," *Modern Language Quarterly* 76: 2 (June 2015): 159–80.

CHAPTER 6

Epigraphs: Stephen Hales, *A Description of Ventilators: Whereby Great Quantities of Fresh Air May with Ease be conveyed into Mines, Goals, Hospitals, Work-Houses and Ships* (London, 1743), t.p. John Dryden, *The Works of John Dryden*, ed. A. B. Chambers and William Frost (Berkeley: University of California Press, 1974), 4: 177.

1. Samuel Richardson, *Clarissa: Preface, Hints of Prefaces, and Postscript*, ed. R. F. Brissenden (Los Angeles: Williams Andrews Clark Memorial Library, 1964), 12. *Hints* "appears to have been put together . . . while [Richardson] was revising the Preface and Postscript to the first edition," but "[l]arge portions . . . have never previously appeared in print" before Brissenden's edition (*Preface, Hints*, i).

2. T. C. Duncan Eaves and Ben D. Kimpel affirm: "when he was an active printer and active writer Richardson read few books but his own." *Samuel Richardson: A Biography* (Oxford: Clarendon Press, 1971), 570. For an index of books Richardson printed, see William M. Sale's *Samuel Richardson: Master Printer* (Ithaca: Cornell University Press, 1950).

3. Sale, *Master Printer*, 75. Whether Richardson was the printer for the Royal Society before he was officially chosen in 1752, Sale explains, "is a question that must remain unanswered, since no official records of the printing for the Society were kept before 1752" (74).

4. Eaves and Kimpel, *Samuel Richardson*, 570.

5. Richardson, *Preface, Hints*, 13–14.

6. Samuel Richardson to Solomon Lowe, January 21, 1749, *Selected Letters of Samuel Richardson*, ed. John Carroll (Oxford: Clarendon Press, 1964), 124.

7. Samuel Richardson, *Clarissa, or, The History of a Young Lady*, ed. Angus Ross (London: Penguin, 1985), 1386. All further references to this edition will be noted parenthetically.

8. See Tita Chico, *Designing Women: The Dressing Room in Eighteenth-Century English Literature and Culture* (Lewisburg: Bucknell University Press, 2005), and Brenda Bean, "Sight and Self-Disclosure: Richardson's Revision of Swift's 'The Lady's Dressing Room,'" *Eighteenth-Century Life* 14 (February 1990): 1–23, which trace literary representations of women's cosmetic artifice from Juvenal and Ovid, respectively, through Swift. For broader synopsis of this terrain, see Will Pritchard, "Masks and Faces: Female Legibility in the Restoration Era," *Eighteenth-Century Life* 24: 3 (Fall 2000): 31–52.

9. For evocations of Richardson's problematic distinction of Clarissa from Sinclair, the prostitutes, or other less virtuous women, see Tassie Gwilliam, *Samuel Richardson's Fictions of Gender* (Stanford: Stanford University Press, 1993); Jocelyn Harris, "Grotesque, Classical and Pornographic Bodies in *Clarissa*," in *New Essays on Samuel Richardson*, ed. Albert J. Rivero (New York: St. Martin's Press, 1996): 101–16; Jocelyn Harris, "Protean Lovelace," *Eighteenth-Century Fiction* 2: 4 (July 1990): 327–46; Terry Castle, *Clarissa's Ciphers: Meaning and Disruption in Richardson's "Clarissa"* (Ithaca: Cornell University Press, 1982), 87–100; Lois A. Chaber, "Christian Form and Anti-Feminism in *Clarissa*," *Eighteenth-Century Fiction* 15: 3 (2003): 507–37; Judith Wilt, "He Could Go No Farther: A Modest Proposal about Lovelace and Clarissa," *PMLA* 92: 1 (January 1977): 19–32.

10. Jonathan Swift, "The Lady's Dressing Room," in *The Poems of Jonathan Swift*, ed. Harold Williams (Oxford: Clarendon Press, 1937), 2: 529. All further references to this edition will be noted parenthetically by line number.

11. John Locke, *An Essay Concerning Human Understanding*, ed. Peter H. Nidditch (Oxford: Oxford University Press, 1975), 2.15.§9, 202–3.

12. Locke, *Essay*, 2.15.§9, 203.

13. With Locke's brief attempt to propose a category of primary ideas, his invocation of clear and distinct ideas of sensible points marks the most Cartesian turn in his *Essay*.

14. David Hume, *A Treatise of Human Nature*, ed. Peter Nidditch, analytical index L. A. Selby-Bigge (Oxford: Oxford University Press, 1978), 225, 228.

15. Hume, *Treatise*, 35.

16. Ibid., 38–39.

17. Ibid., 215.

18. Ibid., 27.

19. Ibid., 40, 41.

20. Ibid., 228.

21. John Bender excerpts Hume's phrase to indicate a general perceptual relativism. See *Ends of Enlightenment* (Stanford: Stanford University Press, 2012): Bender cites "[t]his 'double existence,' to use a phrase from Hume's *Treatise of Human Nature*" to evoke "the importance of multiple scales and points of view to the new methods of presentation in natural science" (60); and, in a biographical register, Hume's division of "the strict thinking of philosophy . . . from the impulses of ordinary life": "Hume sums up this split-focused existence as one in which 'we endeavour to set ourselves at ease . . . by feigning a double existence' " (68). But Hume's passage launches an attack on primary-secondary difference, which he claims as the "monstrous offspring" (*Treatise*, 215) of an empirical commitment to perdurable objects that provoke intermittent, dissimilar ideas: "This hypothesis is the philosophical one of the double existence of perceptions and objects; which pleases our reason, in allowing, that our dependent perceptions are interrupted and different; and at the same time is agreeable to the imagination, in attributing a continu'd existence to something else, which we call *objects*" (ibid.). Hume refers to neither quotidian versus skeptical perception nor "multiple . . . points of view," but to what he claims as the radical incoherence of "interrupted and different" perceptions of empirically unknowable matter.

22. Hume, *Treatise*, 27.

23. The indivisible integrity of Hume's vanishingly small "spot of ink" stands opposed to the printed point claimed by Robert Hooke's *Micrographia* (1665) as the second exhibit in his gallery of microscopically amplified visual objects. Hooke's illustration of the magnified "mark

of a *full stop*, or *period*" reveals that what appears "smothly [*sic*] *engraven*" and "round to the eye" becomes, under the microscope, "like *smutty daubings* on a matt or uneven floor with a blunt extinguisht brand or stick's end." Robert Hooke, *Micrographia: or, Some Physical Descriptions of Minute Bodies* (1665), reprint (New York: Dover, 1961), all citations, 3. Hume refuses magnification as an epistemological supplement to empirically comprehended particles.

24. Lovelace adopts Swift's metaphor of the empty tub that diverts hunted whales: "But when a man talks to a lady upon such subjects, let her be ever so much in *alt*, 'tis strange if he cannot throw out a tub to the whale—if he cannot divert her from resenting one bold thing by uttering two or three full as bold" (412).

25. Jonathan Swift, *A Tale of A Tub and Other Works*, ed. Angus Ross and David Woolley (Oxford: Oxford University Press, 1986), 15.

26. Swift, *Tub*, 60–61. Emphasis Swift's.

27. [George Starkey,] *Secrets Reveal'd: or, An Open Entrance to the Shut-Palace of the King: Containing, The greatest Treasure in Chymistry, Never yet so plainly Discovered* (London: W. Godbid for William Cooper, 1669), Author's preface, n.p. Emphasis Starkey's.

28. Starkey, *Secrets Reveal'd*, 32.

29. [George Starkey,] *A True Light of Alchymy* (London: I. Dawks, 1709), sig. A2.

30. Swift, *Tub*, 72. Michael McKeon designates *Tale of a Tub* "the era's most brilliant account of human culture as a libidinal economy in which the most varied sorts of achievement may be traced to, and explained by reference to, a sexual substance that resides deep in our bodies and that, consisting of vapors, effluvium, wind, gas, exhalation, is neither matter nor spirit but both at once." *The Secret History of Domesticity: Public, Private, and the Division of Knowledge* (Baltimore: Johns Hopkins University Press, 2005), 293. Bracketing Swift's portrayal of female Quaker ministers, "spirit" and "wind" may signal less a "libidinal economy" fueled by sublimated libidinal drive than humanizing metaphysics (and the attendant pretense to ladylike femininity) undone by scatological stink.

31. Swift, *Tub*, 73.

32. Walter Pagel, *Paracelsus: An Introduction to Philosophical Medicine in the Era of the Renaissance* (Basel: Karger, 1982), 6. Pagel clarifies: " 'Bombast' has nothing to do with 'bombastic' in the sense of 'turgid language' or 'tall talk.' It indicates descent from a very old and noble Swabian family which had their original seat at Hohenheim near Stuttgart" (6).

33. Lawrence M. Principe, *The Secrets of Alchemy* (Chicago: University of Chicago Press, 2013), 119. See Principe's synopsis: "It was probably while trying to make the Philosopher's Stone that Hennig Brand discovered the element phosphorus in the 1660s by strongly distilling residues from human urine . . . The use of excrement as a starting material stems from the ancient axiom that the material of the Philosopher's Stone 'is of cheap price and found everywhere' and 'is trodden underfoot' " (119). See William R. Newman and Lawrence M. Principe, *Alchemy Tried in the Fire: Starkey, Boyle, and the Fate of Helmontian Chymistry* (Chicago: University of Chicago Press, 2002), for discussion of Boyle's experimental appropriation of Helmont's "saline spirit of human urine" (278). See also Cathy Cobb, Monty L. Fetterolf, and Harold Goldwhite, *The Chemistry of Alchemy from Dragon's Blood to Donkey Dung: How Chemistry Was Forged* (Amherst, N.Y.: Prometheus Books, 2014), 219–24, to reproduce the Polish alchemist Michael Sendivogius's (c. 1566–c. 1636) extraction of sal ammoniac (ammonium chloride), fixed salt (potassium carbonate and sodium carbonate), and saltpeter (nitrates) from decomposed "barnyard soil" (219) or manure.

34. Swift, *Tub*, 79–80.

35. Ibid., 85.

36. Ibid., 86.

37. Jonathan Swift, *Gulliver's Travels*, ed. Robert DeMaria (London: Penguin, 2001), 168.

38. Judith Butler, *Gender Trouble: Feminism and the Subversion of Identity* (New York: Routledge, 1990), 137.

39. In concert with biological theorists like Anne Fausto-Sterling, Butler now denies the integrity of sex-gender difference to argue that sexed anatomy reflects investiture by social norms, even down to its constituent protein molecules. See Judith Butler, *Bodies That Matter: On the Discursive Limits of "Sex"* (London: Routledge, 1993), and Anne Fausto-Sterling, *Sexing the Body: Gender Politics and the Construction of Sexuality* (New York: Basic Books, 2000). See also Donna J. Haraway, "Situated Knowledges: The Science Question in Feminism and the Privilege of Partial Perspective," in *Simians, Cyborgs, and Women: The Reinvention of Nature* (London: Routledge, 1991): 183–201. Haraway writes: "Difference is theorized *biologically* as situational, not intrinsic, at every level from gene to foraging pattern, thereby fundamentally changing the biological politics of the body" (200).

40. Michael Stolberg, "A Woman Down to Her Bones: The Anatomy of Sexual Difference in the Sixteenth and Early Seventeenth Centuries," *Isis* 94 (2003): 276. For another important refutation of Laqueur, which traces a chemical understanding of sexual difference, see Amy Eisen Cislo, *Paracelsus's Theory of Embodiment: Conception and Gestation in Early Modern Europe* (London: Pickering and Chatto, 2010).

41. Stolberg, "Woman Down to Her Bones," 281.

42. Ibid., 291.

43. Jonathan Swift, "Strephon and Chloe," in *The Poems of Jonathan Swift*, ed. Williams, 2: 584. All further references to this edition will be noted parenthetically by line number.

44. See Carol Houlihan Flynn, *The Body in Swift and Defoe* (Cambridge: Cambridge University Press, 1990). By "[i]nverting Platonic idealism," Swift makes women the ultimate referent for material baseness: "the female body threatens not only to offend, but to scatter reason and swallow up certainty" (90). Nora F. Crow claims Flynn as "one of the few women scholars who see Swift as hostile to women." "Swift and the Woman Scholar," in *Pope, Swift, and Women Writers*, ed. Donald C. Mell (Newark: University of Delaware Press, 1996), 234.

45. Swift, *Gulliver's Travels*, 169–70.

46. James Keill, *An Account of Animal Secretion, The Quantity of Blood In the Human Body, And Muscular Motion* (London: Printed for George Strahan, 1708), 25–26.

47. Keill, *Animal Secretion*, 27.

48. Thomas Twining writes: "the good Pastor Stephen Hales / Weighed moisture in a pair of scales, / To lingering death put Mares and Dogs, / And stripped the Skins from living Frogs. / Nature he loved, her Works intent, / To search or sometimes to torment." Cited in Stephen Hales, *Vegetable Staticks: Or, An Account of some Statical Experiments on the Sap in Vegetables*, foreword M. A. Hoskin (London: Oldbourne Book Co., 1961), xii. See also Alexander Pope's reported commentary on "Dr. Hales": "he commits most of these barbarities with the thought of its being of use to man; but how do we know that we have a right to kill creatures that we are so little above as dogs, for our curiosity, or even for some use to us?" Joseph Spence, *Anecdotes, Observations, and Characters, of Books and Men*, ed. Edmund Malone (London: John Murray, 1820), 60.

49. Keill, *Animal Secretion*, 116–17.

50. Ibid., 104, 105, 105.

51. Ibid., 107.

52. Hales, *Vegetable Staticks*, 4.

53. Ibid., 5.

54. Hales, *Ventilators*, 32.

55. Hales, *Statical Essays: Containing Hæmastaticks; or, An Account of some Hydraulick and Hydrostatical Experiments made on the Blood and Blood-Vessels of Animals* (1733), reprint (New York: Hafner Publishing Co., 1964), 329.

56. Keill, *Animal Secretion*, 8.

57. Ibid., 8, 9.

58. Keill distinguishes his physiological "Oeconomy" from Sydenham's, concluding: "So a Physician ignorant of the Animal Oeconomy, is ignorant of the Structure of the Machine he undertakes to regulate, and the best and exactest Histories of Diseases can never suggest to him any Indication of Cure" (ibid., xiv).

59. Isaac Newton, *Opticks* [1730 ed.], foreword Albert Einstein, intro. Edmund Whittaker, pref. I. Bernard Cohen (New York: Dover, 1979), 389, 388.

60. Newton, *Opticks*, 378.

61. Ibid., 394.

62. Ibid., 380.

63. Ibid., 377.

64. Ibid., both citations, 383.

65. Ibid., 390, 389–90.

66. Ibid., 389.

67. Ibid., 393.

68. Keill, *Animal Secretion*, 2.

69. Ibid., 24.

70. Hales, *Vegetable Staticks*, both citations, xxvii.

71. Ibid., 119. This passage is located (with variant typography and punctuation) in *Opticks*, 396.

72. Keill notes: "Now that the *Blood* contains a great number of Globules of Air is evident from the Quantity it yeilds [*sic*] in the Air Pump" (*Animal Secretion*, 163).

73. Hales, *Vegetable Staticks*, 119, 120.

74. Ibid., 134.

75. Ibid., 140.

76. Ibid., 134.

77. Ibid., 176.

78. Robert Boyle, *New Experiments Physico-Mechanicall, Touching The Spring of the Air, and its Effects*, in *The Works of Robert Boyle*, ed. Michael Hunter and Edward B. Davis (London: Pickering and Chatto, 1999), 1: 159.

79. Hales, *Vegetable Staticks*, 141.

80. Ibid., 153.

81. Hales, *Hæmastaticks*, 324.

82. Hales, *Ventilators*, 42–43.

83. Ibid., 45.

84. Ibid., 129.

85. Ibid., 36.

86. Ibid., 35.

87. Ibid., 38.

88. See Ian Baucom, *Specters of the Atlantic: Finance Capital, Slavery, and the Philosophy of History* (Durham: Duke University Press, 2005), whose deconstructive methodology leaves untouched what Baucom claims as "the matter of nature or history" (287) repressed by sympathetic mind-readers who thereby avoid "the facts of the world" (287). As one perpetrator of "idealist" (286) representation, Baucom characterizes an eighteenth-century "romantic novel" (217) ill-equipped to engage with bodies or history.

89. Hales, *Ventilators*, 50.

90. Ibid., 48.

91. Writing to Lady Bradshaigh (1751), Richardson cites this sentence to describe himself: "I . . . now am fallen into the evil days of indifferent health, and nerves so bad, that my body and mind pull different ways, and so tear me between them!" *Selected Letters*, 185. The internalized antipathy of spirit to body is not, for Richardson, a phenomenon that defines femininity.

92. In *Actions and Objects from Hobbes to Richardson* (Stanford: Stanford University Press, 2010), Jonathan Kramnick shows, by noting Clarissa's references to her ingestion of meals, that she continues to eat through the duration of her life. He concludes: "Clarissa dies without any special causal role for her mind" (229). But a paradox persists: if Clarissa eats, why does she become emaciated and die? This outcome appears to recombine willed self-maintenance and a physiological failure to absorb nutrients.

93. See Carole Pateman, *The Sexual Contract* (Stanford: Stanford University Press, 1988). John Locke's *Two Treatises of Government* (1690) rejects marriage as a prototype for public government: Locke argues that the mandate "*And thy desire shall be to thy Husband, and he shall rule over thee*" (3 Gen. 16) is an instance of private and not public power: "it can be only a Conjugal Power, not Political, the Power that every Husband hath to order the things of private Concernment in his Family." Locke, *Two Treatises of Government*, ed. Peter Laslett (Cambridge: Cambridge University Press, 1988), 174. But to argue that God's curse does not invest Adam with political "Prerogatives and Priviledges," Locke detaches Eve's punishment from any inherent virtue on Adam's part: "*Eve* was laid below him, and so he had accidentally a Superiority over her, for her greater Punishment" (Locke, *Two Treatises*, 172). Locke belatedly cites a "Foundation in Nature" (174) absent from his prior motivation of Eve's inferiority.

94. In these moments, Lovelace assumes the perspective glossed by Frances Ferguson in "Rape and the Rise of the Novel," *Representations* 20 (Fall 1987): 88–112, as his confidence in the retroactively legitimizing power formally stipulated by marriage.

95. In "Knowing Love: The Epistemology of *Clarissa*," *ELH* 74: 4 (Winter 2007), Katherine Binhammer argues that in *Clarissa* women "can have an ontological status which is not sexed" (862). While I agree with Binhammer's claim, I depart from her attribution of it to Clarissa, whose relation to the institution of wifely duty is ambivalent: while Anna mocks Clarissa because "you think more highly of a *husband*'s prerogative, than most people do of the *royal* one" (1152), the integrity of Clarissa's respect for the "*husband*'s prerogative" fuels her refusal to marry Lovelace. Richardson equivocally reconciles intellectual parity and domestic power: "all that I contend for, is, that genius, whether in men or women, should take its course; that, as a ray of the divinity, it should not be suppressed. But I acknowledge that the great and indispensable duties of women are of the domestic kind" (*Selected Letters*, 178).

96. In the most explicit enactment of sexual violation this text offers, Solmes "even snatched my trembling, my struggling hand; and ravished it to his odious mouth" (319). Richardson elaborates the threat of marital rape through the appendage of the "hand" by drawing

out its figural equivocation with forced consent. He frames the unwilled marriage service as an assault: "This Solmes shall never have my vows I am resolved! And I will say nothing but No, as long as I shall be able to speak. And who will presume to look upon such an act of violence as a marriage?" (365) Here Richardson posits the limit case of an "act of violence" in which Clarissa remains conscious to voice her "No."

97. On Lockean understanding as a leveler of sexual difference, see Alice Browne, *The Eighteenth-Century Feminist Mind* (Detroit: Wayne State University Press, 1987); Melissa Butler, "Early Liberal Roots of Feminism: John Locke's Attack on Patriarchy," in *Feminist Interpretations of John Locke*, ed. Nancy Hirschmann and Kirstie McClure (University Park: Pennsylvania State University Press, 2007); Sheryl O'Donnell, "My Idea in Your Mind: John Locke and Damaris Cudworth Masham," in *Mothering the Mind: Twelve Studies of Writers and Their Silent Partners*, ed. Ruth Perry and Martine W. Brownley (New York: Holmes and Meier, 1984); and Kathy Squadrito, "Locke on the Equality of Sexes," *Journal of Social Philosophy* 10: 1 (1979). By contrast, Leo Braudy argues that impenetrability is *Clarissa*'s ruling value: "The response . . . to the threat of penetration, whether physical, mental, or spiritual, is to become impenetrable." Leo Braudy, "Penetration and Impenetrability in *Clarissa*," in *Modern Essays on Eighteenth-Century Literature*, ed. Leopold Damrosch (New York: Oxford University Press, 1988), 274. On penetration as Clarissa's rather than Lovelace's power, see James Grantham Turner, "Lovelace and the Paradoxes of Libertinism," in *Samuel Richardson: Tercentenary Essays*, ed. Margaret Anne Doody and Peter Sabor (Cambridge: Cambridge University Press, 1989).

98. Wilt, "Modest Proposal," 20, 26.

99. In "Reading Rape: Maxist-Feminist Figurations of the Literal," *Diacritics* 13: 4 (Winter 1983), William Warner underscores Castle's and Terry Eagleton's refusal to expose the rape to deconstructive scrutiny: it "is the text's ultra-real event" (19). Warner desists from ascribing the rape any referential refinement, citing it instead as a "vacant negative expression" (27) that connotes "the vexed ambivalence of human desire" (28).

100. On the criminality of women whose amorous motives exclude desire, see Laura Rosenthal, *Infamous Commerce: Prostitution in British Literature and Culture* (Ithaca: Cornell University Press, 2006).

101. John Locke, *Some Thoughts Concerning Education*, ed. Ruth W. Grant and Nathan Tarcov (Indianapolis: Hackett Publishing, 1996), 33.

102. Locke, *Some Thoughts*, 34.

103. See Locke's opposing narrative of the little girl who must be beaten eight times before her will is "suppled" (*Some Thoughts*, 55); the quantitatively restricted application of force preserves her volition while assuring future compliance. With the alternative interminability that would break a child, Locke anticipates Richardson's recurrent—and, one could say, Sadean—attention to how long it would take the whores to break Clarissa.

104. Gwilliam astutely notes: "The rape leaves Lovelace in the impossible position of seeing Clarissa as totally different from all other women, yet absolutely indistinguishable from them. The rape has, in a sense, no ontological status." *Samuel Richardson's Fictions of Gender*, 84.

105. Karen Barad, "Posthumanist Performativity: Toward an Understanding of How Matter Comes to Matter," *Signs* 28: 3 (2003): 815.

106. *Oxford English Dictionary* for "hoyden."

107. Most appraisals of Clarissa's emaciation and death do not affirm their feminist import. See Harris, "Grotesque, Classical and Pornographic Bodies" and Chaber, "Christian Form

and Anti-Feminism in *Clarissa*." For defense of Clarissa's starvation as form of agency, see Donnalee Frega, *Speaking in Hunger: Gender, Discourse, and Consumption in "Clarissa"* (Columbia: University of South Carolina Press, 1998).

108. Clarissa's fitness for intimate duty is relayed when Lovelace witnesses an instant of "delicate transport" (695) inspired by the prospect of reunion with her father.

109. Samuel Richardson, *Pamela; or, Virtue Rewarded*, ed. Thomas Keymer and Alice Wakely (Oxford: Oxford University Press, 2008), 167.

110. Sigmund Freud, *The Interpretation of Dreams*, trans. and ed. James Strachey (New York: Avon, 1965), 328.

111. Richardson, *Pamela*, 196. Emphasis in original.

112. On the appeal of rakes as a primary-secondary problem, see my "Secondary Qualities and Masculine Form in *Clarissa* and *Sir Charles Grandison*," *Eighteenth-Century Fiction* 24: 2 (Winter 2011–12): 195–226.

113. Richardson to Lady Bradshaigh, December 15, 1748, *Selected Letters*, 113.

114. Richardson, *Pamela*, 180.

115. Ibid., 378.

116. Ibid., 385.

EPILOGUE

Epigraphs: Stephen Hales, *Vegetable Staticks* (London: Oldbourne Book Co., 1961), 179–80. Lavoisier et al. translated by Arthur Donovan, *Antoine Lavoisier: Science, Administration, and Revolution* (Oxford: Blackwell, 1993), 160. The passage cites a memoir or presentation read by Lavoisier before the Royal Academy of Sciences on April 18, 1787. See Louis-Bernard Guyton de Morveau, Antoine-Laurent Lavoisier, Claude Louis Berthollet, and Antoine François de Fourcroy, *Méthode de Nomenclature Chimique* (Paris: Chez Cuchet, 1787), 5. To conserve Lavoisier's verb "pénétrer," I would suggest that he and his colleagues tried to "enter the same spirit."

1. Ian Watt, *The Rise of the Novel: Studies in Defoe, Richardson and Fielding* (Berkeley: University of California Press, 1957), 297.

2. See Jane Austen, *Pride and Prejudice*, ed. Robert P. Irvine (Peterborough, Ontario: Broadview Press, 2002), intro., 15. All further citations from this edition are noted parenthetically.

3. Jean-Pierre Poirier, *Lavoisier: Chemist, Biologist, Economist*, trans. Rebecca Balinski (Philadelphia: University of Pennsylvania Press, 1993), xix. In *Antoine Lavoisier*, Donovan cites Lavoisier as "the founder of modern chemistry" (1).

4. Henry Guerlac, *Antoine-Laurent Lavoisier: Chemist and Revolutionary* (New York: Charles Scribner's Sons, 1973), 78.

5. Donovan, *Antoine Lavoisier*, 102.

6. Hales, *Vegetable Staticks*, 107.

7. Guerlac, *Antoine-Laurent Lavoisier*, 85.

8. For more detail about the significance of calcined metals' weight gain in disproving phlogiston, and Lavoisier's publication "Réflexions sur le phlogistique" ("Reflections on Phlogiston"), see Donovan and Guerlac. See also Poirier, *Lavoisier*: "Thus the phenomena of combustion and calcination were explained more simply without phlogiston, which was an imaginary being whose existence was useless." Poirier proceeds: "Actually, the matter of fire described by

Lavoisier was not so different from phlogiston: the sole point distinguishing them was that the former was placed in oxygen, whereas the latter was in the combustible body" (177).

9. Guerlac, *Antoine-Laurent Lavoisier*, 87, 121.

10. Guyton de Morveau et al., *Méthode de Nomenclature Chimique*, 14. This and all subsequent translations are mine.

11. Étienne Bonnot de Condillac, *Essai sur l'origine des connaissances humaines*, in *Oeuvres de Condillac* (Paris: Ch. Houel, 1798), 1: 14. On Condillac's preoccupation with Lockean development, see *Essai sur l'origine des connaissances humaines* (1746) (*Essay on the Origin of Human Knowledge*), *Traité des sensations* (1754) (*Treatise on Sensation*), and *Cours d'Études pour l'instruction du Prince du Parme* (1775) (*Course of Study for the Instruction of the Prince of Parma*). Condillac's *Treatise on Sensation* infuses empirical knowledge into a statue, sense by sense, evoking much of this process from the inside to mark an imagined horizon of empirical phenomenology. See *Course of Study* for Condillac's famous statement, cited by Lavoisier: "To think thus becomes an art, and this art is the art of speaking." Condillac, *Cours d'Études pour l'instruction du Prince du Parme: La grammaire*, in *Oeuvres de Condillac* (Paris: Ch. Houel, 1798), 5: xl.

12. Guyton de Morveau et al., *Méthode de Nomenclature Chimique*, 9.

13. Antoine-Laurent Lavoisier, *Elements of Chemistry* (1790), trans. Robert Kerr, reprint (New York: Dover, 1965), xxxi. This passage occurs in identical form in *Méthode de Nomenclature Chimique*: "Les noms d'*huile de tartre par défaillance*, d'*huile de vitriol*, de *beurres d'arsenic* & d'*antimoine*, de *fleurs de zinc*, &c. sont plus ridicules encore, parce qu'ils sont naître des idées fausses; parce qu'il n'existe, à proprement parler, dans le règne minéral, & sur-tout dans le règne métallique, ni beurre, ni huile, ni fleurs" (24).

14. Guyton de Morveau et al., *Méthode de Nomeclature Chimique*, 103. Despite the identical stem "chim-," "chimérique" (chimerical) and "chimie" (chemistry) claim divergent etymologies: the former Greek stem refers a she-goat or monster, the latter to magic or black magic.

15. Ibid., 18.

16. Ibid., 31–33. Translated with the assistance of Natania Meeker. See Lavoisier's *Elements*, 51–52, for a shorter version of this passage. I am grateful to William N. West for assistance with Lavoisier's Greek.

17. Guerlac, *Antoine-Laurent Lavoisier*, 93.

18. Ibid., 116.

19. Lavoisier, *Elements*, 66.

20. Henry George Liddell and Robert Scott, *A Greek-English Lexicon*. Revised and augmented throughout by Sir Henry Stuart Jones with the assistance of Roderick McKenzie (Oxford: Clarendon Press, 1940). Accessed online at: http://www.perseus.tufts.edu/hopper/text?doc=Perseus%3Atext%3A1999.04.0057%3Aalphabetic+letter%3D*o%3Aentry+group%3D50%3Aentry%3Do%29cu%2Fs2. Emphasis removed from original.

21. Poirier, *Lavoisier*, 188. Guerlac notes: "Only when his investigations dealt with the reactive or combinatorial behavior of substances . . . did Lavoisier speak of doing chemistry." *Antoine-Laurent Lavoisier*, 61.

22. For example, Austen converted Richardson's *Sir Charles Grandison* (1753–54) into a play. See *Jane Austen's 'Sir Charles Grandison'*, transcribed and ed. Brian Southam (Oxford: Oxford University Press, 1981), 4–11, on the play's authorship; Southam disputes its attribution to Austen's niece Jane Anna Elizabeth Austen.

23. Slavoj Žižek, *The Sublime Object of Ideology* (London: Verso, 1989), 63–64. Emphasis in original.

24. On *Pride and Prejudice* and judgment, see Vivasvan Soni, "Committing Freedom: The Cultivation of Judgment in Rousseau's *Emile* and Austen's *Pride and Prejudice*," *The Eighteenth Century: Theory and Interpretation* 51: 3 (Fall 2010): 363–87.

25. Samuel Richardson, *Clarissa; or, The History of a Young Lady*, ed. Angus Ross (London: Penguin, 1985), 184.

26. The virtue of Frances Burney's marriageable men—Lord Orville of *Evelina* (1778); Mortimer Delvile of *Cecilia* (1782); Edgar Mandelbert of *Camilla* (1796); and Albert Harleigh of *The Wanderer* (1814)—is perceptually evident. See my "How the Wanderer Works: Reading Burney and Bourdieu," *ELH* 68 (Winter 2001): 965–89, and "Burney's Conservatism: Masculine Value and 'the ingenuous Cecilia,'" in *Recognizing the Romantic Novel*, ed. Jillian Heydt-Stevenson and Charlotte Sussman (Liverpool: Liverpool University Press, 2008), on Burney's representations of the perceptual and ontological status of men's and women's social virtue.

27. For formal, feminist, and ethnographic analysis of the romance plot in late twentieth-century America, a plot that pivots on a silent, often brutal man's sudden extroversion of hidden gentleness, see Janice Radway, *Reading the Romance: Women, Patriarchy, and Popular Literature* (Chapel Hill: University of North Carolina Press, 1984).

28. John Gregory, *A Father's Legacy to His Daughters* (Dublin: John Colles, 1774), notes of a woman who elicits marital attention: "When you perceive it, it excites your gratitude; this gratitude rises into a preference, and this preference perhaps at last advances to some degree of attachment" (37).

29. Austen affirms the redundancy of Elizabeth's gratitude to a plot which already proves her readiness to wed, as Darcy attests: "I am not indebted for my present happiness to your eager desire of expressing your gratitude. I was not in a humour to wait for any opening of your's" (379). For a reading that recognizes the conservatism of Elizabeth's plot while affirming the feminine happiness it permits, see Claudia L. Johnson, *Jane Austen: Women, Politics, and the Novel* (Chicago: University of Chicago Press, 1988). A more deterministic "chemical" path for the marriage plot, predicated not on relational contingency but ineluctable attraction, is charted by Johann Wolfgang von Goethe's *Elective Affinities* (1809).

30. *Pride and Prejudice* was originally called *First Impressions*. See Robert Irvine, intro., 15.

31. Watt, *Rise of the Novel*, 29.

32. Ibid., all citations, 295.

33. The first formula, voiced by Lizzy, is iterated from the vantage of Mr. Bennet: "I know that you could be neither happy nor respectable, unless you truly esteemed your husband; unless you looked up to him as a superior. Your lively talents would place you in the greatest danger in an unequal marriage" (375). Here Austen confronts the problem of wifely superiority posed by *Clarissa*. Again *Pride and Prejudice* elicits a better outcome, not because Darcy is "superior" but because he is good enough to sustain Lizzy's "esteem[]" and thereby conserve the patriarchal warrant for marriage in Lizzy's empirical experience of it.

34. Watt, *Rise of the Novel*, 11.

35. John Bender, *Ends of Enlightenment* (Stanford: Stanford University Press, 2012), both citations, 100.

BIBLIOGRAPHY

Aaron, Richard Ithamar. *John Locke*. Oxford: Oxford University Press, 1937.

Aaron, Richard Ithamar, and J. Gibb. *An Early Draft of Locke's Essay, Together with Excerpts from his Journals*. Oxford: Clarendon Press, 1936.

Alaimo, Stacy, and Susan Hekman, eds. *Material Feminisms*. Bloomington: Indiana University Press, 2008.

Alexander, Peter. *Ideas, Qualities, and Corpuscles: Locke and Boyle on the External World*. Cambridge: Cambridge University Press, 1985.

[Anonymous.] *L'Ecole des filles ou La Philosophie des dames*. Paris: Éditions Allia, 1997.

Anstey, Peter R. "Boyle on Seminal Principles." *Studies in History and Philosophy of Biological and Biomedical Sciences* 33 (2002): 597–630.

———. *John Locke and Natural Philosophy*. Oxford: Oxford University Press, 2011.

———, ed. *John Locke: Critical Assessments*. 4 vols. New York: Routledge, 2006.

———, ed. *The Philosophy of John Locke: New Perspectives*. London: Routledge, 2003.

———. *The Philosophy of Robert Boyle*. London: Routledge, 2000.

Appiah, Anthony. "The Uncompleted Argument: Du Bois and the Illusion of Race." In *"Race," Writing, and Difference*, ed. Henry Louis Gates Jr., 21–37. Chicago: University of Chicago Press, 1986.

Aravamudan, Srinivas. *Enlightenment Orientalism: Resisting the Rise of the Novel*. Chicago: University of Chicago Press, 2012.

Arbuthnot, John. *An Essay Concerning the Effects of Air on Human Bodies*. London: J. Tonson, 1733.

Aristotle. *The Metaphysics*. Trans. Hugh Lawson-Tancred. London: Penguin, 1998.

———. *On Generation and Corruption*. In *The Complete Works of Aristotle*, ed. Jonathan Barnes. Vol. 1. Princeton: Princeton University Press, 1984.

Armstrong, Nancy. *Desire and Domestic Fiction: A Political History of the Novel*. Oxford: Oxford University Press, 1987.

———. "The Other Side of Modern Individualism: Locke and Defoe." In *Individualism: The Cultural Logic of Modernity*. Lanham, Md.: Lexington Books, 2011.

Atherton, Margaret. "Corpuscles, Mechanism and Essentialism in Berkeley and Locke." *Journal of the History of Philosophy* 29 (1991): 47–67.

———. "'Ideas in the Mind, Qualities in Bodies': Some Distinctive Features of Locke's Account of Primary and Secondary Qualities." In *Minds, Ideas, and Objects*, ed. Phillip D. Cummins and Günter Zöller. North American Kant Society Studies in Philosophy, vol. 2. Atascadero, Calif.: Ridgeview, 1992.

Aubin, Penelope. *A Collection of Entertaining Histories and Novels . . . By Mrs. Penelope Aubin*. London, 1739.

———. *The Life and Amorous Adventures of Lucinda, an English Lady.* London: Printed for E. Bell et al., 1722.

———. *The Life of Charlotta Du Pont, An English Lady.* London: A. Bettesworth, 1723.

———. *The Life of Madam de Beaumount.* London: Printed for E. Bell et al., 1721.

———. *The Noble Slaves: or, The Lives and Adventures of Two Lords and Two Ladies.* London: Printed for E. Bell et al., 1722.

———. *The Strange Adventures of the Count de Vinevil and his Family.* London: Printed for E. Bell et al., 1721.

Austen, Jane. *Jane Austen's "Sir Charles Grandison."* Ed. Brian Southam. Oxford: Oxford University Press, 1981.

———. *Pride and Prejudice.* Ed. Robert P. Irvine. Peterborough, Ontario: Broadview Press, 2002.

Ayers, Michael. "The Foundations of Knowledge and the Logic of Substance: The Structure of Locke's General Philosophy." In *Locke's Philosophy*, ed. G. A. J. Rogers. Oxford: Oxford University Press, 1994.

———. *Locke, Volume I: Epistemology.* London: Routledge, 1991.

———. *Locke, Volume II: Ontology.* London: Routledge, 1991.

———. "Locke's Logical Atomism." In *John Locke: Critical Assessments*, ed. Richard Ashcraft. Vol. 4. London: Routledge, 1991.

Backscheider, Paula R. *Daniel Defoe: His Life.* Baltimore: Johns Hopkins University Press, 1989.

Bacon, Francis. *The Advancement of Learning.* Ed. Stephen Jay Gould. New York: Modern Library, 2001.

———. *The New Organon.* Ed. Lisa Jardine and Michael Silverthorne. Cambridge: Cambridge University Press, 2000.

Balibar, Étienne. "Is There a 'Neo-racism'?" in *Race, Nation, Class: Ambiguous Identities*, ed. Balibar and Immanuel Wallerstein. London: Verso, 1991.

Barad, Karen. *Meeting the Universe Halfway: Quantum Physics and the Entanglement of Matter and Meaning.* Durham: Duke University Press, 2007.

———. "Posthumanist Performativity: Toward an Understanding of How Matter Comes to Matter." *Signs* 28: 3 (2003): 801–31.

Battestin, Martin. *The Providence of Wit: Aspects of Form in Augustan Literature and the Arts.* Charlottesville: University Press of Virginia, 1989.

Battestin, Martin, and Ruth Battestin. *Henry Fielding: A Life.* London: Routledge, 1993.

Baucom, Ian. *Specters of the Atlantic: Finance Capital, Slavery, and the Philosophy of History.* Durham: Duke University Press, 2005.

Beach, Adam R. "African Slaves, English Slave Narratives, and Early Modern Morocco." *Eighteenth-Century Studies* 46: 3 (2013): 333–48.

Bean, Brenda. "Sight and Self-Disclosure: Richardson's Revision of Swift's 'The Lady's Dressing Room.'" *Eighteenth-Century Life* 14 (February 1990): 1–23.

Bell, Walter George. *The Great Plague in London in 1665.* London: John Lane, 1924.

Bender, John. *Ends of Enlightenment.* Stanford: Stanford University Press, 2012.

———. *Imagining the Penitentiary: Fiction and the Architecture of Mind in Eighteenth-Century England.* Chicago: University of Chicago Press, 1989.

Bennett, Jane. *Vibrant Matter: A Political Ecology of Things.* Durham: Duke University Press, 2010.

Bensaude-Vincent, Bernadette, and Isabelle Stengers. *A History of Chemistry*. Cambridge, Mass.: Harvard University Press, 1996.

Berkeley, George. *An Essay Towards a New Theory of Vision*. Dublin, 1709.

Berlant, Lauren. *The Anatomy of National Fantasy: Hawthorne, Utopia, and Everyday Life*. Chicago: University of Chicago Press, 1991.

Berryman, Sylvia. *The Mechanical Hypothesis in Ancient Greek Natural Philosophy*. Cambridge: Cambridge University Press, 2009.

Binhammer, Katherine. "Knowing Love: The Epistemology of *Clarissa*." *ELH* 74: 4 (Winter 2007): 859–79.

Birch, Thomas. *The History of the Royal Society of London*. 4 vols. London, 1756–57.

Blackwell, Mark. *The Secret Life of Things: Animals, Objects, and It-Narratives in Eighteenth-Century England*. Lewisburg: Bucknell University Press, 2007.

Bolton, Martha Brandt. "Locke on the Semantic and Epistemic Role of Simple Ideas of Sensation." *Pacific Philosophical Quarterly* 85: 3 (2004): 301–21.

———. "The Real Molyneux Question and the Basis of Locke's Answer." In *Locke's Philosophy*, ed. G. A . J. Rogers. Oxford: Oxford University Press, 1994.

Bowers, Toni. *Force or Fraud: British Seduction Stories and the Problem of Resistance, 1660–1760*. Oxford: Oxford University Press, 2011.

Boyle, Robert. *The Early Essays and Ethics of Robert Boyle*. Ed. John T. Harwood. Carbondale: Southern Illinois University Press, 1991.

———. *Selected Philosophical Papers of Robert Boyle*. Ed. M. A. Stewart. Indianapolis: Hackett Publishing, 1991.

———. *The Works of Robert Boyle*. Ed. Michael Hunter and Edward B. Davis. 14 vols. London: Pickering and Chatto, 1999.

Braudy, Leo. "Penetration and Impenetrability in *Clarissa*." In *Modern Essays on Eighteenth-Century Literature*, ed. Leopold Damrosch. New York: Oxford University Press, 1988.

Browne, Alice. *The Eighteenth-Century Feminist Mind*. Detroit: Wayne State University Press, 1987.

Butler, Judith. *Bodies That Matter: On the Discursive Limits of "Sex."* London: Routledge, 1993.

———. *Gender Trouble: Feminism and the Subversion of Identity*. New York: Routledge, 1990.

Butler, Melissa. "Early Liberal Roots of Feminism: John Locke's Attack on Patriarchy." In *Feminist Interpretations of John Locke*, ed. Nancy Hirschmann and Kirstie McClure. University Park: Pennsylvania State University Press, 2007.

Castle, Terry. *Clarissa's Ciphers: Meaning and Disruption in Richardson's "Clarissa."* Ithaca: Cornell University Press, 1982.

Chaber, Lois A. "Christian Form and Anti-Feminism in *Clarissa*." *Eighteenth-Century Fiction* 15: 3 (2003): 507–37.

Chalmers, Alan. "Experimental versus Mechanical Philosophy in the Work of Robert Boyle: A Reply to Anstey and Pyle." *Studies in History and Philosophy of Science* 33 (2002): 191–97.

———. "The Lack of Excellence of Boyle's Mechanical Philosophy." *Studies in History and Philosophy of Science* 24 (1993): 541–63.

Charleton, Walter. *Natural History of Nutrition, Life and Voluntary Motion*. London, 1659.

———. *A Ternary of Paradoxes. Of The Magnetic Cure of Wounds. Nativity of Tartar in Wine. Image of God in Man*. London, 1650.

Chetwood, William. *The Voyages and Adventures of Captain Robert Boyle, In several Parts of the World. Intermix'd with the Story of Mrs. Villars, an English Lady with whom He made his Surprizing Escape from Barbary*. London: Printed for John Watts, 1726.

———. *The Voyages, Dangerous Adventures And imminent Escapes of Captain Richard Falconer*. London: Printed for W. Chetwood, 1720.

———. *The Voyages, Travels and Adventures, of William Owen Gwin Vaughan, Esq*. London: Printed for John Watts, 1736.

Cheyne, George. *An Essay on Regimen. Together with Five Discourses, Medical, Moral, and Philosophical*. London, 1740.

Chico, Tita. *Designing Women: The Dressing Room in Eighteenth-Century English Literature and Culture*. Lewisburg: Bucknell University Press, 2005.

Chiles, Katy L. *Transformable Race: Surprising Metamorphoses in the Literature of Early America*. Oxford: Oxford University Press, 2014.

Cislo, Amy Eisen. *Paracelsus's Theory of Embodiment: Conception and Gestation in Early Modern Europe*. London: Pickering and Chatto, 2010.

Clericuzio, Antonio. *Elements, Principles, and Corpuscles: A Study of Atomism and Chemistry in the Seventeenth Century*. Dordrecht: Kluwer, 2000.

———. "From van Helmont to Boyle: A Study of the Transmission of Helmontian Chemical and Medical Theories in Seventeenth-Century England." *British Journal for the History of Science* 26: 3 (1993): 303–34.

———. "A Redefinition of Boyle's Chemistry and Corpuscular Philosophy." *Annals of Science* 47 (1990): 561–89.

———. "Robert Boyle and the British Helmontians." In *Alchemy Revisited: Proceedings of the International Conference of the History of Alchemy at the University of Groningen*, ed. Z. R. W. M. von Martels, 192–99. Leiden: Brill, 1990.

Cobb, Cathy, Monty L. Fetterolf, and Harold Goldwhite. *The Chemistry of Alchemy from Dragon's Blood to Donkey Dung: How Chemistry Was Forged*. Amherst, N.Y.: Prometheus Books, 2014.

Colley, Linda. *Captives*. New York: Pantheon Books, 2002.

Comenius, John Amos. *The Great Didactic of John Amos Comenius*. Trans. and intro. M. W. Keating. London: Adam and Charles Black, 1896.

———. *Orbis Sensualium Pictus . . . Joh. Amos Comenius's Visible World*. Trans. Charles Hoole. London, 1685.

———. *A Patterne of Universall Knowledge, In a plaine and true Draught*. Trans. Jeremy Collier. London, 1651.

Condillac, Étienne Bonnot de. *Oeuvres de Condillac, revues, corrigées par l'Auteur*. 23 vols. Paris: Ch. Houel, 1798.

Cook, Harold J. *The Decline of the Old Medical Regime in Stuart London*. Ithaca: Cornell University Press, 1986.

———. "The New Philosophy and Medicine in Seventeenth-Century England." In *Reappraisals of the Scientific Revolution*, ed. D. C. Lindberg and Richard S. Westman. Cambridge: Cambridge University Press, 1990.

———. "The Society of Chemical Physicians, the New Philosophy, and the Restoration Court." *Bulletin of the History of Medicine* 61: 1 (Spring 1987): 61–77.

Cowper, William. *The Anatomy of Humane Bodies*. London, 1698.

Crow, Nora F. "Swift and the Woman Scholar." In *Pope, Swift, and Women Writers*, ed. Donald C. Mell, 222–38. Newark: University of Delaware Press, 1996.

Curley, E. M. "Locke, Boyle, and the Distinction between Primary and Secondary Qualities." *Philosophical Review* 81: 4 (1972): 438–64.

Curran, Andrew S. *The Anatomy of Blackness: Science and Slavery in an Age of Enlightenment.* Baltimore: Johns Hopkins University Press, 2011.

Davidson, A. I., and N. Hornstein. "The Primary/ Secondary Quality Distinction: Berkeley, Locke, and the Foundations of Corpuscularian Science." *Dialogue* 23 (1984): 281–303.

Davis, Angela Y. *Women, Race, and Class*. New York: Vintage, 1983.

Davis, Audrey B. *Circulation Physiology and Medical Chemistry in England, 1650–1680*. Lawrence: Coronado Press, 1973.

Davis, Robert C. *Christian Slaves, Muslim Masters*. London: Palgrave, 2003.

Dear, Peter. *Revolutionizing the Sciences: European Knowledge and Its Ambitions, 1500–1700.* Princeton: Princeton University Press, 2001.

Debus, Allen G. *The Chemical Philosophy*. 2 vols. New York: Science History Publications, 1977.

———. *Chemistry, Alchemy and the New Philosophy, 1500–1700*. London: Variorum Reprints, 1987.

———. *Chemistry and Medical Debate: Van Helmont to Boerhaave*. Canton, Mass.: Science History Publications, 2001.

———. *The English Paracelsians*. London: Oldbourne, 1965.

———, ed. *Science, Medicine, and Society in the Renaissance: Essays to Honor Walter Pagel.* New York: Neale Watson Academic Publications, 1972.

Defoe, Daniel. *Due Preparations for the Plague, as well for Soul as Body*. Ed. George A. Aitken. London: J. M. Dent and Co., 1895.

———. *The History of the Principal Discoveries and Improvements, In the Several Arts and Sciences*. London: Printed for W. Mears . . . F. Clay . . . and D. Browne, 1727.

———. *A Journal of the Plague Year*. Ed. Cynthia Wall. London: Penguin, 2003.

———. *Roxana: The Fortunate Mistress*. Ed. John Mullan. Oxford: Oxford University Press, 1996.

Delbourgo, James. "The Newtonian Slave Body: Racial Enlightenment in the Atlantic World." *Atlantic Studies: Literary, Cultural and Historical Perspectives* 9: 2 (2012): 185–207.

Democritus. Fragment 589. In *The Pre-Socratic Philosophers*. Trans. and ed. G. S. Kirk and J. E. Raven. Cambridge: Cambridge University Press, 1957.

Derrida, Jacques. *Margins of Philosophy*. Trans. Alan Bass. Sussex: Harvester Press, 1982.

Descartes, René. *The Philosophical Writings of Descartes*. Trans. John Cottingham, Robert Stoothoff, and Dugald Murdoch. 2 vols. Cambridge: Cambridge University Press, 1985.

Dewhurst, Kenneth. *John Locke (1632–1716) Physician and Philosopher: A Medical Biography. With an Edition of the Medical Notes in his Journals*. London: Wellcome Historical Medical Library, 1963.

———. "Locke and Sydenham on the Teaching of Anatomy." *Medical History* 2 (1958): 3–8.

———. "Locke's Contribution to Boyle's Researches on the Air and on Human Blood." *Notes and Records of the Royal Society* 17 (1962): 198–206.

Digby, Kenelm. *Of Bodies, and of Mans Soul* (1669). Reprint. Kessinger Publishing's Rare Mystical Reprints, n. d.

Dijksterhuis, E. J. *The Mechanization of the World Picture: Pythagoras to Newton*. Trans. C. Dikshoorn. Oxford: Clarendon Press, 1961.

Dobbs, Betty Jo Teeter. *The Foundations of Newton's Alchemy or "The Hunting of the Greene Lyon."* Cambridge: Cambridge University Press, 1975.

Dobbs, Betty Jo Teeter, and Margaret C. Jacob. *Newton and the Culture of Newtonians*. Atlantic Highlands, N.J.: Humanities Press, 1995.

Donovan, Arthur. *Antoine Lavoisier: Science, Administration, and Revolution*. Oxford: Blackwell, 1993.

Downing, Lisa. "Are Corpuscles Unobservable in Principle for Locke?" *Journal of the History of Philosophy* 30: 1 (1992): 33–52.

———. "Locke's Newtonianism and Lockean Newtonianism." *Perspectives on Science* 5 (1997): 285–310.

———. "The Status of Mechanism in Locke's *Essay*." *Philosophical Review* 107: 3 (1998): 381–414.

———. "The Uses of Mechanism: Corpuscularianism in Drafts A and B of Locke's *Essay*." In *Late Medieval and Early Modern Corpuscular Matter Theories*, ed. Christoph Lüthy, John E. Murdoch, and William R. Newman, 33–52. Leiden: Brill, 2001.

Dryden, John. *The Works of John Dryden*. Ed. A. B. Chambers and William Frost. 4 vols. Berkeley: University of California Press, 1974.

Du Châtelet, Émilie. *Institutions de Physique*. Paris, 1740.

Eagleton, Terry. "Clarissa Harlowe and Her Times." *Essays in Criticism* 5: 4 (October 1955): 315–40.

———. *The Rape of Clarissa: Writing, Sexuality, and Class Struggle in Samuel Richardson*. Minneapolis: University of Minnesota Press, 1982.

Eaves, T. C. Duncan, and Ben D. Kimpel. *Samuel Richardson: A Biography*. Oxford: Clarendon Press, 1971.

Ellis, Markman. *The Politics of Sensibility: Race, Gender and Commerce in the Sentimental Novel*. Cambridge: Cambridge University Press, 1996.

Emerton, Norma E. *The Scientific Reinterpretation of Form*. Ithaca: Cornell University Press, 1984.

Equiano, Olaudah. *The Interesting Narrative of the Life of Olaudah Equiano, or Gustavus Vassa, the African*. Ed. Werner Sollors. New York: Norton, 2001.

Fausto-Sterling, Anne. *Sexing the Body: Gender Politics and the Construction of Sexuality*. New York: Basic Books, 2000.

Ferguson, Frances. *Pornography, the Theory: What Utilitarianism Did to Action*. Chicago: University of Chicago Press, 2004.

———. "Rape and the Rise of the Novel." *Representations* 20 (Fall 1987): 88–112.

Fielding, Henry. *An Apology for the Life of Mrs. Shamela Andrews*. By Mr. Conny Keyber. London: A. Dodd, 1741.

———. *The Author's Farce; and the Pleasures of the Town. Written by Scriblerus Secundus*. London: J. Roberts, 1730.

———. *The Charge to the Jury: or, the Sum of the Evidence, on the Trial of A.B.C.D. and E.F.* London: M. Cooper, 1745.

———. *Contributions to The Champion and Related Writings*, ed. W. B. Coley. Oxford: Clarendon Press, 2003.

———. *The History of the Adventures of Joseph Andrews* and *An Apology for the Life of Mrs. Shamela Andrews*. Ed. Douglas Brooks-Davies. Intro. Thomas Keymer. Oxford: Oxford University Press, 2008.

———. *The Life of Mr. Jonathan Wild the Great*. Ed. Hugh Amory and Claude Rawson. Notes Linda Bree. Oxford: Oxford University Press, 2003.
———. *Miscellanies by Henry Fielding, Esq.* Ed. Henry Knight Miller, vol. 1. Oxford: Wesleyan University Press, 1972.
———. "Some Papers Proper to be Read before the R—l Society, Concerning the Terrestrial Chrysipus, Golden-Foot or Guinea." In *Miscellanies by Henry Fielding, Esq.*, vol. 1. Ed. Henry Knight Miller. Oxford: Wesleyan University Press, 1972.
———. *Tom Jones*. Ed. John Bender and Simon Stern. Oxford: Oxford University Press, 1996.
———. *Tom Thumb. A Tragedy. As it is Acted at the Theatre in the Hay-Market. Written by Scriblerus Secondus*. 2nd ed. London: J. Roberts, 1730.
Flynn, Carol Houlihan. *The Body in Swift and Defoe*. Cambridge: Cambridge University Press, 1990.
Frank, Robert G. *Harvey and the Oxford Physiologists: A Study of Scientific Ideas and Social Interaction*. Berkeley: University of California Press, 1980.
Frega, Donnalee. *Speaking in Hunger: Gender, Discourse, and Consumption in "Clarissa."* Columbia: University of South Carolina Press, 1998.
Freud, Sigmund. *The Interpretation of Dreams*. Trans. and ed. James Strachey. New York: Avon, 1965.
Fried, Michael. *Absorption and Theatricality: Painting and Beholder in the Age of Diderot*. Chicago: University of Chicago Press, 1980.
Friedman, Michael. "Explanation and Scientific Understanding." *Journal of Philosophy* 71 (1974): 5–19.
Gal, Ofer, and Raz Chen-Morris. *Baroque Science*. Chicago: University of Chicago Press, 2013.
Gaukroger, Stephen. "The Role of Natural Philosophy in the Development of Locke's Empiricism." *British Journal for the History of Philosophy* 17 (2009): 55–83.
Geber. *The "Summa Perfectionis" of Pseudo-Geber*. Trans. and ed. William R. Newman. Leiden: E. J. Brill, 1991.
Gelbert, Nina Rattner. "The Intellectual Development of Walter Charleton." *Ambix* 18 (1971): 149–68.
Gibson, Alexander George. *The Physician's Art; An Attempt to Expand John Locke's Fragment, De Arte Medica*. Oxford: Clarendon Press, 1933.
Gilman, Ernest B. *Plague Writing in Early Modern England*. Chicago: University of Chicago Press, 2009.
———. "The Subject of the Plague." *Journal for Early Modern Cultural Studies* 10.2 (Fall/Winter 2010): 23–44.
Goldberg, Jonathan. *The Seeds of Things: Theorizing Materiality and Sexuality in Renaissance Representations*. New York: Fordham University Press, 2009.
Goliniski, Jan. *Making Natural Knowledge*. Cambridge: Cambridge University Press, 1998.
[Graunt, John.] *London's Dreadful Visitation: Or, a Collection of All the Bills of Mortality For this Present Year: Beginning the 20th of December 1664. and ending the 19th of December following*. London, 1665.
Gregory, John. *A Father's Legacy to His Daughters*. Dublin: John Colles, 1774.
Grosrichard, Alan. *The Sultan's Court: European Fantasies of the East*. Trans. Liz Heron. London: Verso, 1998.

Grosz, Elizabeth. "Irigaray, Ethics, and the Arts: Ontologies, Topologies, Temporalities." Talk delivered May 5, 2015, Northwestern University.

Guerlac, Henry. *Antoine-Laurent Lavoisier: Chemist and Revolutionary*. New York: Charles Scribner's Sons, 1973.

———. "Can There Be Colors in the Dark? Physical Color Theory before Newton." *Journal of the History of Ideas* 47 (1986): 3–20.

Guyton de Morveau, Louis-Bernard, Antoine-Laurent Lavoisier, Claude Louis Berthollet, and Antoine François de Fourcroy. *Méthode de Nomenclature Chimique*. Paris: Chez Cuchet, 1787.

Gwilliam, Tassie. *Samuel Richardson's Fictions of Gender*. Stanford: Stanford University Press, 1993.

Habermas, Jürgen. *The Structural Transformation of the Public Sphere: An Inquiry into a Category of Bourgeois Society*. Trans. Thomas Burger. Cambridge, Mass.: MIT Press, 1989.

Hacking, Ian. *Representing and Intervening: Introductory Topics in the Philosophy of Natural Science*. Cambridge: Cambridge University Press, 1983.

Hales, Stephen. *A Description of Ventilators: Whereby Great Quantities of Fresh Air May With Ease be conveyed into Mines, Goals, Hospitals, Work-Houses and Ships*. London, 1743.

———. *Statical Essays: Containing Hæmastaticks; or, An Account of some Hydraulick and Hydrostatical Experiments made on the Blood and Blood-Vessels of Animals* (1733). Reprint. New York: Hafner Publishing Co., 1964.

———. *Vegetable Staticks: Or, An Account of some Statical Experiments on the Sap in Vegetables*. Foreword M. A. Hoskin. London: Oldbourne Book Co., 1961.

Hall, A. Rupert. *All Was Light: An Introduction to Newton's* Opticks. Oxford: Clarendon Press, 1993.

Hall, A. Rupert, and Marie Boas Hall. "Newton's Theory of Matter." *Isis* 51 (1960): 131–44.

Hall, Marie Boas. "An Early Version of Boyle's *Sceptical Chemist*." *Isis* 45 (1954): 153–68.

———. "The Establishment of Mechanical Philosophy." *Osiris* 10 (1952): 412–541.

———. *Robert Boyle and Seventeenth-Century Chemistry*. Cambridge: Cambridge University Press, 1958.

Haraway, Donna. *Simians, Cyborgs, and Women: The Reinvention of Nature*. London: Routledge, 1991.

Harris, Jocelyn. "Grotesque, Classical and Pornographic Bodies in *Clarissa*." In *New Essays on Samuel Richardson*, ed. Albert J. Rivero, 101–16. New York: St. Martin's Press, 1996.

———. "Protean Lovelace." *Eighteenth-Century Fiction* 2: 4 (July 1990): 327–46.

Harrison, John, and Peter Laslett. *The Library of John Locke*. Oxford: Oxford University Press, 1971.

Hatfield, Glenn W. *Henry Fielding and the Language of Irony*. Chicago: University of Chicago Press, 1968.

Haywood, Eliza. *Fantomina: or, Love in a Maze*. In *Popular Fiction by Women, 1660–1730: An Anthology*. Ed. Paula Backscheider and John Richetti. Oxford: Oxford University Press, 1996.

———. *Love in Excess*. Ed. David Oakleaf. Peterborough, Ontario: Broadview Press, 2000.

Henry, John. "Occult Qualities and the Experimental Philosophy: Active Principles in Pre-Newtonian Matter Theory." *History of Science* 24 (1986): 335–81.

———. *The Scientific Revolution and the Origins of Modern Science.* New York: St. Martin's Press, 1997.

Hill, James. "Locke's Account of Cohesion and its Philosophical Significance." *British Journal for the History of Philosophy* 12 (2004): 611–30.

———. "Primary Qualities, Secondary Qualities and Locke's Impulse Principle." *British Journal for the History of Philosophy* 17: 1 (2009): 85–98.

———. "Was Locke An Atomist?" *Locke Studies* 5 (2005): 75–101.

Hobbes, Thomas. *Leviathan.* Ed. A. P. Martinich. Peterborough, Ontario: Broadview Press, 2002.

Holmes, Frederic L. "Do We Understand Historically How Experimental Knowledge is Acquired?" *History of Science* 30 (1992): 119–36.

———. *Eighteenth-Century Chemistry as an Investigative Enterprise.* Berkeley: University of California Press, 1989.

Hooke, Robert. *Micrographia: or, Some Physiological Descriptions of Minute Bodies* (1665). Reprint. New York: Dover, 1961.

hooks, bell. *Ain't I a Woman?: Black Women and Feminism.* Boston: South End Press, 1981.

Howell, A. C. "Res et Verba: Words and Things." *ELH* 13 (1946): 131–42.

Hume, David. *A Treatise of Human Nature.* Ed. Peter Nidditch. Analytical index L. A. Selby-Bigge. Oxford: Oxford University Press, 1978.

Hunt, Lynn. *Inventing Human Rights: A History.* New York: Norton, 2007.

Hunter, J. Paul. *Before Novels: The Cultural Contexts of Eighteenth-Century English Fiction.* New York: Norton, 1990.

———. *Occasional Form: Henry Fielding and the Chains of Circumstance.* Baltimore: The Johns Hopkins University Press, 1975.

Hunter, Michael. *Boyle: Between God and Science.* New Haven: Yale University Press, 2009.

———. "Boyle versus the Galenists: A Suppressed Critique of Seventeenth-Century Medical Practice and Its Significance." *Medical History* 41 (1997): 322–61.

———. *Robert Boyle by Himself and His Friends.* London: William Pickering, 1994.

———, ed. *Robert Boyle Reconsidered.* Cambridge: Cambridge University Press, 1994.

Irigaray, Luce. *Speculum of the Other Woman.* Trans. Gillian C. Gill. Ithaca: Cornell University Press, 1985.

Jackson, Reginald. "Locke's Distinction between Primary and Secondary Qualities." *Mind* 38 (1929): 56–76.

James, Susan. *Passion and Action: The Emotions in Seventeenth-Century Philosophy.* Oxford: Oxford University Press, 1997.

Jameson, Fredric. *The Political Unconscious: Narrative as a Socially Symbolic Act.* Ithaca: Cornell University Press, 1981.

Jardine, Lisa. *Ingenious Pursuits: Building the Scientific Revolution.* London: Little, Brown, 1999.

Jarvis, R. C. "The Death of Walpole: Henry Fielding and a Forgotten Cause Célèbre." *Modern Language Review* 41: 2 (April 1946): 113–30.

Johnson, Claudia L. *Jane Austen: Women, Politics, and the Novel.* Chicago: University of Chicago Press, 1988.

Johnson, Monte Ransome. *Aristotle on Teleology.* Oxford: Clarendon Press, 2005.

Johnson, William. *Agyrto-mastik Or, some brief animadversions upon two late treatises: one of Master George Thomsons, entituled Galeno-Pale* etc. London, 1665.

Kaplan, Barbara Beigun. *"Divulging of Useful Truths in Physick": The Medical Agenda of Robert Boyle.* Baltimore: Johns Hopkins University Press, 1993.

Kargon, Robert H. *Atomism in England from Hariot to Newton*. Oxford: Clarendon Press, 1966.
Keele, Kenneth D. "The Sydenham-Boyle Theory of Morbific Particles." *Medical History* 18 (1974): 240–48.
Keill, James. *An Account of Animal Secretion, The Quantity of Blood In the Human Body, And Muscular Motion*. London: Printed for George Strahan, 1708.
Kirby, Vicki. *Telling Flesh: The Substance of the Corporeal*. London: Routledge, 1997.
Kozaczka, Edward J. "Penelope Aubin and Narratives of Empire." *Eighteenth-Century Fiction* 25: 1 (Fall 2012): 199–226.
Kramnick, Jonathan Brody. *Actions and Objects from Hobbes to Richardson*. Stanford: Stanford University Press, 2010.
Kroll, Richard W. F. *The Material Word: Literate Culture in the Restoration and Early Eighteenth Century*. Baltimore: Johns Hopkins University Press, 1991.
Kroll, Richard, Richard Ashcraft, and Perez Zagorin., eds. *Philosophy, Science, and Religion in England, 1640–1700*. Cambridge: Cambridge University Press, 1992.
Kuhn, Thomas S. "Robert Boyle and Structural Chemistry in the Seventeenth Century." *Isis* 43: 1 (April 1952): 12–36.
Lamb, Jonathan. *The Things Things Say*. Princeton: Princeton University Press, 2011.
Laqueur, Thomas. *Making Sex: Body and Gender from the Greeks to Freud*. Cambridge, Mass.: Harvard University Press, 1990.
Laslett, Peter. "The Foundation of the Royal Society and the Medical Profession in England." *British Medical Journal* 2: 5193 (1960): 165–69.
———. "The Recovery of Locke's Library." In *The Philosophical Canon in the 17th and 18th Centuries: Essays in Honour of John W. Yolton*, ed. G. A. J. Rogers and Sylvana Tomaselli. Rochester, N.Y.: University of Rochester Press, 1996.
Lasswitz, Kurd. "The Experimental Origin of Chemical Atomic and Molecular Theory before Boyle." *Chymia* 2 (1949): 65–80.
Latour, Bruno. *We Have Never Been Modern*. Trans. Catherine Porter. Cambridge, Mass.: Harvard University Press, 1993.
Lavoisier, Antoine-Laurent. *Elements of Chemistry* (1790). Trans. Robert Kerr. Reprint. New York: Dover, 1965.
Lémery, Nicolas. *Cours de chimie* (1675). Reprint. Paris: Les Introuvables/ Éditions d'aujourd'hui, 1981.
Lennon, Thomas M. "Locke and the Logic of Ideas." *History of Philosophy Quarterly* 18 (2001): 139–54.
Levinson, Marjorie. "What Is New Formalism?" *PMLA* 122: 2 (2007): 558–68.
Lewis, Jayne Elizabeth. *Air's Appearance: Literary Atmosphere in British Fiction, 1660–1794*. Chicago: University of Chicago Press, 2012.
Locke, John. *An Essay Concerning Human Understanding*. Ed. Peter H. Nidditch. Oxford: Oxford University Press, 1975.
———. *An Examination of P. Malebranche's Opinion of Seeing all things in God*. In *Posthumous Works of Mr. Locke*. London, 1706.
———. *Some Thoughts Concerning Education* and *Of the Conduct of the Understanding*. Ed. Ruth W. Grant and Nathan Tarcov. Indianapolis: Hackett Publishing, 1996.
———. *Two Treatises of Government*. Ed. Peter Laslett. Cambridge: Cambridge University Press, 1988.

Long, Edward. *The History of Jamaica. Or, General Survey of the Antient and Modern State of that Island.* 2 vols. London: T. Lowndes, 1774.
Lubey, Kathleen. *Excitable Imaginations: Eroticism and Reading in Britain, 1650–1750.* Lewisburg: Bucknell University Press, 2012.
Lucretius. *The Nature of Things.* Trans. Frank O. Copley. New York: Norton, 1977.
Luhning, H. R. "A Crafted Debut: Haywood's *Love in Excess* and the Literary Marketplace." *Lumen* 28 (2009): 97–110.
Lüthy, Christoph, John E. Murdoch, and William R. Newman, eds. *Late Medieval and Early Modern Corpuscular Matter Theories.* Leiden: Brill, 2001.
Lynch, Deidre. *The Economy of Character: Novels, Market Culture, and the Business of Inner Meaning.* Chicago: University of Chicago Press, 1998.
Mace, Nancy A. *Henry Fielding's Novels and the Classical Tradition.* Newark: University of Delaware Press, 1996.
Mackenzie, Henry. *The Man of Feeling.* Ed. Maureen Harkin. Peterborough, Ontario: Broadview Press, 2005.
Mackie, J. L. *Problems from Locke.* Oxford: Clarendon Press, 1976.
Macpherson, Sandra. *Harm's Way: Tragic Responsibility and the Novel Form.* Baltimore: Johns Hopkins University Press, 2010.
Makdisi, Saree. *Making England Western: Occidentalism, Race, and Imperial Culture.* Chicago: University of Chicago Press, 2014.
Mandelbaum, Maurice. *Philosophy, Science, and Sense Perception.* Baltimore: Johns Hopkins University Press, 1964.
Markley, Robert. *Fallen Languages: Crises of Representation in Newtonian England, 1660–1740.* Ithaca: Cornell University Press, 1993.
Matar, Nabil. *Britain and Barbary, 1589–1689.* Gainesville: University Press of Florida, 2005.
———. *Turks, Moors, and Englishmen in the Age of Discovery.* New York: Columbia University Press, 1999.
McCann, Edwin. "Locke's Philosophy of Body." In *The Cambridge Companion to Locke,* ed. Vere Chappell, 56–88. Cambridge: Cambridge University Press, 1994.
McKeon, Michael. *The Origins of the English Novel, 1600–1740.* Baltimore: Johns Hopkins University Press, 1987.
———. *The Secret History of Domesticity: Public, Private, and the Division of Knowledge.* Baltimore: Johns Hopkins University Press, 2005.
McMullin, Ernan. "Structural Explanation." *American Philosophical Quarterly* 15 (1978): 139–47.
Meillassoux, Quentin. *After Finitude: An Essay on the Necessity of Contingency.* Trans. Ray Brassier. London: Bloomsbury Academic, 2012.
Meinel, Christoph. "Early Seventeenth-Century Atomism: Theory, Epistemology, and the Insufficiency of Experiment." *Isis* 79 (1988): 68–103.
———. "Theory or Practice? The Eighteenth-Century Debate on the Scientific Status of Chemistry." *Ambix* 30 (1983): 121–32.
Metzger, Hélène. *La méthode philosophique en histoire des sciences.* Tours: Librairie Arthème Fayard, 1987.
Meyerson, Emile. *Identity and Reality.* New York: Gordon and Breach, 1989.
Michaels, Walter Benn. *Our America: Nativism, Modernism, and Pluralism.* Durham: Duke University Press, 1997.

———. "Race into Culture: A Critical Genealogy of Cultural Identity." *Critical Inquiry* 18: 4 (Summer 1992): 655–85.

Miller, D. A. *The Novel and the Police*. Berkeley: University of California Press, 1989.

Milton, J. R. "Locke at Oxford." In *Locke's Philosophy: Content and Context*, ed. G. A. J. Rogers, 29–48. Oxford: Clarendon Press, 1994.

———. "Locke's Life and Times." In *The Cambridge Companion to Locke*, ed. Vere Chappell, 5–25. Cambridge: Cambridge University Press, 1994.

Milton, John. *Complete Poems and Major Prose*. Ed. Merritt Y. Hughes. New York: Macmillan, 1957.

Mitchell, John, and Peter Collinson. *An Essay upon the Causes of the Different Colours of People in Different Climates; By John Mitchell, M. D. Communicated to the Royal Society by Mr. Peter Collinson, F. R. S. Philosophical Transactions of the Royal Society* (London, 1744), 43: 102–150.

Moi, Toril. *Simone de Beauvoir: The Making of an Intellectual Woman*. Oxford: Oxford University Press, 1994.

Molineux, Catherine. *Faces of Perfect Ebony: Encountering Atlantic Slavery in Imperial Britain*. Cambridge, Mass.: Harvard University Press, 2012.

Moote, A. Lloyd, and Dorothy C. Moote. *The Great Plague: The Story of London's Most Deadly Year*. Baltimore: Johns Hopkins University Press, 2004.

Moran, Bruce. *Distilling Knowledge: Alchemy, Chemistry, and the Scientific Revolution*. Cambridge, Mass.: Harvard University Press, 2005.

Morgan, Jennifer L. *Laboring Women: Reproduction and Gender in New World Slavery*. Philadelphia: University of Pennsylvania Press, 2004.

Morton, Charles. *Compendium Physicae*. Intro. Samuel Eliot Morison. In *Publications of the Colonial Society of Massachusetts*, vol. 33. Boston: Published by the Society, 1940.

Nadler, Stephen M. *Arnauld and the Cartesian Philosophy of Ideas*. Princeton: Princeton University Press, 1989.

A New Canting Dictionary. London: Booksellers of London and Westminster, 1725.

Newman, Lex, ed. *The Cambridge Companion to Locke's* Essay Concerning Human Understanding. Cambridge: Cambridge University Press, 2007.

———. "Locke on Sensitive Knowledge and the Veil of Perception—Four Misconceptions." *Pacific Philosophical Quarterly* 85: 3 (2004): 273–300.

Newman, William R. "The Alchemical Sources of Robert Boyle's Corpuscular Philosophy." *Annals of Science* 53 (1996): 567–85.

———. *Atoms and Alchemy: Chymistry and the Experimental Origins of the Scientific Revolution*. Chicago: University of Chicago Press, 2006.

———. "The Background to Newton's Chymistry." In *The Cambridge Companion to Newton*, ed. I. Bernard Cohen and George E. Smith, 358–69. Cambridge: Cambridge University Press, 2002.

———. *Gehennical Fire: The Lives of George Starkey, an American Alchemist in the Scientific Revolution*. Cambridge, Mass.: Harvard University Press, 1994.

———. *Promethean Ambitions: Alchemy and the Quest to Perfect Nature*. Chicago: University of Chicago Press, 2004.

Newman, William R., and Lawrence M. Principe. *Alchemy Tried in the Fire: Starkey, Boyle, and the Fate of Helmontian Chymistry*. Chicago: University of Chicago Press, 2002.

———. "Alchemy vs. Chemistry: The Etymological Origins of a Historiographic Mistake." *Early Science and Medicine* 3: 1 (February 1998): 32–65.

Newton, Isaac. *The Correspondence of Isaac Newton*. Ed. H. W. Turnbull. 3 vols. Cambridge: Royal Society at the University Press, 1960.

———. "Dibner MS. 101 B SCDIRB." *The Chymistry of Isaac Newton*. Ed. William R. Newman. http://purl.dlib.indiana.edu/iudl/newton/ALCH00081.

———. *Opticks; or A Treatise of the Reflections, Refractions, Inflections & Colours of Light* (1730). Foreword Albert Einstein, intro. Edmund Whittaker, pref. I. Bernard Cohen, and analytical table of contents Duane H. D. Roller. New York: Dover, 1979.

Nicholson, Watson. *The Historical Sources of Defoe's "Journal of the Plague Year."* Boston: Stratford Co., 1919.

Novak, Maximillian E. *Daniel Defoe: Master of Fictions*. Oxford: Oxford University Press, 2003.

Nussbaum, Felicity. "Between 'Oriental' and 'Blacks So Called,' 1688–1788." In *The Postcolonial Enlightenment: Eighteenth-Century Colonialism and Postcolonial Theory*, ed. Daniel Carey and Lynn Festa, 137–66. Oxford: Oxford University Press, 2009.

Nutton, Vivian. "The Reception of Fracastoro's Theory of Contagion. The Seed That Fell Among Thorns?" *Osiris* (2nd series) 6 (1990): 196–234.

———. "The Seeds of Disease: An Explanation of Contagion and Infection from the Greeks to the Renaissance." *Medical History* 27 (1983): 1–34.

O'Donnell, Sheryl. "My Idea in Your Mind: John Locke and Damaris Cudworth Masham." In *Mothering the Mind: Twelve Studies of Writers and Their Silent Partners*, ed. Ruth Perry and M. W. Brownley. New York: Holmes and Meier, 1984.

Osler, Margaret J. "Biography, Culture and Science: The Formative Years of Robert Boyle." *History of Science* 22 (1989): 151–79.

———. "The Intellectual Sources of Boyle's Philosophy of Nature: Gassendi's Voluntarism and Boyle's Physico-Theological Project." In *Philosophy, Science, and Religion in England 1640–1700*, ed. Richard Kroll, Richard Ashcroft, and Perez Zagorin, 178–98. Cambridge: Cambridge University Press, 1992.

———. John Locke and the Changing Ideal of Scientific Knowledge." *Journal of the History of Ideas* 31 (1970): 3–16.

———, ed. *Rethinking the Scientific Revolution*. Cambridge: Cambridge University Press, 2000.

Pagel, Walter. *Joan Baptista Van Helmont: Reformer of Science and Medicine*. Cambridge: Cambridge University Press, 1982.

———. *Paracelsus: An Introduction to Philosophical Medicine in the Era of the Renaissance*. 2nd rev. ed. Basel: Karger, 1982.

Park, Julie. *The Self and It: Novel Objects in Eighteenth-Century England*. Stanford: Stanford University Press, 2010.

Pateman, Carole. *The Sexual Contract*. Stanford: Stanford University Press, 1988.

Paulson, Ronald. *Satire and the Novel in Eighteenth-Century England*. New Haven: Yale University Press, 1967.

Plato. *Timaeus*. Trans. Robin Waterfield. Oxford: Oxford University Press, 2008.

Poirier, Jean-Pierre. *Lavoisier: Chemist, Biologist, Economist*. Trans. Rebecca Balinski. Philadelphia: University of Pennsylvania Press, 1993.

Porter, Stephen. *The Great Plague*. Gloucestershire: Amberley, 2009.

Potter, Elizabeth. *Gender and Boyle's Law of Gases*. Bloomington: Indiana University Press, 2001.

Pratt, Mary Louise. *Imperial Eyes: Travel Writing and Transculturation*. London: Routledge, 1992.

Principe, Lawrence M. *The Aspiring Adept: Robert Boyle and His Alchemical Quest.* Princeton: Princeton University Press, 1998.

———, ed. *Chymists and Chymistry: Studies in the History of Alchemy and Early Modern Chemistry.* Sagamore Beach, Mass.: Science History Publications, 2007.

———. *The Secrets of Alchemy.* Chicago: University of Chicago Press, 2013.

Principe, Lawrence M., and William R. Newman. "Some Problems with the Historiography of Alchemy." In *Secrets of Nature: Astrology and Alchemy in Early Modern Europe,* ed. William R. Newman and Anthony Grafton. Cambridge, Mass.: MIT Press, 2001.

Pritchard, Will. "Masks and Faces: Female Legibility in the Restoration Era." *Eighteenth-Century Life* 24 (Fall 2000): 31–52.

Pyle, Andrew. *Atomism and Its Critics.* Bristol: Thoemmes Press, 1995.

———. "Boyle on Science and the Mechanical Philosophy: A Reply to Chalmers." *Studies in History and Philosophy of Science* 33 (2002): 175–90.

Radway, Janice. *Reading the Romance: Women, Patriarchy, and Popular Literature.* Chapel Hill: University of North Carolina Press, 1984.

Rattansi, P. M. "The Helmontian-Galenist Controversy in Restoration England." *Ambix* 12: 1 (1964): 1–23.

———. "Newton's Alchemical Studies." In *Science, Medicine and Society in the Renaissance: Essays to Honor Walter Pagel,* ed. Allen G. Debus, 167–82. New York: Neale Watson Academic Publications, 1972.

Ribble, Frederick G., and Anne G. Ribble. *Fielding's Library: An Annotated Catalogue.* Charlottesville: Bibliographical Society of the University of Virginia, 1996.

Richardson, Samuel. *Clarissa. Or, the history of a young lady . . . In eight volumes.* London, 1751.

———. *Clarissa, or, The History of a Young Lady.* Ed. Angus Ross. London: Penguin, 1985.

———. *Clarissa: Preface, Hints of Prefaces, and Postscript.* Ed. and intro. R. F. Brissenden. Los Angeles: Williams Andrews Clark Memorial Library, 1964.

———. *Pamela; or, Virtue Rewarded.* Ed. T. C. Duncan Eaves and Ben D. Kimpel. Boston: Houghton Mifflin, 1971.

———. *Pamela; or, Virtue Rewarded.* Ed. Thomas Keymer and Alice Wakely. Intro. Thomas Keymer. Oxford: Oxford University Press, 2001.

———. *Selected Letters of Samuel Richardson.* Ed. John Carroll. Oxford: Clarendon Press, 1964.

———. *Sir Charles Grandison.* Ed. Jocelyn Harris. Otago, New Zealand: Otago University Print, 2001.

Rickless, Samuel C. "Locke on Primary and Secondary Qualities." *Pacific Philosophical Quarterly* 78 (1997): 297–319.

Riskin, Jessica. *Science in the Age of Sensibility: The Sentimental Empiricists of the French Enlightenment.* Chicago: University of Chicago Press, 2002.

Rogers, G. A. J. "The Intellectual Setting and Aims of the *Essay.*" In *The Cambridge Companion to Locke's "Essay Concerning Human Understanding,"* ed. Lex Newman, 7–32. Cambridge: Cambridge University Press, 2007.

———, ed. *Locke's Philosophy: Content and Context.* Oxford: Oxford University Press, 1994.

Rogers, G. A. J., and Sylvana Tomaselli, eds. *The Philosophical Canon in the 17th and 18th Centuries: Essays in Honour of John W. Yolton.* Rochester, N.Y.: University of Rochester Press, 1996.

Romanell, Patrick. *John Locke and Medicine: A New Key to Locke.* Buffalo, NY: Prometheus Books, 1984.

Rosenthal, Laura. *Infamous Commerce: Prostitution in British Literature and Culture*. Ithaca: Cornell University Press, 2006.

Sabor, Peter, and Thomas Keymer. *"Pamela" in the Marketplace: Literary Controversy and Print Culture in Eighteenth-Century Britain and Ireland*. Cambridge: Cambridge University Press, 2005.

Sabra, A. I. *Theories of Light from Descartes to Newton*. Cambridge: Cambridge University Press, 1981.

Sale, William M. *Samuel Richardson: Master Printer*. Ithaca: Cornell University Press, 1950.

Sargent, Rose-Mary. *The Diffident Naturalist: Robert Boyle and the Philosophy of Experiment*. Chicago: University of Chicago Press, 1995.

Sartre, Jean-Paul. *Being and Nothingness*. Trans. and intro. Hazel E. Barnes. New York: Washington Square Press, 1956.

Schmidgen, Wolfram. *Eighteenth-Century Fiction and the Law of Property*. Cambridge: Cambridge University Press, 2002.

———. *Exquisite Mixture: The Virtues of Impurity in Early Modern England*. Philadelphia: University of Pennsylvania Press, 2012.

Sennert, Daniel. *Thirteen Books of Natural Philosophy*. London: Peter Cole, 1660.

Sepper, Dennis L. *Newton's Optical Writings: A Guided Study*. New Brunswick, N.J.: Rutgers University Press, 1994.

Shackelford, Jole. *A Philosophical Path for Paracelsian Medicine: The Ideas, Intellectual Context, and Influence of Petrus Severinus: 1540/2–1602*. Copenhagen: Museum Tusculanum Press, 2004.

———. "Seeds with a Mechanical Purpose: Severinus' Semina and Seventeenth-Century Matter Theory." In *Reading the Book of Nature: The Other Side of the Scientific Revolution*, ed. Allen G. Debus and Michael T. Walton, 15–44. Kirksville, Mo.: Sixteenth Century Journal Publishers, 1998.

Shapin, Steven, and Simon Schaffer. *Leviathan and the Air-Pump: Hobbes, Boyle, and the Experimental Life*. Princeton: Princeton University Press, 1985.

Shapiro, Alan E. *Fits, Passions, and Paroxysms: Physics, Method, and Chemistry and Newton's Theories of Colored Bodies and Fits of Easy Reflection*. Cambridge: Cambridge University Press, 1993.

———. "Newton's Optics and Atomism." In *The Cambridge Companion to Newton*, ed. I. Bernard Cohen and George E. Smith, 227–55. Cambridge: Cambridge University Press, 2002.

Sharp, L. "Walter Charleton's Early Life, 1620–1659, and Relationship to Natural Philosophy in Mid-Seventeenth Century England." *Annals of Science* 30 (1973): 311–40.

Slack, Paul. *The Impact of Plague in Tudor and Stuart England*. London: Routledge, 1985.

Smith, Pamela. *The Body of the Artisan: Art and Experience in the Scientific Revolution*. Chicago: University of Chicago Press, 2004.

Soni, Vivasvan. "Committing Freedom: The Cultivation of Judgment in Rousseau's *Emile* and Austen's *Pride and Prejudice*." *Eighteenth Century: Theory and Interpretation* 51: 3 (Fall 2010): 363–87.

———. "Judging, Inevitably: Aesthetic Judgment and Novelistic Form in Fielding's *Joseph Andrews*." *Modern Language Quarterly* 76: 2 (June 2015): 159–80.

Sparrow, Tom. *The End of Phenomenology: Metaphysics and the New Realism*. Edinburgh: Edinburgh University Press, 2014.

Spence, Joseph. *Anecdotes, Observations, and Characters, of Books and Men*. Ed. Edmund Malone. London: John Murray, 1820.
Squadrito, Kathy. "Locke on the Equality of Sexes." *Journal of Social Philosophy* 10: 1 (1979): 6–11.
Starkey, George. *Alchemical Laboratory Notebooks and Correspondence*. Ed. William R. Newman and Lawrence M. Principe. Chicago: Chicago University Press, 2004.
———. *An Epistolar Discourse to the Learned and Deserving Author of Galeno-pale*. London: R. Wood, 1665.
———. *Natures Explication and Helmont's Vindication*. London, 1657.
———. *Pyrotechny Asserted and Illustrated*. London, 1658.
———. [Philalethes, pseud.] *Ripley Reviv'd: Or, An Exposition upon Sir George Ripley's Hermetico-Poetical Works*. London: Tho. Ratcliffe and Nat. Thompson, for William Cooper, 1678.
———. [Philalethes, pseud.] *Secrets Reveal'd: or, An Open Entrance to the Shut-Palace of the King. Containing, The greatest Treasure in Chymistry, Never yet so plainly Discovered*. London: W. Godbid for William Cooper, 1669.
———. *A Smart Scourge for a Silly, Sawcy Fool*. London, 1665.
———. *A True Light of Alchymy*. London: I. Dawks, 1709.
Stolberg, Michael. "A Woman Down to Her Bones: The Anatomy of Sexual Difference in the Sixteenth and Early Seventeenth Centuries." *Isis* 94 (2003): 274–99.
Stubbe, Henry. *The Lord Bacon's Relation of the Sweating-Sickness examined, in a Reply to George Thomson, Pretender to Physick and Chymistry, together with a Defence of Phlebotomy*. London, 1671.
Swaminathan, Srividhya, and Adam R. Beach, eds. *Invoking Slavery in the Eighteenth-Century British Imagination*. Burlington, Vt.: Ashgate, 2013.
Swift, Jonathan. *Gulliver's Travels*. Ed. Robert DeMaria. London: Penguin, 2001.
———. *The Poems of Jonathan Swift*. Ed. Harold Williams. 2 vols. Oxford: Clarendon Press, 1937.
———. *A Tale of A Tub and Other Works*. Ed. Angus Ross and David Woolley. Oxford: Oxford University Press, 1986.
Sydenham, Thomas. *Collections of Acute Diseases. The Second and Third Part*. London, 1688.
Thackray, Arnold. *Atoms and Powers*. Cambridge, Mass.: Harvard University Press, 1970.
Thomas, Sir Henry. "The Society of Chymical Physitians." In *Science, Medicine, and History*, ed. E. A. Underwood, vol. 2, 55–71. London, 1953.
Thompson, Helen. "Burney's Conservatism: Masculine Value and 'the ingenuous Cecilia.'" In *Recognizing the Romantic Novel*, ed. Jillian Heydt-Stevenson and Charlotte Sussman, 99–125. Liverpool: Liverpool University Press, 2008.
———. "How the Wanderer Works: Reading Burney and Bourdieu." *ELH* 68 (Winter 2001): 965–89.
———. *Ingenuous Subjection: Compliance and Power in the Eighteenth-Century Domestic Novel*. Philadelphia: University of Pennsylvania Press, 2005.
———. "'In *Idea*, a thousand nameless Joys': Secondary Qualities in Arnauld, Locke, and Haywood's *Lasselia*." *Eighteenth Century: Theory and Interpretation* 48: 3 (Fall 2007): 225–44.
———. "Secondary Qualities and Masculine Form in *Clarissa* and *Sir Charles Grandison*." *Eighteenth-Century Fiction* 24: 2 (Winter 2011–12): 195–226.

Thomson, George. *A Check Given to the Insolent Garrulity of Henry Stubbe etc.* London, 1671.

———. *A Gag for Johnson That Published Animadversions upon Galeno-pale.* London: R. Wood, 1665.

———. *Galeno-pale; or, a Chymical Trial of the Galenists.* London: R. Wood, 1665.

———. *Loimologia: A Consolatory Advice, And some brief Observations Concerning the Present Pest.* London: L. Chapman, 1665.

———. *Loimotomia: or the Pest Anatomized.* London: Nathaniel Crouch, 1666.

Thorndike, Lynn. *A History of Magic and Experimental Science.* 8 vols. New York: Columbia University Press, 1958.

Tiffany, Daniel. *Toy Medium: Materialism and Modern Lyric.* Berkeley: University of California Press, 2000.

Turner, James Grantham. "Lovelace and the Paradoxes of Libertinism." In *Samuel Richardson: Tercentenary Essays,* ed. Margaret Anne Doody and Peter Sabor, 70–88. Cambridge: Cambridge University Press, 1989.

Twigg, Graham. *The Black Death: A Biological Reappraisal.* London: Batsford Academic and Educational, 1984.

Van Helmont, John Baptista. *Oriatrike or, Physick Refined.* Trans. John Chandler. London, 1662.

Vickers, Ilse. *Defoe and the New Sciences.* Cambridge: Cambridge University Press, 1996.

Wahrman, Dror. *The Making of the Modern Self: Identity and Culture in Eighteenth-Century England.* New Haven: Yale University Press, 2004.

Wall, Cynthia Sundberg. *The Prose of Things: Transformations of Description in the Eighteenth Century.* Chicago: University of Chicago Press, 2006.

Walmsley, Peter. *Locke's* Essay *and the Rhetoric of Science.* Lewisburg: Bucknell University Press, 2003.

Walton, Michael T. "Boyle and Newton on the Transmutation of Water and Air, from the Root of Helmont's Tree." *Ambix* 27 (1980): 11–18.

Warner, William. *Licensing Entertainment: The Elevation of Novel Reading in Britain, 1684–1750.* Berkeley: University of California Press, 1998.

———. "Reading Rape: Marxist-Feminist Figurations of the Literal." *Diacritics* 13: 4 (Winter 1983): 12–32.

Watt, Ian. *The Rise of the Novel: Studies in Defoe, Richardson, and Fielding.* Berkeley: University of California Press, 1957.

Wear, Andrew. *Knowledge and Practice in English Medicine, 1550–1680.* Cambridge: Cambridge University Press, 2000.

Webster, Charles. "Alchemical and Paracelsian Medicine." In *Health, Medicine, and Mortality in the Sixteenth Century,* ed. Charles Webster, 301–34. Cambridge: Cambridge University Press, 1979.

———. "The College of Phyicians: 'Solomon's House' in Commonwealth England." *Bulletin for the History of Medicine* 41 (1967): 393–412.

———. "English Medical Reformers of the Puritan Revolution: A Background to the 'Society of Chymical Physitians.'" *Ambix* 14: 1 (February 1967): 16–41.

———. *From Paracelsus to Newton: Magic and the Making of Modern Science.* Mineola, N.Y.: Dover, 2005.

———. *The Great Instauration: Science, Medicine and Reform, 1626–1660.* London: Duckworth, 1975.

———. *Paracelsus: Medicine, Magic and Mission at the End of Time*. New Haven: Yale University Press, 2008.

———. "Water as the Ultimate Principle of Nature: The Background to Boyle's *Sceptical Chymist*." *Ambix* 13 (1966): 96–107.

Welham, Debbie. "The Particular Case of Penelope Aubin." *Journal for Eighteenth-Century Studies* 31: 1 (2008): 63–76.

———. "The Political Afterlife of Resentment in Penelope Aubin's *The Life and Amorous Adventures of Lucinda* (1721)." *Women's Writing* 20: 1 (2013): 49–63.

West, Muriel. "Notes on the Importance of Alchemy to Modern Science in the Writings of Francis Bacon and Robert Boyle." *Ambix* 9 (1961): 102–14.

Westfall, Richard S. *The Construction of Modern Science: Mechanisms and Mechanics*. Cambridge: Cambridge University Press, 1977.

———. *The Life of Isaac Newton*. Cambridge: Cambridge University Press, 1994.

———. "Newton and the Hermetic Tradition," in *Science, Medicine, and Society in the Renaissance*, ed. Allen G. Debus, 183–92. New York: Neale Watson Academic Publications, 1972.

Wheeler, Roxann. *The Complexion of Race: Categories of Difference in Eighteenth-Century British Culture*. Philadelphia: University of Pennsylvania Press, 2000.

Willis, Thomas. *A Plain and Easie Method for Preserving [by God's Blessing] those that are Well from the Infection of the Plague . . . Written in the Year 1666*. London: W. Crook, 1691.

Wilson, Catherine. *Epicureanism at the Origins of Modernity*. Oxford: Clarendon Press, 2008.

———. *The Invisible World: Early Modern Philosophy and the Invention of the Microscope*. Princeton: Princeton University Press, 1995.

Wilson, Elizabeth A. *Psychosomatic: Feminism and the Neurological Body*. Durham: Duke University Press, 2004.

Wilson, Margaret D. *Ideas and Mechanism: Essays on Early Modern Philosophy*. Princeton: Princeton University Press, 1999.

———. "Superadded Properties: The Limits of Mechanism in Locke." *American Philosophical Quarterly* 16: 2 (1979): 143–50.

Wilt, Judith. "He Could Go No Farther: A Modest Proposal About Lovelace and Clarissa." *PMLA* 91: 1 (January 1977): 19–32.

Wittig, Monique. *The Straight Mind and Other Essays*. Foreword Louise Turcotte. Boston: Beacon Press, 1992.

Wojcik, Jan W. *Robert Boyle and the Limits of Reason*. Cambridge: Cambridge University Press, 1997.

Wolfe, Colin T., and Ofer Gal, eds. *The Body as Object and Instrument of Knowledge: Embodied Empiricism in Early Modern Science*. Dordrecht: Springer, 2010.

Wood, Neal. *The Politics of Locke's Philosophy: A Social Study of "An Essay Concerning Human Understanding."* Berkeley: University of California Press, 1983.

Woodward, Hezekiah. *A Light to Grammar*. London, 1641.

Yolton, John W. *John Locke and the Way of Ideas*. Oxford: Clarendon Press, 1968.

———. *Locke and the Compass of Human Understanding: A Selective Commentary on the "Essay."* Cambridge: Cambridge University Press, 1970.

———. *Perceptual Acquaintance from Descartes to Reid*. Minneapolis: University of Minnesota Press, 1984.

———. *Thinking Matter: Materialism in Eighteenth-Century Britain*. Minneapolis: University of Minnesota Press, 1983.

Young, John T. *Faith, Medical Alchemy and Natural Philosophy: Johann Moriaen, Reformed Intelligencer and the Hartlib Circle*. Brookfield, Vt.: Ashgate, 1998.

Zach, Wolfgang. "Mrs. Aubin and Richardson's Earliest Literary Manifesto (1739)." *English Studies* 62 (1981): 271–85.

Zitin, Abigail. "Thinking Like an Artist: Hogarth, Diderot, and the Aesthetics of Technique." *Eighteenth-Century Studies* 46: 4 (Summer 2013): 555–70.

Žižek, Slavoj. *The Sublime Object of Ideology*. London: Verso, 1989.

INDEX

ACKNOWLEDGMENTS

When I encountered Robert Boyle in Chem 11 during my freshman year of college, I could never have dreamed that he would occupy so many of my future years. I am grateful to my chemistry professors at Amherst College, especially Patricia O'Hara, who let me work in her biochemistry laboratory while she respected where my double major in Chemistry and English would take me, and David Hansen, who made organic chemistry an ongoing revelation of the productive powers of form. More fundamentally, I thank my father, Edward Thompson, for venturing out of his own laboratory to demonstrate acid indicators for show-and-tell in my first-grade class. My memories of that magenta litmus allow me to hazard some sympathy with Boyle's fascination at color's vanishing and reapparition.

Over the long evolution of this book, my biggest debt is to my friend and colleague Vivasvan Soni, whose involvement with these pages has been unstinting. His arrival at Northwestern has transformed my understanding of what intellectual, practical, and concretely utopian community can be. I can envision neither this book nor my professional life without the incentive of his own convictions.

For support in many registers, I owe much to Ruth Mack, whose responses to this project have made me see what it could do. Natania Meeker's formidably enriching affirmation sustained me well before I knew what the book was about. In countless ASECS and MLA hotel rooms, Jennifer Snead has offered commiseration and comedy, insight and abiding inspiration. I thank Sarah Ellenzweig for her generative and necessary engagement. Jayne Lewis provided an incomparably luminous and empowering appraisal of this project. Although I have cut Clarissa's *ouroboros* from these pages, mnemonics of its metaphysics persist, not least in the felt resonance of these multifarious debts.

For his history of chymical matter theory and experiment, I am indebted to William Newman, whose *Atoms and Alchemy* radically expanded my appreciation of the corpuscle. Wolfram Schmidgen's rigorous feedback made this a better book. Helen Deutsch, Jonathan Lamb, and Robert Markley enabled my access to fellowship support, for which I am deeply in their debt. I also thank Jonathan Lamb for the opportunity to participate in the Mellon Seminar at Vanderbilt University in 2006 and 2007. For invitations to share work, I thank Emily Hodgson Anderson, Nancy Armstrong, Alison Conway, Mary Helen McMurran, Robert Mitchell, Jesse Molesworth, Rebecca Spang, and Linda Zerilli. Adam Frank invited me to the University of British Columbia, which brought me close to the Pacific Ocean at a revisionary late moment. Among those whose encouragement, insight, and comradeship have enabled this book in vital ways, I thank Toni Bowers, Tim Campbell, Tita Chico, Andrew Clark, Andrew Curran, Peter DeGabriele, Paul Farber, Marcie Frank, Lisa Freeman, William Galperin, Kristen Girten, Corrinne Harol, Heather Keenleyside, Jess Keiser, Paul Kelleher, Theresa Kelley, Julie Park, Nush Powell, John Richetti, Rivka Swenson, Dorothy Wang, and Sharif Youssef. Helen Tartar's interest moved me forward.

At Northwestern, Nathan Mead has provided critical help with the images as well as forbearing assistance during my tenure as director of Graduate Studies. Nick Davis has buoyed my spirits with his discernment, exuberance, and inimitable care. I thank Laurie Shannon, departmental chair at the end stages of this book, for perfectly timed and galvanizing encouragement. I am grateful to Doris Garraway for staunch friendship at home and abroad.

For invaluable institutional support, I am grateful for two short-term Northwestern University Research Grants Committee awards, which enabled me to pursue research in Paris at the Bibliothèque Nationale and Bibliothèque de l'Arsenal. Northwestern's Alice Kaplan Institute for the Humanities provided a fellowship year at a decisive early moment in the development of the book. I am deeply indebted to the Newberry Library, Chicago, for a year-long NEH fellowship. I thank the staff at the Newberry, especially John Powell, for making image permissions so easy; for equally felicitous help, I thank the Lilly Library, Bloomington, Indiana. A portion of Chapter 3 was originally published as "'It Was Impossible to Know These People': Secondary Qualities and the Form of Character in *A Journal of the Plague Year*," *The Eighteenth Century: Theory and Interpretation* 54: 2 (Summer 2013). I thank the editors for permission to reprint that material in revised form here. I am indebted to my editor at Penn Press, Jerry Singerman, for his unfailing integrity and confidence in this project.

Now to more intimate thanks. For the miracle of vegetable growth, which coincided with deep revisions to the manuscript, I thank the Chicago Park

District and Ruby Garden for my community plot. I owe this rediscovery of the joy of gardening to my mother, Deborah Thompson, who planted the seed. It is impossible to imagine the long process of this book and its attendant context without my late cats, Armida and Rinaldo, whose hearts were close to mine during the longest nights of my life. I am enlivened every day by the tenacious devotion to play of their successor, Fantomina. No doubt the second book means different things for different authors. In my experience, I learned that I could not hope to write it, let alone love it, if I did not try to love myself. For saving help with that endeavor, I am grateful for the abundant dedication of my therapist Carol. I thank my stepdaughter, Leah, for her loving generosity as we learn and grow together. Finally, I must acknowledge my husband, Jeffrey Sklansky, who has come to this book relatively late in the day. By giving it, and me, the incredible gift of his attention, he has transmuted my sense of what we might be together. In gratitude for love whose powers I continue to fathom, I give my book to him.

Lightning Source UK Ltd.
Milton Keynes UK
UKOW03n2159201216
290484UK00002B/25/P

9 780812 248722